Atomic Masses of the Elements and Their Symbols

Element	Symbol	Atomic number	Atomic mass (amu)	Element	Symbol	Atomic number	Atomic mass (amu)
actinium	Ac	89	[227]	mendelevium	Md	101	[258]
aluminum	Al	13	26.98	mercury	Hg	80	200.59
americium	Am	95	[243]	molybdenum	Mo	42	95.94
antimony	Sb	51	121.76	neodymium	Nd	60	144.24
argon	Ar	18	39.95	neon	Ne	10	20.18
arsenic	As	33	74.92	neptunium	Np	93	[237]
astatine	At	85	[210]	nickel	Ni	28	58.69
barium	Ba	56	137.33	niobium	Nb	41	92.91
berkelium	Bk	97	[247]	nitrogen	N	7	14.01
beryllium	Be	4	9.012	nobelium	No	102	[259]
bismuth	Bi	83	208.98	osmium	Os	76	190.23
bohrium	Bh	107	[264]	oxygen	O	8	16.00
boron	B	5	10.81	palladium	Pd	46	106.42
bromine	Br	35	79.91	phosphorus	P	15	30.97
cadmium	Cd	48	112.41	platinum	Pt	78	195.08
calcium	Ca	20	40.08	plutonium	Pu	94	[244]
californium	Cf	98	[251]	polonium	Po	84	[209]
carbon	C	6	12.01	potassium	K	19	39.10
cerium	Ce	58	140.12	praseodymium	Pr	59	140.91
cesium	Cs	55	132.91	promethium	Pm	61	[145]
chlorine	Cl	17	35.45	protactinium	Pa	91	231.04
chromium	Cr	24	52.00	radium	Ra	88	[226]
cobalt	Co	27	58.93	radon	Rn	86	[222]
copper	Cu	29	63.55	rhenium	Re	75	186.21
curium	Cm	96	[247]	rhodium	Rh	45	102.91
darmstadtium	Ds	110	[281]	roentgenium	Rg	111	[272]
dubnium	Db	105	[262]	rubidium	Rb	37	85.47
dysprosium	Dy	66	162.50	ruthenium	Ru	44	101.07
einsteinium	Es	99	[252]	rutherfordium	Rf	104	[261]
erbium	Er	68	167.26	samarium	Sm	62	150.36
europium	Eu	63	151.96	scandium	Sc	21	44.96
fermium	Fm	100	[257]	seaborgium	Sg	106	[266]
fluorine	F	9	19.00	selenium	Se	34	78.96
francium	Fr	87	[223]	silicon	Si	14	28.09
gadolinium	Gd	64	157.25	silver	Ag	47	107.87
gallium	Ga	31	69.72	sodium	Na	11	22.99
germanium	Ge	32	72.64	strontium	Sr	38	87.62
gold	Au	79	196.97	sulfur	S	16	32.07
hafnium	Hf	72	178.49	tantalum	Ta	73	180.95
hassium	Hs	108	[277]	technetium	Tc	43	[98]
helium	He	2	4.002	tellurium	Te	52	127.60
holmium	Ho	67	164.93	terbium	Tb	65	158.93
hydrogen	H	1	1.0079	thallium	Tl	81	204.38
indium	In	49	114.82	thorium	Th	90	232.04
iodine	I	53	126.90	thulium	Tm	69	168.93
iridium	Ir	77	192.22	tin	Sn	50	118.71
iron	Fe	26	55.85	titanium	Ti	22	47.87
krypton	Kr	36	83.80	tungsten	W	74	183.84
lanthanum	La	57	138.91	uranium	U	92	238.03
lawrencium	Lr	103	[262]	vanadium	V	23	50.94
lead	Pb	82	207.21	xenon	Xe	54	131.29
lithium	Li	3	6.941	ytterbium	Yb	70	173.04
lutetium	Lu	71	174.97	yttrium	Y	39	88.91
magnesium	Mg	12	24.31	zinc	Zn	30	65.41
manganese	Mn	25	54.94	zirconium	Zr	40	91.22
meitnerium	Mt	109	[268]				

Note: The names of elements 112–118 are provisional; brackets [ ] denote the most stable isotope of a radioactive element.
Online at: http://www.iupac.org/publications/pac/2003/pdf/7508x1107.pdf

About the Authors

Ira Blei was born and raised in Brooklyn, New York, where he attended public schools and graduated from Brooklyn College with B.S. and M.A. degrees in chemistry. After receiving a Ph.D. degree in physical biochemistry from Rutgers University, he worked for Lever Brothers Company in New Jersey, studying the effects of surface-active agents on skin. His next position was at Melpar Incorporated, in Virginia, where he founded a biophysics group that researched methods for the detection of terrestrial and extraterrestrial microorganisms. In 1967, Ira joined the faculty of the College of Staten Island, City University of New York, and taught chemistry and biology there for three decades. His research has appeared in the *Journal of Colloid Science,* the *Journal of Physical Chemistry*, and the *Archives of Biophysical and Biochemical Science*. He has two sons, one an engineer working in Berkeley, California, and the other a musician who lives and works in San Francisco. Ira is outdoors whenever possible, overturning dead branches to see what lurks beneath or scanning the trees with binoculars in search of new bird life, and has recently served as president of Staten Island's local Natural History Club.

George Odian is a tried and true New Yorker, born in Manhattan and educated in its public schools, including Stuyvesant High School. He graduated from The City College with a B.S. in chemistry. After a brief work interlude, George entered Columbia University for graduate studies in organic chemistry, earning M.S. and Ph.D. degrees. He then worked as a research chemist for 5 years, first at the Thiokol Chemical Company in New Jersey, where he synthesized solid rocket propellants, and subsequently at Radiation Applications Incorporated in Long Island City, where he studied the use of radiation to modify the properties of plastics for use as components of space satellites and in water-desalination processes. George returned to Columbia University in 1964 to teach and conduct research in polymer and radiation chemistry. In 1968, he joined the chemistry faculty at the College of Staten Island, City University of New York, and has been engaged in undergraduate and graduate education there for three decades. He is the author of more than 60 research papers in the area of polymer chemistry and of a textbook titled *Principles of Polymerization,* now in its fourth edition, with translations in Chinese, French, Korean, and Russian. George has a son, Michael, who is an equine veterinarian practicing in Maryland. Along with chemistry and photography, one of George's greatest passions is baseball. He has been an avid New York Yankees fan for more than five decades.

Ira Blei and George Odian arrived within a year of each other at the College of Staten Island, where circumstances eventually conspired to launch their collaboration on a textbook. Both had been teaching the one-year chemistry course for nursing and other health science majors for many years, and during that time they became close friends and colleagues. It was their habit to have intense, ongoing discussions about how to teach different aspects of the chemistry course, each continually pressing the other to enhance the clarity of his presentation. Out of those conversations developed their ideas for this textbook.

An Introduction to General Chemistry

Connecting Chemistry to Your Life

SECOND EDITION

Ira Blei

George Odian

College of Staten Island
City University of New York

W. H. Freeman and Company · New York

Senior Acquisitions Editor: Clancy Marshall
Senior Marketing Manager: Krista Bettino
Developmental Editor: Donald Gecewicz
Publisher: Craig Bleyer
Media Editor: Victoria Anderson
Associate Editor: Amy Thorne
Photo Editor: Patricia Marx
Photo Researcher: Elyse Rieder
Design Manager: Diana Blume
Project Editor: Jane O'Neill
Illustrations: Fine Line Illustrations
    and Imagineering Media Services, Inc.
Illustration Coordinator: Bill Page
Production Coordinator: Julia DeRosa
Composition: Schawk, Inc.
Printing and Binding: RR Donnelley

Library of Congress Control Number: 2005935009

ISBN 0-7167-7073-3
EAN 9780716743750

Printed in the United States of America

First printing

W. H. Freeman and Company
41 Madison Avenue
New York, NY 10010
Houndmills, Basingstoke RG21 6XS, England
www.whfreeman.com

Contents in Brief

Contents

Preface

An *Introduction to General Chemistry: Connecting Chemistry to Your Life* is designed to be used in a one-semester course in general chemistry. Often, this course precedes a one-semester course in organic and biochemistry or serves as a precursor to other chemistry courses. Our book was written for students who intend to pursue careers as nurses, dieticians, physician's assistants, physical therapists, or environmental scientists.

Goals of This Book

Our chief objective in writing both editions of this book is to promote a better understanding of chemical *principles*—the comprehensive laws that explain how matter behaves—through the use of real-world examples. Students who merely memorize today's scientific information without understanding the basic underlying principles will not be prepared for the demands of the future. On the other hand, students who have a clear understanding of basic physical and chemical phenomena will have the tools that they need to understand new facts and ideas and will be able to incorporate new knowledge into their professional practices in appropriate and meaningful ways.

The other central goal of our book is to introduce students to how the human body works at the level of molecules and ions—that is, to the chemistry of physiological function. Throughout the book, we take every opportunity to illustrate chemical principles with specific examples of biomolecules and with real-world applications having physiological or medical contexts. Our focus in the coverage of general chemistry is on providing clear explication of fundamental chemical principles to prepare students for the exploration of topics in organic chemistry and biochemistry.

In the process of exploring and using chemical principles, we emphasize two major themes throughout: (1) the ways in which intermolecular and intramolecular forces influence the properties of substances and (2) the relations between molecular structures within the body and their physiological function.

New to This Edition

- In response to reviewer recommendations for more coverage of reactions, we added in-depth coverage to Chapter 4, "Chemical Calculations." Chapter 10, "Chemical and Biological Effects of Radiation," has been enhanced by additional discussion of the basics of the electromagnetic spectrum as well as more information on X-rays and their applications in the medical field.

- Because visuals are so important to chemistry as a discipline and to chemistry textbooks, we have taken particular care with the illustrations in this new edition. Chapter 3 is enhanced by several revised illustrations as well as a new figure illustrating electronegativity, one of the central concepts of chemistry.

- In line with the second major goal of this textbook—showing students how the human body works at the level of molecules and ions—we changed the Pictures of Health that appear in most chapters. Each Picture of Health combines a photograph of an actual person with a drawing of the body and its processes in action, thus showing students how "macroscopic" everyday activities relate to the molecular and ionic activity that goes on within the body. We think that the Picture of Health feature will engage students and that each Picture of Health helps to visually reinforce the concepts described in words in the main text. At the same time, the range of activities shown—from eating cotton candy to farming to playing tennis—highlights chemistry's central role in life.

- We know that students rely on a textbook for review and for test preparation. For that reason, we changed the format of the Summary at the end of each chapter. The new format—a list of short bulleted paragraphs—will make it easier for a student to identify the most important concepts in each chapter.

- We enhanced the more conceptual questions in each chapter. The Expand Your Knowledge category within the Exercise sets will show the students how to synthesize the concepts in the chapter—getting the students to think more like chemists.

- There are three kinds of boxes in this textbook: Chemistry in Depth, Chemistry Within Us, and Chemistry Around Us. Each of these kinds of boxes is designed to give the student more information and an awareness of the myriad applications of chemistry. To enhance the role of these boxes in the classroom and to reinforce their purpose, we added "box exercises" to the Expand Your Knowledge category in the Exercises at the ends of chapters. The box exercises relate to the boxes and the applications in them, and these exercises will draw student's attention to this interesting feature. Look for the flask icons 🔺🔺🔺 in the Exercise sections. Further, we added new applications or updated information to many of these boxes—demonstrating the dynamism of chemistry and its constant effects on our lives.

- Finally, the design of the new edition brightens the Concept Checklists, making them easier for students to find. The various lists of rules (such as the rules for naming certain compounds) are now that much easier to find, too, inasmuch as they follow a similar checklist format. We wanted our readers to be able to navigate our book easily, and its clean and logical design will help them to do so.

Pedagogical Features

The features of this book are **applications, problem-solving strategies, visualization,** and **learning tools,** in a real-world context to connect chemistry to students' lives.

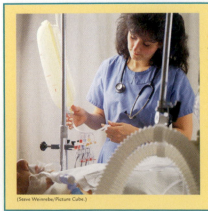

Chemistry in Your Future

When a head-injury patient arrives at the hospital, a key concern of the health-care team is to prevent or reduce excess fluid collection around the brain. Such swelling, or edema, is a natural response to injury but, when the injured organ is the brain, the added fluid pressure can cause severe damage. One way to deal with this problem is to administer an intravenous solution of mannitol, a water-soluble compound having no biological activity. Its only physiological effect is to increase the osmotic pressure of the filtrate in the kidneys' tubules, thus increasing the amount of fluid disposed of in the urine. What is osmotic pressure, and how does it affect body fluids? After reading this chapter on solutions, you will know.

(Steve Weinrebe/Picture Cube.)

Making Connections with Applications

Students are motivated to learn a subject if they are convinced of its fundamental importance and personal relevance. Examples of the relevance of chemical concepts are woven into the text and emphasized through several key features.

Chemistry in Your Future A scenario at the beginning of each chapter describes a typical workplace situation that illustrates a practical, and usually professional, application of the contents of that chapter. A link to the book's Web site leads the student to further practical information.

A Picture of Health This completely revised series of drawings and photographs shows how chapter topics apply to human physiology and health.

Three Categories of Boxes A total of 35 boxed essays, divided into three categories, broaden and deepen the reader's understanding of basic ideas. Icons in the exercise sets reinforce the use of these practical essays.

Chemistry Within Us These boxes describe applications of chemistry to human health and well-being.

Chemistry Around Us These boxes describe applications of chemistry to our everyday life (including commercial products) and to biological processes in organisms other than humans.

Chemistry in Depth These boxes provide a more detailed description of selected topics, ranging from chromatography to the quantitative aspects of equilibrium systems.

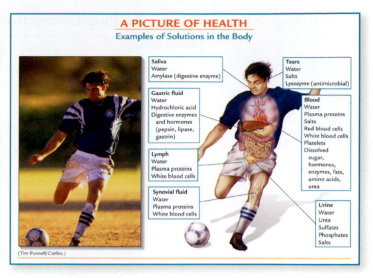

Making Connections Through Problem Solving

Learning to work with chemical concepts and developing problem-solving skills are integral to understanding chemistry. We help students develop these skills.

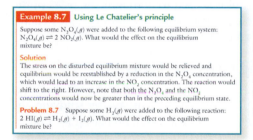

In-Chapter Examples Nearly 140 in-chapter examples with step-by-step solutions, each followed by a similar in-chapter problem, allow students to verify and practice their skills.

End-of-Chapter Exercises More than 700 end-of-chapter exercises are divided into three categories:

- **Paired Exercises** are arranged according to chapter sections; each odd-numbered paired exercise is followed by an even-numbered exercise of the same type.

- **Unclassified Exercises** do not reference specific chapter sections but test the student's overview of chapter concepts.

- **Expand Your Knowledge Exercises** challenge students to expand their problem-solving skills by applying them to more-complex questions or to questions that require the integration of material from different chapters.

Answers to Odd-Numbered Exercises are supplied at the end of the book. Step-by-step solutions to the odd-numbered exercises are supplied in the *Student Solutions Manual*. Step-by-step solutions to even-numbered exercises are supplied in the *Instructor's Resource Manual*. Step-by-step solutions to in-chapter problems are supplied in the *Study Guide*.

Expand Your Knowledge

Note: The icons  denote exercises based on material in boxes.

9.76 The ion-product of water increases as temperature increases. Is the dissociation of water an exothermic or an endothermic reaction?

9.77 What is a conjugate acid–base pair?

9.78 What is the meaning of pH?

9.79 Suppose the capacity of your stomach to be 2.00 L and its pH to be 1.50. How many grams of the antacid sodium bicarbonate, $NaHCO_3$, would be necessary to bring the pH of your stomach to 7.00?

9.80 How would you prepare a buffer to maintain the pH of an aqueous solution as close to pH 5.00 as possible? (Consult Table 9.5.)

9.81 Calculate the pH of a 0.010 M aqueous solution of sodium acetate at 25°C (see Box 9.1).

9.82 Calculate the pH of a 0.010 M aqueous solution of ammonium chloride at 25°C (see Box 9.1).

9.83 Use the ionization constant in Table 9.4 to calculate the osmotic pressure of a 0.0050 M aqueous solution of formic acid at 25°C.

what species are present in the solution, and what is the relation of the pH to the pK_a of acetic acid?

9.86 Would the pH at the equivalence point of the titration in Exercise 9.85 be 7.00? Explain your answer.

9.87 Arrange the pH values of the following 0.010 M aqueous solutions in the order highest to lowest: Na_3PO_4, Na_2HPO_4, NaH_2PO_4.

9.88 Your laboratory instructor provides you with four solutions. She tells you that they are equimolar solutions of monoprotic weak acids. You are to measure the pH of each solution and correlate them with their dissociation constants, smallest to largest. The list with their pH values is (1) 6.65; (2) 3.41; (3) 4.82; (4) 2.85.

9.89 What can you say about the relative basic strengths of the anions produced by the step-by-step dissociation of diprotic or triprotic acids such as carbonic acid or phosphoric acid?

9.90 Write equations that describe the acid–base reactions of each of the following pairs of Brønsted–Lowry acids and bases. (a) H_2O and NH_2^-, (b) H_2O and HS^-, (c) $HCNO$ and NH_3, (d) HNO_2 and CH_3COO^-.

Content from image boxes:

Example 8.7 Using Le Chatelier's principle

Suppose some $N_2O_4(g)$ were added to the following equilibrium system: $N_2O_4(g) \rightleftharpoons 2\ NO_2(g)$. What would be the effect on the equilibrium mixture be?

Solution

The stress on the disturbed equilibrium mixture would be relieved and equilibrium would be reestablished by a reduction in the N_2O_4 concentration, which would lead to an increase in the NO_2 concentration. The reaction would shift to the right. However, note that both the N_2O_4 and the NO_2 concentrations would now be greater than in the preceding equilibrium state.

Problem 8.7 Suppose some $H_2(g)$ were added to the following reaction: $2\ HI(g) \rightleftharpoons H_2(g) + I_2(g)$. What would be the effect on the equilibrium mixture be?

A PICTURE OF HEALTH
Examples of Solutions in the Body

Saliva
Water
Amylase (digestive enzyme)

Gastric fluid
Water
Hydrochloric acid
Digestive enzymes and hormones (pepsin, lipase, gastrin)

Lymph
Water
Plasma proteins
White blood cells

Synovial fluid
Water
Plasma proteins
White blood cells

Tears
Water
Salts
Lysozyme (antimicrobial)

Blood
Water
Plasma proteins
Salts
Red blood cells
White blood cells
Platelets
Dissolved sugar, hormones, enzymes, fats, amino acids, urea

Urine
Water
Urea
Sulfates
Phosphates
Salts

(Tim Pannell/Corbis.)

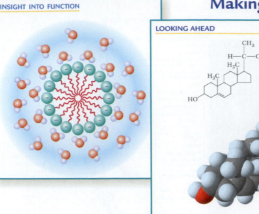

Making Connections Through Visualization

Illustrations Illustrations and tables have been carefully chosen or designed to support the text and are carefully labeled for clarity. Special titles on certain illustrations—Insight into Properties, Insight into Function, and Looking Ahead—emphasize the use of secondary attractive forces and molecular structure as unifying themes throughout the book.

Ball-and-Stick and Space-Filling Molecular Models

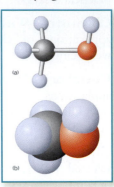

Molecular structures of compounds, especially organic compounds, offer students considerable interpretive challenge. Throughout the book, two-dimensional molecular structures are supported by generous use of ball-and-stick and space-filling molecular models to aid in the visualization of three-dimensional structures of molecules.

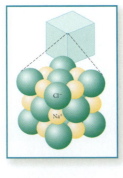

Functional Use of Color Color is used functionally and systematically in schematic illustrations and equations to draw attention to key changes or components and to differentiate one key component from another. For example, in molecular models, the carbon, hydrogen, oxygen, and nitrogen atoms are consistently illustrated in black, white, red, and blue, respectively. In structural representations of chemical reactions, color is used to highlight the parts of the molecule undergoing change. The strategic use of color makes complex chemical concepts less daunting and easier to understand.

Making Connections by Using Learning Tools

Learning Objectives Each chapter begins with a list of learning objectives that preview the skills and concepts that students will master by studying the chapter. Students can use the list to gauge their progress in preparing for exams.

Concept Checklists The narrative is punctuated with short lists serving to highlight or summarize important concepts. They provide a periodic test of comprehension in a first reading of the chapter, as well as an efficient means of reviewing the chapter's key points.

> ✓ When the attractive forces between different molecules are similar, solutions will form.
> ✓ When the attractive forces between different molecules are different, solutions will not form.
>
> *Concept checklist*

Rules Rules for nomenclature, balancing reaction equations, and other important procedures are highlighted so that students can find them easily when studying or doing homework.

>> Chapter 8 describes the role of kinetic energy in a chemical reaction.

Cross-References Cross-referencing in the text and margins alerts students to upcoming topics, suggests topics to review, and draws connections between material in different parts of the book.

Chapter Summaries Serving as a brief study guide, the Summary at the end of each chapter points out the major concepts presented in each section of the chapter.

Key Words Important terms are listed at the end of each chapter and keyed to the pages on which their definitions appear.

Organization

An Introduction to General Chemistry: Connecting Chemistry to Your Life

To understand the molecular basis of physiological functioning, students must have a thorough grounding in the fundamental concepts of general chemistry. Chapter 1 describes the qualitative and quantitative tools of chemistry. It is followed by a consideration of atomic and molecular structure and chemical bonding in Chapters 2 and 3. In Chapter 4, the major types of chemical reactions are presented, along with the quantitative methods for describing the mass relations in those reactions. Chapters 5 and 6 consider the physical properties of molecules and the nature of the interactions between them. Chapter 7 examines the properties of solutions, particularly diffusion and osmotic phenomena. A study of chemical kinetics and equilibria, in Chapter 8, also considers some basic aspects of enzyme function. Chapter 9 treats acids and bases, critical for an understanding of physiological function. Chapter 10 deals with the effects of the interaction of radiation with biological systems and with the use of radiation in medical diagnosis and therapy.

Flexibility for Chemistry Courses

We recognize that all introductory courses are not alike. For that reason, we offer this text in three versions, so you can choose the option that is right for you:

- *General, Organic, and Biochemistry* (ISBN 0-7167-4375-2)—the comprehensive 26-chapter text
- *An Introduction to General Chemistry* (ISBN 0-7167-7073-3)—10 chapters that cover the core concepts in general chemistry
- *Organic and Biochemistry* (ISBN 0-7167-7072-5)—16 chapters that cover organic and biochemistry plus two introductory chapters that review general chemistry

For further information on the content in each of these versions, please visit our Web site: http://www.whfreeman.com/bleiodian2e.

Supplements

A mouse icon in the margins of the textbook indicates that a resource on the book's companion Web site (www.whfreeman.com/bleiodian2e) accompanies that section of the book. Animations, simulations, videos, and more resources found on the book's companion site help to bring the book to life. Its practice tools such as interactive quizzes help students review for exams.

For Students

Student Solutions Manual, by Mark D. Dadmun of the University of Tennessee–Knoxville, contains complete solutions to the odd-numbered end-of-chapter exercises.

Study Guide, by Marcia L. Gillette of Indiana University, Kokomo, provides reader friendly reinforcement of the concepts covered in the textbook. Includes chapter outlines, hints, practice exercises with answers, and more.

General, Organic, and Biochemistry Laboratory Manual, Second Edition, by Sara Selfe of Edmonds Community College.

For Instructors

Instructor's Resource Manual, by Mark D. Dadmun of the University of Tennessee–Knoxville, contains complete solutions to the even-numbered end-of-chapter exercises, chapter outlines, and chapter overviews.

New! Enhanced Instructor's Resource CD-ROM To help instructors create lecture presentations, Web sites, and other resources, this CD-ROM allows instructors to search and export the following resources by key term or chapter: all text images; animations, videos, PowerPoint, and more found on the Web site; and the printable electronic Instructor's Manual (available in Microsoft

For Students (continued)

Web Site, www.whfreeman.com/bleiodian2e, offers a number of features for students and instructors including online study aids such as quizzes, molecular visualizations, chapter objectives, chapter summaries, Web review exercises, flashcards, Web-linked exercises, molecules in the news, and a periodic table.

For Instructors (continued)

Word format), which can be fully edited and includes answers to even-numbered end-of-chapter questions.

Test Bank, by Margaret G. Kimble of Indiana University–Purdue University, contains more than 2500 multiple-choice, fill-in-the-blank, and short-answer questions, available in both print and electronic formats.

More than 200 **Overhead Transparencies.**

Instructor's Web Site, which is password-protected, contains student resources, laboratory information, and PowerPoint files.

Course Management Systems (WebCT, Blackboard) As a service to adopters, electronic content will be provided for this textbook, including the instructor and student resources in either WebCT or Blackboard formats.

Acknowledgments

We are especially grateful to the many educators who reviewed the manuscript and offered helpful suggestions for improvement. For the first edition, we thank the following persons:

Brad P. Bammel, Boise State University; George C. Bandik, University of Pittsburgh; Bruce Banks, University of North Carolina, Greensboro; Lorraine C. Brewer, University of Arkansas; Martin L. Brock, Eastern Kentucky University; Steven W. Carper, University of Nevada, Las Vegas; John E. Davidson, Eastern Kentucky University; Geoffrey Davies, Northeastern University; Marie E. Dunstan, York College of Pennsylvania; James I. Durham, Blinn College; Wes Fritz, College of DuPage; Patrick M. Garvey, Des Moines Area Community College; Wendy Gloffke, Cedar Crest Community College; T. Daniel Griffiths, Northern Illinois University; William T. Haley, Jr., San Antonio College; Edwin F. Hilinski, Florida State University; Vincent Hoagland, Sonoma State University; Sylvia T. Horowitz, California State University, Los Angeles; Larry L. Jackson, Montana State University; Mary A. James, Florida Community College, Jacksonville; James Johnson, Sinclair Community College; Morris A. Johnson, Fox Valley Technical College; Lidija Kampa, Kean College; Paul Kline, Middle Tennessee State University; Robert Loeschen, California State University, Long Beach; Margaret R. R. Manatt, California State University, Los Angeles; John Meisenheimer, Eastern Kentucky University; Frank R. Milio, Towson University; Michael J. Millam, Phoenix College; Renee Muro, Oakland Community College; Deborah M. Nycz, Broward Community College; R. D. O'Brien, University of Massachusetts; Roger Penn, Sinclair Community College; Charles B. Rose, University of Nevada, Reno; William Schloman, University of Akron; Richard Schwenz, University of Northern Colorado; Michael Serra, Youngstown State College; David W. Seybert, Duquesne University; Jerry P. Suits, McNeese State University; Tamar Y. Susskind, Oakland Community College; Arrel D. Toews, University of North Carolina, Chapel Hill; Steven P. Wathen, Ohio University; Garth L. Welch, Weber State University; Philip J. Wenzel, Monterey Peninsula College; Thomas J. Wiese, Fort Hays State University; Donald H. Williams, Hope College; Kathryn R. Williams, University of Florida; William F. Wood, Humboldt State University; Les Wynston, California State University, Long Beach.

We also wish to thank the students of George C. Bandik, University of Pittsburgh; Sharmaine Cady, East Stroudsburg University; Wes Fritz, College of DuPage; Wendy Gloffke, Cedar Crest Community College; Paul Kline, Middle Tennessee State University; Sara Selfe, Edmonds Community College; Jerry P. Suits, McNeese State University; and Arrel D. Toews, University of North Carolina, Chapel Hill, whose comments on the text and exercises provided invaluable guidance in the book's development.

For the second edition, we thank the following persons:

Kathleen Antol, Saint Mary's College; Clarence (Gene) Bender, Minot State University–Bottineau; Verne L. Biddle, Bob Jones University; John J. Blaha, Columbus State Community College; Salah M. Blaih, Kent State University, Trumbull; Laura Brand, Cossatot Community College; R. Todd Bronson, College of Southern Idaho; Charmita Burch, Clayton State University; Sharmaine Cady, East Stroudsburg University; K. Nolan Carter, University of Central Arkansas; Jeannie T. B. Collins, University of Southern Indiana; Thomas G. Conally, Alamance Community College; Loretta T. Dorn, Fort Hays State University; Daniel Freeman, University of South Carolina; Laura DeLong Frost, Georgia Southern University; Edwin J. Geels, Dordt College; Marcia L. Gillette, Indiana University, Kokomo; James K. Hardy, University of Akron; Harvey Hopps, Amarillo College; Shell L. Joe, Santa Ana College; James T. Johnson, Sinclair Community College; Margaret G. Kimble, Indiana University-Purdue University, Fort Wayne; Richard Kimura, California State University, Stanislaus; Robert R. Klepper, Iowa Lakes Community College; Edward A. Kremer, Kansas City, Kansas Community College; Jeanne L. Kuhler, Southern Illinois University; Darrell W. Kuykendall, California State University, Bakersfield; Jennifer Whiles Lillig, Sonoma State University; Robert D. Long, Eastern New Mexico University; David H. Magers, Mississippi College; Janet L. Marshall, Raymond Walters College–University of Cincinnati; Douglas F. Martin, Penn Valley Community College; Craig P. McClure, University of Alabama at Birmingham; Ann H. McDonald, Concordia University Wisconsin; Robert P. Metzger, San Diego State University; K. Troy Milliken, Waynesburg College; Qui-Chee A. Mir, Pierce College; Cynthia Molitor, Lourdes College; John A. Myers, North Carolina Central University; E. M. Nicholson, Eastern Michigan University; Naresh Pandya, Kapiolani Community College; John W. Peters, Montana State University; David Reinhold, Western Michigan University; Elizabeth S. Roberts-Kirchhoff, University of Detroit, Mercy; Sara Selfe, Edmonds Community College; David W. Smith, North Central State College; Sharon Sowa, Indiana University of Pennsylvania; Koni Stone, California State University, Stanislaus; Erach R. Talaty, Wichita State University; E. Shane Talbott, Somerset Community College; Ana M. Q. Vande Linde, University of Wisconsin–Stout; Thomas J. Wiese, Fort Hays State University; John Woolcock, Indiana University of Pennsylvania.

Special thanks are due to Irene Kung, University of Washington; Stan Manatt, California Institute of Technology; and Mark Wathen, University of Northern Colorado, who checked calculations for accuracy for the first edition; and Mark D. Dadmun and Marcia L. Gillette, who checked calculations for accuracy for the second edition.

Finally, we thank the people of W. H. Freeman and Company for their constant encouragement, suggestions, and conscientious efforts in bringing this second edition of our book to fruition. Although most of these people are listed on the copyright page, we would like to add some who are not and single out some who are listed but deserve special mention. We want to express our deepest thanks to Clancy Marshall for providing the opportunity, resources, and enthusiastic support for producing this second edition; to Jane O'Neill and Patricia Zimmerman for their painstaking professionalism in producing a final manuscript and published book in which all can feel pride; and to Moira Lerner (first edition) and Donald Gecewicz (second edition) whose creativity, cheerful encouragement, and tireless energy were key factors in the manuscript's evolution and preparation.

The authors welcome comments and suggestions from readers at: irablei@bellsouth.net; odian@mail.csi.cuny.edu.

PART 1

GENERAL CHEMISTRY

Living organisms are highly organized, with each successive level of organization more complex than the last. Atoms and small molecules are bonded together into molecules of great size, which are then organized into microscopic structures and cells. Cells are then organized into macroscopic tissues and organs, organs into organ systems and organisms. A simple illustration of this theme begins with the fact that our lives depend upon the oxygen in the air. And, although we live in a sea of air, there are times when we must carry it with us—just as the scuba diver on the cover of this book is doing. Oxygen travels a long and tortuous path from the air in our lungs to the most distant cells, and breathing air is only the first step in its journey through the blood to all the cells of our body. The illustration at the right provides a case in point. Red blood cells (top), which carry oxygen to all parts of our bodies, are able to do so because of the special structure of the protein called hemoglobin (center right) that they contain; and the key components of these large proteins are smaller molecules called heme, which contain a form of iron (Fe), to which oxygen becomes attached. Part 1 begins the story of how the properties of simple atoms and molecules lead to the construction of this complex machinery of life.

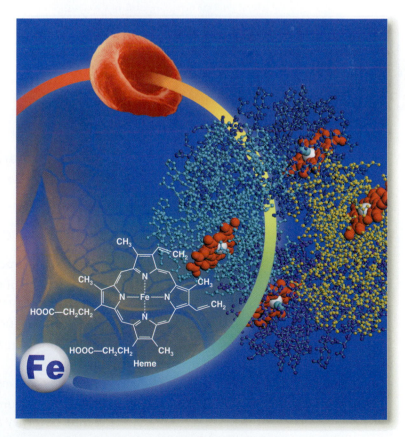

CHAPTER 1

THE LANGUAGE OF CHEMISTRY

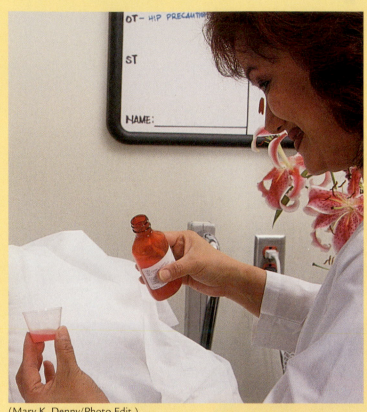

(Mary K. Denny/Photo Edit.)

Chemistry in Your Future

You arrive for your shift at the skilled-nursing facility and read on a patient's chart that the doctor has prescribed a 100-mg dose of Colace. The pharmacy sends up a bottle of the medication in syrup form, containing 20 mg of medicine in each 5 mL of syrup. How many milliliters of the syrup do you give to your patient? A simple calculating technique that you learned in Chemistry helps you find the answer.

For more information on this topic and others in this chapter, go to www.whfreeman.com/bleiodian2e

Learning Objectives

- Describe the characteristics of elements, compounds, and mixtures.
- Name the units of the metric system and convert them into the units of other systems.
- Describe the relation between uncertainty and significant figures.
- Use scientific notation in expressing numbers and doing calculations.
- Use the unit-conversion method in solving problems.
- Define mass, volume, density, temperature, and heat, and describe how they are measured.

Chemistry is the study of matter and its transformations, and no aspect of human activity is untouched by it. The discoveries of chemistry have transformed the food that we eat, the homes that we live in, and the manufactured objects that we use in our daily lives. In addition to explaining and transforming the chemical world outside of our bodies, chemists have developed a detailed understanding of the chemistry within us, the underlying **physiological function.** By physiological function, we mean a function of a living organism or of an individual cell, tissue, or organ of which it is composed. Today, students preparing for careers in any of the life sciences must learn the basic principles of chemistry to acquire a meaningful understanding of biology. If you are one of those students, the purpose of this book is to provide you, first, with a firm grounding in chemical science and, second, with a broad understanding of the physiological processes underlying the lives of cells and organisms.

The practical results of chemical research have greatly changed the practice of medicine. As recently as 70 years ago, families were regularly devastated when children and young adults died from bacterial infections such as diphtheria and scarlet fever. Entire hospitals were once dedicated to the care of patients with tuberculosis, and mental wards were filled with patients suffering from tertiary syphilis. That our experience is so different today is a result of the development of antibacterial drugs such as the sulfonamides, streptomycin, and penicillin. Medical professionals are no longer forced to stand by as disease takes its toll. Armed with a powerful pharmacological arsenal, they have some confidence in their ability to cure those formerly deadly infections.

Since the early 1950s, when the chemical structure of deoxyribonucleic acid (DNA) was described by James Watson and Francis Crick, the pace of accomplishment in the understanding of life processes has been truly phenomenal. The Watson and Crick model of DNA structure was rapidly followed by further developments that allowed biologists and chemists to treat chromosomes (the molecules of inheritance, which dictate the development of living things) literally as chemical compounds. In one of the more interesting and promising of these new approaches, pharmacology and genetics have been combined to study how a person's genetic inheritance can affect the body's response to drugs. A person's genetic makeup may be the key to creating personalized drugs with greater efficacy and safety. In addition to direct medical applications, basic research into the chemistry and biology of DNA has led to the development of new pharmacological products, such as human insulin produced in bacteria.

Parts 1 and 2 of this book, "General Chemistry" and "Organic Chemistry," will provide you with the tools that you need to understand and enjoy Part 3, "Biochemistry." At times you may feel impatient with the pace of the work. Your impatience is understandable because it is difficult to see an immediate connection between elementary chemical concepts and the biochemistry of DNA, but a good beginning will get us there. The present chapter launches our exploration of the chemistry underlying physiological processes with introductory remarks about the composition of matter, conventions for reporting measurements and doing calculations in chemistry, and descriptions of basic physical and chemical properties commonly studied in the laboratory.

1.1   THE COMPOSITION OF MATTER

Humans have been practicing chemistry for hundreds of thousands of years, probably since the first use of fire. Chemical processes—processes that transform the identity of substances—are at the heart of cooking, pottery making, metallurgy, the concoction of herbal remedies, and countless other long-time human pursuits. But these early methods were basically recipes developed in a hit-or-miss fashion over periods of thousands of years. The science of chemistry is only about 300 years old. Its accomplishments are the result of

1.1 CHEMISTRY IN DEPTH

The Scientific Method

The scientific method is basically a common-sense approach to establishing knowledge. What we present here is a distillation of the efforts of many minds and thousands of years of thoughtful curiosity about the world around us. The bottom line for all scientific inquiry is the idea of cause and effect. This idea simply means that whatever effect or observation one can make must have its origin in an identifiable cause. Scientific inquiry has progressed rapidly in the past few hundred years. With that progress came an understanding that there is a significant difference between asking why an event took place and asking how it took place. Asking ten people why an event took place could result in ten different explanations. However, asking ten people how the event took place most often resulted in only one explanation. A how explanation trumps a why explanation because it is useful; that is, it provides a road map for future study.

The elements of the scientific method are the observation of demonstrable facts, the creation of hypotheses to explain or account for those facts, and experimental testing of hypotheses. As more tests validate a hypothesis, more confidence is placed in it until, finally, it may become a theory. An important aspect of this method is the willingness to discard or modify a hypothesis when it is not supported by experiment. A hypothesis is only as good as its last exposure to a rigorous test.

Repeated observations of natural processes can also lead to the development of what are called laws—concise statements of the behavior of nature with no explanation of that behavior, to which there is no exception. For example, Newton's law of gravity says nothing about the mechanism underlying the law but merely asserts its universality. These ideas are illustrated by a flow diagram in Figure 1.1. Let's see how the scientific method worked in a real situation.

In 1928, it was discovered that a nonpathogenic strain of pneumococcus could be transformed into a virulent strain by exposure to chemical extracts of the virulent strain. Call this discovery a fact or an observation. The bacteria is *Diplococcus pneumoniae*, and the virulent strain causes pneumonia. The biological process was called transformation. The material in these extracts responsible for the transmittance of inheritance was called "transforming principle," but its chemical nature was unknown.

To uncover the chemical identity of the transforming principle, scientists required a hypothesis, a guess or hunch regarding what that transforming principle might be. Most biochemists at that time believed that inheritance was carried by proteins, and that became the first hypothesis proposed. It could be readily tested because proteins could be inactivated by heat and destroyed by enzymes such as trypsin and pepsin (the stomach enzyme that degrades proteins). The transforming principle survived all experiments devised to inactivate or destroy proteins in the transforming cell extracts. This fact established that the transforming principle could not be a protein, and that hypothesis had to be discarded. An alternative testable hypothesis was proposed—that the transforming substance could be DNA. The transforming principle was exposed to an enzyme that could degrade only DNA and no other substance. The result was the complete inactivation of the transforming principle. This result was the first indication that the transforming principle was DNA and that DNA was probably the universal carrier of genetic information. Since that time, many other experimental discoveries have supported the original hypothesis. Because of all the subsequent experimental support of the idea that DNA is the molecule that carries genetic information, it now has the status of a theory, a hypothesis in which scientists have a high degree of confidence.

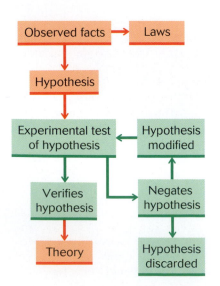

quantitative methods of investigation and experimentation—that is, of systematic measurement and calculation. The general approach, called the scientific method, is discussed in Box 1.1 and diagrammed in Figure 1.1.

The science of chemistry began with the recognition that, to develop an understanding of chemical processes, one must first study the properties of **pure substances.** The notion of purity is not a simple one and requires more than a simple definition. At this point, however, let's simply say that the early chemists were familiar with certain substances—mercury, for example—that appeared to be neither adulterated by nor mixed with anything else. These substances were therefore called pure, and their characteristics served as a model for determining the purity of other, more complicated substances.

Some methods for obtaining pure materials are illustrated on the following page in Figures 1.2 (filtration) and 1.3 (distillation) and on page 6 in Box 1.2 and Figure 1.4 (chromatography). They were found to have unique and consistent

Figure 1.1 A flowchart illustrating the scientific method.

physical and chemical properties. **Physical properties** include the temperature at which a substance melts (changes from a solid to a liquid) or freezes (changes from liquid to solid), color, and densities (Figure 1.5 on the following page). A pure substance undergoes physical changes (freezing, melting, evaporation, and condensation), illustrated in Figure 1.6 on page 7, without losing its identity.

A **chemical property** is the ability of a pure substance to chemically react with other pure substances. In a chemical reaction (Figure 1.7 on page 7), substances lose their chemical identities and form new substances with new physical and chemical properties.

When chemists applied various separation methods to the materials around them and studied the physical and chemical properties of the resulting substances, they discovered that most familiar materials were **mixtures;** that is, they consisted of two or more pure substances in varying proportions. Some mixtures—salt and pepper, for example—are visibly discontinuous; the different components are easy to distinguish. Such a mixture is **heterogeneous.** Other mixtures—sugar and water, for example—have a uniform appearance throughout. The eye cannot distinguish one component from another, even under the strongest microscope. They are called **solutions** and are described as **homogeneous.**

The pepper–salt mixture can be separated into its components by, first, the addition of water. The salt will dissolve in the water, whereas the pepper will remain a solid. Next, the mixture is poured through a filter as illustrated in Figure 1.2; the pepper remains on the filter, and the dissolved salt passes through. Finally, the salt–water solution is separated into its components by allowing the water to evaporate, which leaves the salt behind.

In contrast with pure substances, whose properties are consistent and predictable, mixtures have properties that are variable and depend on the proportions of the components. Consider the mixture of sugar in water. You can dissolve one, two, or more teaspoonsful in a cup of water, and the appearance of the mixture remains the same (in other words, the mixture is a solution and homogeneous). Yet you know from experience that the property known as sweetness increases as the sugar content of the mixture increases.

Figure 1.2 Filtration is used to separate liquids from solids. The filter paper retains the solid because the particles of the solid are too large to pass through the pores, or openings, in the paper. Micropore filters, which have pore sizes small enough to retain bacteria, are used to produce sterile water, sterile pharmaceutical preparations, and bacteria-free bottled beer. (Chip Clark.)

》》 Much of chemistry is concerned with solutions, as you'll see in Chapters 7 and 9.

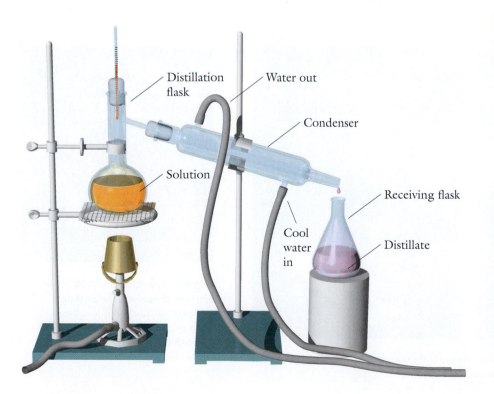

Figure 1.3 A distillation apparatus. If two liquids are to be separated, the liquid with the lower boiling temperature will vaporize at a lower temperature and leave the distillation flask before the higher-boiling liquid. The vaporized liquid leaves the flask and enters the condenser, a long glass tube surrounded by a glass jacket through which cold water is circulated. There the cooled vapor condenses to a liquid and is collected in the receiving flask. A solution of a solid also can be separated by this technique, in which case the solid remains in the distillation flask.

1.2　CHEMISTRY IN DEPTH

Chromatography

Chromatography is a separation technique in which a mixture of substances in a pure liquid called the developer moves past a solid substance that remains stationary. Each component of the mixture interacts to a different extent with the stationary substance and therefore moves at a different rate. The developer does not interact with the stationary substance and acts as a neutral medium, allowing the components of the mixture to interact with the stationary solid. Just as athletes running at different rates will become separated from one another, the different components also become separated. In the earliest application of this method, the green pigments of plants (the chlorophylls) were separated as their liquid mixture flowed down a column packed with solid calcium carbonate. The colored (chroma means color) components moved down the column at different rates. Those that interacted most strongly with the solid lagged behind those that interacted weakly. Eventually, the various components cleanly separated. This method is called column chromatography.

Paper chromatography and thin-layer chromatography (TLC) are two related methods for separating substances in solution. In both, a drop of the mixture is placed on a strip of filter paper or on a thin layer of solid (such as silica gel or aluminum oxide deposited on a plastic strip) and allowed to dry. The strip is placed upright in a small pool of developer and acts as a wick, drawing the liquid along with the mixture of substances along the solid. After sufficient time, the strip is removed, and the solvent is allowed to evaporate. The components interacting least strongly with the solid will have moved farthest along the solid strip, leaving behind those interacting most strongly with the solid. If the compounds possess color, they will appear as a series of spots at different positions along the strip. If the compounds are colorless, additional treatment is necessary to locate them. Some of these treatments use radioactivity and are discussed in Chapter 10. Chromatography is used not only to separate homogeneous mixtures of substances, but also to identify unknown substances by comparison of their chromatographic characteristics with those of known pure compounds under identical conditions.

Paper chromatography and thin-layer chromatography have proved invaluable in separating the products of biochemical reaction products and identifying complex substances with very similar chemical properties. In particular, TLC (see Figure 1.4) is used extensively in the pharmaceutical industry as a quality-control check in the manufacture of complex substances such as penicillin and steroid hormones.

Concept check　　✓　　A mixture is composed of at least two pure substances and is either homogeneous (visibly continuous) or heterogeneous (visibly discontinuous).

Figure 1.4　Thin-layer chromatography can separate complex mixtures and allow the identification of each compound. (Chip Clark.)

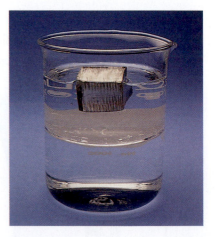

Figure 1.5　Lithium is an element that is less dense than water or oil, and oil is less dense than water. The oil floats on water, and lithium floats on the oil. (Chip Clark.)

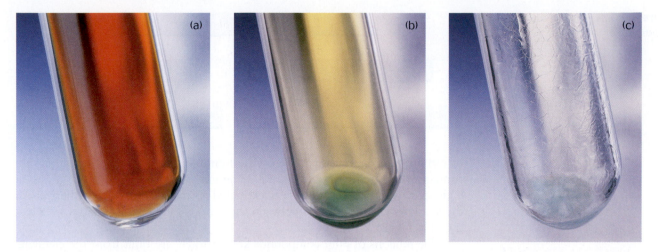

Figure 1.6 The substance nitric oxide (NO_2) can exist in three states, (a) gas, (b) liquid, and (c) solid, and can be reversibly transformed from one state into another without losing its chemical identity. (Richard Megna/Fundamentals Photographs.)

Chemists studying the chemical properties of pure substances found that some of the substances could be decomposed, by chemical means, into simpler pure substances. Furthermore, they found that those simpler substances could not be further decomposed. Decomposable pure substances are called **compounds** (Figure 1.8), and those that cannot be further decomposed are called **elements** (Figure 1.9). An element is a substance that can neither be separated chemically into simpler substances nor be created by combining simpler substances.

When elements combine to form compounds, they always do so in fixed proportions. For example, glucose, also called dextrose, is a chemical combination

Figure 1.7 The solid metallic element iron reacts vigorously with the gaseous element chlorine to form the new solid substance iron chloride. (Chip Clark.)

Figure 1.8 Chemical compounds found in the kitchen. (Richard Megna/Fundamentals Photographs.)

Figure 1.9 Some common elements. *Clockwise from left:* the red-brown liquid bromine, the silvery liquid mercury, and the solids iodine, cadmium, red phosphorus, and copper. (W. H. Freeman photograph by Ken Karp.)

Figure 1.10 Analysis of the composition of matter.

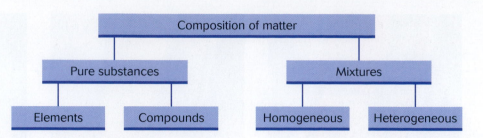

of carbon, oxygen, and hydrogen. One hundred grams of glucose will always contain 40.00 grams of carbon, 53.33 grams of oxygen, and 6.67 grams of hydrogen, whether the glucose is extracted from rose hips or synthesized in the laboratory. The relations between the various categories of matter are illustrated in Figure 1.10.

The millions of pure compounds known today are built from the elements whose names and symbols can be found in the table inside the back cover of this book. Most of the symbols that chemists use to represent elements are derived from the first letter of the capitalized name of the element. However, when the names of two or more elements begin with the same first letter, the symbol for the more recently discovered element is usually formed by adding the second letter of the name, in lower case. For example, the symbol for carbon is C; for calcium, Ca; for cerium, Ce. Other symbols are formed from the elements' Latin, Arabic, or German names. For example, the symbol for potassium is K, after *kalium,* the element's Latin name of Arabic origin. Other examples are W for tungsten, whose German name is *Wolfram,* and Fe for iron, whose Latin name is *ferrum.* These examples and additional ones are listed in Table 1.1.

TABLE 1.1 | **Common Names of the Elements Whose Symbols Are Derived from Latin, German, Greek, or Arabic Names**

Common name	Symbol	Symbol origin
silver	Ag	*argentum*
gold	Au	*aurum*
iron	Fe	*ferrum*
mercury	Hg	*hydrargyrum*
potassium	K	*kalium*
sodium	Na	*natrium*
lead	Pb	*plumbum*
tin	Sn	*stannum*
antimony	Sb	*stibium*
tungsten	W	*Wolfram*

Concept checklist

✓ There are only two kinds of pure substances: elements and compounds.

✓ An element can neither be decomposed into simpler pure substances nor be created by combining simpler substances.

✓ Elements combine to form compounds, which are substances containing fixed proportions of their constituent elements. The composition of a given compound is always the same, regardless of where or how the substance may have formed.

✓ A compound can be decomposed, by chemical means, into simpler pure substances.

✓ The physical and chemical properties of compounds are always different from those of the elements from which they were formed.

✓   Elements are identified by symbols derived from their English, Latin, Arabic, Greek, or German names.

1.2   MEASUREMENT AND THE METRIC SYSTEM

It is far more common to use quantitative rather than qualitative language to describe properties of matter. After sulfur is described as a yellow powder, there are virtually no other descriptive qualities that can help differentiate pure sulfur from all other elements. On the other hand, by carefully measuring sulfur's quantitative properties—its melting point, density, specific heat, and coefficient of expansion, as well as the exact composition of its compounds with oxygen, chlorine, and so forth—we soon compile a profile that is unique. The key concept of the preceding sentence is **measure,** and the meaning of that word is best expressed by a common dictionary definition:

> the size, capacity, extent, volume, or quantity of anything, especially as determined by comparison with some standard or **unit.**

As this definition suggests, to measure anything, we need a standard system of units. We also need a device designed to allow comparison of the object being measured with the standard or reference unit. The story describing Noah building his ark illustrates two ways of using numbers: (1) Noah *counts* the animals that he is going to take and (2) he *measures* the dimensions of the ark (in cubits, a unit no longer in use but a unit nonetheless). A number resulting from counting is considered exact, but a number resulting from a measurement will always have a degree of uncertainty, depending on the device used for making the measurement. This idea will be considered more fully in Section 1.3.

The measurement system used in science and technology is called the **metric system.** The newest version of this system is called the **Système International d'Unités,** abbreviated as the **SI system.** The units defined by this system are found in Table 1.2. All other units are derived from these fundamental units. Following are examples of **derived units:**

$$\text{Area} = m^2$$
$$\text{Volume} = m^3$$
$$\text{Density} = kg/m^3$$
$$\text{Velocity} = m/s$$

TABLE 1.2   Fundamental Units of the Modern Metric System

Fundamental quantity	Unit name	Symbol
length	meter	m
mass	kilogram	kg
temperature	kelvin	K
time	second	s
amount of substance*	mole	mol
electric current	ampere	A
luminous intensity	candela	cd

*The mole is a chemical quantity that will be considered in Chapter 4.

The first five units in Table 1.2 are those with which we will be concerned in chemistry. The SI system is widely used in the physical sciences because it greatly simplifies the kinds of calculations that are most common in those fields. Because certain older, non-SI units continue to be used in clinical and chemical laboratories, we will also use them in many of our quantitative calculations. A few of them are given in Table 1.3.

TABLE 1.3	Non-SI Units in Common Use			
Quantity	Unit	Symbol	SI definition	SI equivalent
length	Ångström	Å	10^{-10} m	0.1 nanometer
volume	liter	L	10^{-3} m^3	1 decimeter^3
energy	calorie	cal	$(kg \cdot m^2 \cdot s^{-2})$*	4.184 joules

*A centered dot ($\cdot$) is used to denote multiplication in derived units.

The great convenience of the metric system is that all basic units are multiplied or divided by multiples of ten, which makes mathematical manipulation very simple, often as simple as moving a decimal point. It also simplifies the calibration of measuring instruments: all basic units are subdivided into tenth, hundredth, or thousandth parts of those units. The multiples of ten are denoted by prefixes, all of Greek or Latin origin, and are listed in Table 1.4. They are combined with any of the basic metric units to denote quantity or size.

TABLE 1.4	Names Used to Express Metric Units in Multiples or Parts of Ten	
Multiple or part of ten	Prefix	Symbol
1,000,000,000	1 giga-	G
1,000,000	1 mega-	M
1,000	1 kilo-	k
100	1 hecto-	h
10	1 deka-	da
0.1	1 deci-	d
0.01	1 centi-	c
0.001	1 milli-	m
0.000001	1 micro-	μ
0.000000001	1 nano-	n
0.000000000001	1 pico-	p

Example 1.1   Using metric system prefixes

Express (a) 0.005 second (s) in milliseconds (ms); (b) 0.02 meter (m) in centimeters (cm); (c) 0.007 liter (L) in milliliters (mL).

Solution
(a) Use Table 1.4 to find the relation between the prefix and the base unit. Milli represents 0.001 of a unit, so

$$0.001 \text{ s} = 1 \text{ ms}$$

therefore

$$0.005 \text{ s} = 5 \text{ ms}$$

(b) In Table 1.4,

$$0.01 \text{ m} = 1 \text{ cm}$$

therefore

$$0.02 \text{ m} = 2 \text{ cm}$$

(c) In Table 1.4,

$$0.001 \text{ L} = 1 \text{ mL}$$

therefore

$$0.007 \text{ L} = 7 \text{ mL}$$

Problem 1.1   Express (a) 2 ms in seconds (s); (b) 5 cm in meters (m); (c) 100 mL in liters (L).

Many of you are familiar with the English system of weights and measures—pounds (lb), inches (in.), yards (yd), and so forth. Section 1.5 will illustrate a formal mathematical procedure for converting units from that or any system of units into any other. This procedure, called the unit-conversion method, also forms the basis for the general method of problem solving that we will use throughout this book. It relies on the use of equivalences—so-called conversion factors—such as those found in Table 1.5 (page 19). First, however, let us look at some of the practical aspects of taking a measurement.

1.3 MEASUREMENT, UNCERTAINTY, AND SIGNIFICANT FIGURES

It is unlikely that a series of measurements of the same property of the same object made by one or more persons will all result in precisely the same value. This inevitable variability is not the result of mistakes or negligence. No matter how carefully each measurement is made, there is no way to avoid small differences between measurements. These differences arise because, no matter how fine the divisions of a measuring device may be, when a measure falls between two such divisions, an estimate, or "best guess," must be made. This unavoidable estimate is called the **uncertainty** or **variability.** All measurements are made with the assumption that there is a correct, or true, value for the quantity being measured. The difference between that true value and the measured value is called the **error.**

You may have already encountered this difficulty yourself in your chemistry laboratory, which is no doubt equipped with several types of balances for measuring mass (a property related to how much an object weighs; see Section 1.8). Let's assume that a balance has a variability of about 1 gram. This means that, every time a mass of, say, 4 g is placed on this balance, the reading will be slightly different but will probably fall within 1 g of the actual mass (no higher than 5 g and no lower than 3 g). Thus, if we decide to measure 4 g of a substance with this balance, we must take account of its variability and report the mass as 4 ± 1 g (4 plus or minus 1 gram). If, instead, we used a balance with a variability of 0.1 g, the measured value of the mass would be written 4 ± 0.1 g. For a third balance, with a variability of 0.001 g, we would report the mass as 4 ± 0.001 g. Finally, we could use an analytical balance with a variability of 0.0001 g and report the mass as 4 ± 0.0001 g.

Although masses are often reported with accompanying variabilities, as just illustrated, scientists also use a simpler system that takes advantage of a concept called the **significant figure.** This system eliminates the need for a $\pm$ notation. It indicates the uncertainty by means of the number of digits instead.

4 ± 1 g = <u>4</u> g	1 significant figure
4 ± 0.1 g = <u>4.0</u> g	2 significant figures
4 ± 0.001 g = <u>4.000</u> g	4 significant figures
4 ± 0.0001 g = <u>4.0000</u> g	5 significant figures

In this way, degrees of uncertainty are communicated through the numbers of significant figures—here one, two, four, and five significant figures, respectively. For the purpose of counting significant figures, zero can have different meanings, depending on its location within a number:

We have seen that the last digit in a reported value is an estimate. Therefore, a reported measurement of 4.130 g indicates an uncertainty of $\pm$ 0.001 g and thus contains four significant figures.

- A trailing zero, as in 4.130, is significant.

In a report recording a measured value of 35.06 cm, the last digit is assumed to be an estimate, but the zero after the decimal and before the last digit is considered to be an accurate part of the measurement and is therefore significant. There are four significant figures in the number.

- A zero within a number, as in 35.06 cm, is significant.

A report lists a liquid volume of 0.082 L. In this case, the zeroes are acting as decimal place holders, and the measurement contains only two significant figures. The insignificance of the zeroes becomes clear when you realize that 0.082 L can also be written as 82 mL.

- A zero before a digit, as in 0.082, is not significant.

A report such as 20 cm is ambiguous. It could be interpreted as meaning "approximately 20" (say, 20 ± 10) or it might be understood as 20 ± 1. It might also mean 20 cm exactly. The number of significant figures in 20 cm is unclear.

- A number ending in zero with no decimal point, as in 20, is ambiguous.

Ambiguities of this last type can be prevented by the use of scientific, or exponential notation.

1.4 SCIENTIFIC NOTATION

Scientific notation, or **exponential notation,** is a convenient method for preventing ambiguity in the reporting of measurements and for simplifying the manipulation of very large and very small numbers. To express a number such as 233 in scientific notation, we write it as a number between 1 and 10 multiplied by 10 raised to a whole-number power: 2.33×10^2. The number between 1 and 10 (in our example, the number 2.33) is called the **coefficient,** and the whole-number exponent of 10 (in our example, 10^2) is called the **exponential factor.** A key rule to remember in using scientific notation is that any number raised to the zero power is equal to 1. Thus, $10^0 = 1$.

The following examples illustrate numbers rewritten in scientific notation:

3 is written:	3×10^0
24 is written:	2.4×10^1
346 is written:	3.46×10^2
2537 is written:	2.537×10^3

- For any number greater than 1, the decimal is moved to the left to create a coefficient between 1 and 10.

- Next, an exponential factor is created whose positive power is equal to the number of places that the decimal point was moved to the left.

To express numbers smaller than 1 (decimal numbers) in scientific notation, use the following algebraic rule describing the reciprocal of any quantity (units also are expressed in this way):

$$1/X = X^{-1}$$

Therefore, $1/8 = 8^{-1}$, and $1/kg = kg^{-1}$.

To express the number 0.2 in scientific notation, we transform it into a whole-number coefficient between 1 and 10, multiplied by an exponential factor that is decreased by the same power of ten:

$$0.2 = 2 \times 0.1 = 2 \times \frac{1}{10^1} = 2 \times 10^{-1}$$

The number 0.365 is therefore written 3.65×10^{-1}. The numbers 0.046 and 0.00753 are written 4.6×10^{-2} and 7.53×10^{-3}, respectively.

A PICTURE OF HEALTH
Ranges of Measurement in the Body

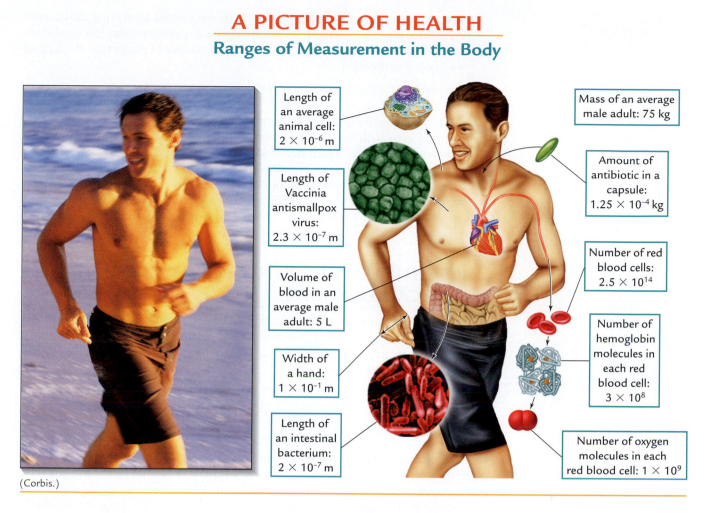

Length of an average animal cell: 2×10^{-6} m

Length of Vaccinia antismallpox virus: 2.3×10^{-7} m

Volume of blood in an average male adult: 5 L

Width of a hand: 1×10^{-1} m

Length of an intestinal bacterium: 2×10^{-7} m

Mass of an average male adult: 75 kg

Amount of antibiotic in a capsule: 1.25×10^{-4} kg

Number of red blood cells: 2.5×10^{14}

Number of hemoglobin molecules in each red blood cell: 3×10^{8}

Number of oxygen molecules in each red blood cell: 1×10^{9}

(Corbis.)

- For any number smaller than 1, the decimal is moved to the right to create a coefficient between 1 and 10.

- Next, an exponential factor is created with a negative power equal to the number of places that the decimal point was moved to the right.

Example 1.2   Expressing a number in scientific notation

Express the number 0.00964 in scientific notation.

Solution
The first task is to create a coefficient between 1 and 10. This task is accomplished by moving the decimal point three places to the right. The result is 9.64. Moving the decimal to the right three places is equivalent to multiplying the decimal number by ten three times, as in the following three steps. To retain the value of the number, each time the decimal number on the left-hand side is multiplied by 10, the result on the right-hand side is reduced by a power of 10.

$$0.00964 = 0.0964 \times 10^{-1}$$
$$0.0964 = 0.964 \times 10^{-1}$$
$$0.964 = 9.64 \times 10^{-1}$$

Instead of taking three separate steps, the transformation into scientific notation is done in one step:

$$0.00964 = 9.64 \times 10^{-3}$$

The coefficient (a number between 1 and 10) was created by moving the decimal point three places to the right. To retain the actual numerical value, the coefficient must be multiplied by 10 raised to the negative number of places that the decimal was moved to the right.

Problem 1.2   Express the number 0.0007068 in scientific notation.

In the discussion of significant figures, it was pointed out that a number ending in zero with no decimal point in the number (21,600, for example) is ambiguous. Scientific notation provides a way to express the number without ambiguity. If the last zero in 21,600 is significant, the number has five significant figures and should be written 2.1600×10^4. If both zeroes are not significant, the number has three significant figures and should be written 2.16×10^4. If a number containing zeroes loses those zeroes when the number is expressed in scientific notation, they were not significant.

1.5   THE USE OF SCIENTIFIC NOTATION IN CALCULATIONS

Although the rules of standard scientific notation require the coefficient to be a number between 1 and 10, for calculations requiring addition or subtraction, it is useful to write numbers by using nonstandard coefficients. For example, 4573 can be written as

$$4573 = 4573 \times 10^0$$
$$4573 = 457.3 \times 10^1$$
$$4573 = 45.73 \times 10^2$$
$$4573 = 4.573 \times 10^3$$

Each of these forms expresses the same numerical value. Every time the value of the coefficient was changed, the value of the exponential factor was adjusted to preserve the original numerical value of 4573. Because we can vary the coefficient and exponential factor of a number without changing the number's value, scientific notation simplifies additions and subtractions of numbers having different exponential factors.

Example 1.3   Adding numbers written in scientific notation

Perform the addition

$$(3.63 \times 10^{-2}) + (4.85 \times 10^{-3})$$

Solution

The more arduous solution would be to convert these expressions into full decimal notation, then add,

$$\begin{array}{r} 0.0363 \\ +0.00485 \\ \hline 0.04115 \end{array}$$

and, after that, convert back into scientific notation: 4.12×10^{-2} (the "rounding" of such numbers will be explained in Section 1.6). But most scientists use the more convenient approach of modifying the expressions so that the exponents are equal:

$$(3.63 \times 10^{-2}) + (0.485 \times 10^{-2})$$

Now, with each coefficient multiplied by the same exponential factor, the addition (or subtraction) takes the form

$$(3.63 + 0.485) \times 10^{-2} = 4.12 \times 10^{-2}$$

Problem 1.3   What is the result of subtracting the following numbers?

$$7.953 \times 10^{-4} - 6.42 \times 10^{-5}$$

To multiply numbers written in scientific notation, we multiply the coefficients and add the exponents.

Example 1.4 **Multiplying numbers written in scientific notation**

Multiply 3.4×10^3 by 2.8×10^{-2}.

Solution

$$(3.40 \times 2.80) \times 10^{[3 + (-2)]} = 9.52 \times 10^1 = 95.2$$

Problem 1.4 Multiply 4.2×10^5 by 0.64×10^{-4}.

To divide numbers written in scientific notation, divide the coefficients and subtract the exponent of 10 in the denominator from the exponent of 10 in the numerator.

Example 1.5 **Dividing numbers written in scientific notation**

Divide 2.8×10^5 by 4.0×10^2.

Solution

$$\frac{2.8 \times 10^5}{4.0 \times 10^2} = 0.70 \times 10^{[5 - (+2)]} = 0.70 \times 10^3 = 7.0 \times 10^2$$

Problem 1.5 Divide 3.45×10^4 by 7.2×10^{-2}.

✓ To multiply in scientific notation, multiply the coefficients and add the exponents.

✓ To divide in scientific notation, divide the coefficients and subtract the exponent of 10 in the denominator from the exponent of 10 in the numerator.

Concept checklist

1.6 CALCULATIONS AND SIGNIFICANT FIGURES

You may well ask, "What's the point? Why should we be responsible for learning about significant figures?" The answer is that it may not necessarily be useful to you, but it could mean a great deal to the next person who has to make calculations based on your report. Did you mean that this patient can receive 1 mL of standard morphine sulfate solution, or 1.0 mL? After all, 1 mL can mean anything from 0.5 mL to 1.4 mL. Which is it to be? Well, let's find out.

Calculations that are numerically correct can sometimes lead to unrealistic results. For example, how should we report the area of a square whose dimensions have been measured as 8.5 in. on a side? Mathematically,

$$\text{Area of a square} = \text{side} \times \text{side} = 8.5 \text{ in.} \times 8.5 \text{ in.}$$
$$= 72.25 \text{ in.}^2$$

Multiplication of 2 two-digit numbers always yields a number with more than two digits. However, information regarding the size of an object can be obtained only by measurement, not by an arithmetic operation. The results of multiplications and divisions using measured quantities are reported according to the following rule:

> The number of significant figures in a number resulting from multiplication or division may not exceed the number of significant figures in the least well known value used in the calculation.

In the preceding example of the area of the 8.5 in. $\times$ 8.5 in. square, the length of a side is known to two significant figures, and therefore the area of the square (length $\times$ length) cannot be known with any greater accuracy. Should we report it as 72 or 73 in.^{2}?

To reduce the number of significant figures and determine the value of the final significant digit, we commonly use a practice called **rounding.** The rules of rounding stipulate that, if the digit after the one that we want to retain is 5 or greater, we increase the value of the digit that we want to retain by 1 and drop the trailing digits. If its value is 4 or less, we leave unchanged the value of the digit that we want to retain and drop the trailing digits.

Note that rounding takes place after the calculation has been completed. That is, the calculation is done by using as many digits as possible. Only the final result is rounded. In determining the area of the 8.5-in. square, because the least well known measurement has only two significant figures, we should round the calculated result of 72.25 and report an area of 72 in.^{2}.

A more perplexing situation might be encountered if we needed to know the area of a rug required to fit a room 74 in. by 173 in. The calculated area is 12,802 in.^{2}. The least accurately known measurement possesses two significant figures, and so the area must be expressed with that number of significant figures as well. The value of the area is reported by first converting the value into scientific notation and then rounding to two significant figures:

$$12{,}802 \text{ in.}^2 = 1.2802 \times 10^4 \text{ in.}^2 = 1.3 \times 10^4 \text{ in.}^2$$

The same considerations hold for division.

Example 1.6   Multiplying and dividing measured quantities

Velocity is defined as $\dfrac{\text{distance}}{\text{time}}$. What velocity must an automobile be driven to cover 639 km in 9.5 hours (h)?

Solution

$$\text{Velocity} = \frac{\text{distance}}{\text{time}} = \left(\frac{639 \text{ km}}{9.5 \text{ h}} \right) = 67.3 \, \frac{\text{km}}{\text{h}} = 67 \, \frac{\text{km}}{\text{h}}$$

Because the least well known quantity, 9.5 h, has two significant figures, it is necessary to round the resultant calculated value to two significant figures.

Problem 1.6   Calculate the volume of a cube that is 8.5 cm on a side. (Volume of a cube 1 cm on a side = 1 cm $\times$ 1 cm $\times$ 1 cm.)

A somewhat different approach is required for addition and subtraction. In both these situations, the number of figures after the decimal point decides the final answer. The final sum or difference cannot have any more figures after the decimal point than are contained in the least well known quantity in the calculation. All significant figures are retained while doing the calculation, and the final result is rounded.

Example 1.7   Adding measured quantities

Add the following measured quantities: 24.62 g, 3.7 g, 93.835 g.

Solution

The least well known of these quantities has only one significant figure after the decimal point, and so the final sum cannot contain any more than that. We add all the values and round off after the sum has been calculated, as follows:

$$
\begin{array}{r}
24.62 \text{ grams} \\
3.7 \text{ grams} \\
\underline{93.835 \text{ grams}} \\
122.155 \text{ grams} = 122.2 \text{ grams}
\end{array}
$$

Problem 1.7   Add the following quantities: 1.9375, 34.23, 4.184.

The same considerations apply to subtractions.

Example 1.8   **Subtracting measured quantities**

Calculate the result of the following subtraction:

$$5.753 \text{ grams} - 2.32 \text{ grams}$$

Solution

The least well known quantity has two significant figures after the decimal point, and so the result cannot contain any more than that. As in addition, we round off after having done the subtraction.

$$
\begin{array}{r}
5.753 \text{ grams} \\
-\ 2.32\ \ \text{grams} \\
\hline
3.433 \text{ grams} = 3.43 \text{ grams}
\end{array}
$$

Problem 1.8   What is the result of the following subtraction?

$$94.935 \text{ m} - 7.6 \text{ m}$$

- The number of significant figures in the result of a multiplication or division may not exceed the number of significant figures found in the least well known value used in the calculation.

- The number of figures after the decimal point in the result of an addition or subtraction may not exceed the number of significant figures after the decimal point in the least well known quantity being used.

Rules for determining significant figures

- If the digit after the one to be retained is 5 or greater, increase the value of the digit to be retained by 1 and drop the trailing digits.

- If the digit after the one to be retained is 4 or less, leave unchanged the value of the digit to be retained and drop the trailing digits.

- Only a final result is rounded. All digits are retained until a calculation is complete.

Rules for rounding numbers

1.7   THE USE OF UNITS IN CALCULATIONS: THE UNIT-CONVERSION METHOD

All of the quantities that you will be working with when you do chemical calculations will have units—for example, mL, cal, and so forth. The method used in solving problems with quantities having units is called the **unit-conversion method.** It is also referred to as the factor-label method, the unit-factor method, or dimensional analysis.

The underlying principle in this problem-solving strategy is the conversion of one type of unit into another by the use of a **conversion factor.**

$$\text{Unit}_1 \times \text{conversion factor} = \text{unit}_2$$

The conversion factor has the form of a ratio that allows cancellation of unit_1 and its replacement with unit_2. The units are quantities that are treated according to the rules of algebra.

Say that unit_1 is g/L and that the required unit_2 is g/mL. To get the desired unit, we must multiply unit_1 by a factor that will allow L to be canceled out so that the result is the required unit along with the correct numerical value. The calculation is

$$\frac{\text{g}}{\text{L}} \times \frac{\text{L}}{1000 \text{ mL}} = \frac{\text{g}}{1000 \text{ mL}}$$

Earlier in this chapter, Example 1.1 asked us to convert 0.001 s into milliseconds, which we accomplished by using the definition 0.001 second (s) = 1 millisecond (ms). Let's now see how the unit-conversion method takes this kind of information and uses it to solve problems.

>> We begin using unit conversions in the very next chapter to understand the nature of matter.

1.8 TWO FUNDAMENTAL PROPERTIES OF MATTER: MASS AND VOLUME

Having established the basis for quantitative description, we are ready to examine two of the basic properties of matter and some of the ways in which they are measured in the laboratory.

Mass

Mass is a measure of the quantity of matter. It is a useful property in the study of matter because it remains constant in the presence of environmental changes such as fluctuations in temperature and pressure. Because an object's mass is determined by weighing, it is common to speak of mass and weight interchangeably; however, they are not the same thing. A mass has weight because it is under the influence of a gravitational field. The gravitational field at the top of Mt. Everest is weaker than that at sea level by about 0.2%; so a 150-lb astronaut at sea level weighs 149.8 lb at the mountain's summit. Astronauts aboard the space shuttle in outer space are nearly weightless because there is little effect of gravity on a mass so far from Earth. Nevertheless, an astronaut's mass at the summit of Mt. Everest, at sea level on Earth, and in outer space is the same. The SI unit of mass is the kilogram.

Mass is measured relative to that of a standard mass. In weighing, an unknown mass is balanced against a known, or reference, mass. Thus, mass measuring devices, both mechanical ones and those that are a combination of mechanical and electronic devices, have come to be known as "balances."

A balance operates on the same principle as the playground seesaw, or teetertotter—a plank supported at its center by a wedge-shaped fulcrum. If equal weights are placed equal distances from the fulcrum, the plank is held in balance, with each end equidistant from the ground. In a laboratory balance, weights are added or removed until they equal that of the unknown mass, and balance is achieved. Figure 1.11 illustrates an older beam type and the newer electronic type of balance used in the chemical laboratory.

Example 1.12 **Converting measures of mass**

Express a mass of 76 g in units of kilograms.

Solution

Using the unit-conversion method, first write the given quantity, and then multiply by the appropriate conversion factor:

$$76 \text{ g} \left(\frac{1 \text{ kg}}{1000 \text{ g}} \right) = 0.076 \text{ kg}$$

Problem 1.12 Express a mass of 2.87 kg in grams.

Figure 1.11 A modern automatic analytical balance *(left)* and an older laboratory beam balance *(right)*. The mass of a sample on one pan of the beam balance is determined by balancing it with a known reference weight on the other pan. (Chip Clark.)

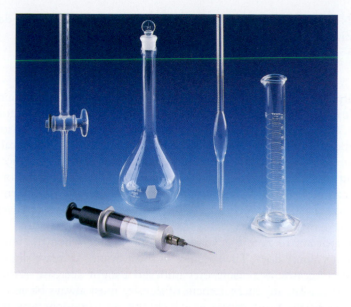

Figure 1.12 Devices for measuring liquid volumes: (*back row, left to right*) a buret, a volumetric flask, a pipet, and a graduated cylinder; (*in front*) a syringe. (Richard Megna/Fundamentals Photographs.)

Volume

Volume is the amount of space that a sample occupies. Figure 1.12 illustrates the devices commonly used in the laboratory to measure liquid volume. The one most commonly used in the beginning chemistry laboratory is the graduated cylinder. A graduated cylinder has an error of about 1%, which means plus or minus about 0.1 mL for a 10-mL graduated cylinder and plus or minus about 1 mL for a 100-mL graduated cylinder. When less error is desired, volumetric (volume-measuring) flasks and delivery pipets are used for fixed volumes and burets are used for variable quantities. These devices will contain or measure out their stated volumes repeatedly to within 0.1%. The syringe, also shown in Figure 1.12, has an error of about 5% to 6%; so a 5-mL drug injection may be in error of plus or minus about 0.3 mL. That size of error is acceptable because of the convenience and versatility of the delivery system.

All volumetric containers are calibrated in milliliters. One milliliter is equal to 1 cm³ (a cubic centimeter, or cc), and there are 1000 mL in 1.0 L. There are 10 cm × 10 cm × 10 cm = 1000 cm³ (mL) in 1.0 L; so, in SI units, $1 \text{ L} = 1 \times 10^{-3} \text{ m}^3$. To get an idea of the dimensions of a liter container, consider Figure 1.13, which shows an adult who is 2 meters tall, a child who is 1 meter tall, and a 1-liter cube, which is 10 centimeters "tall."

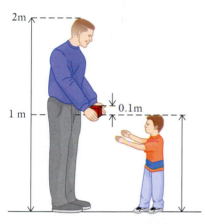

Figure 1.13 The dimensions of a liter container in the shape of a cube are 0.1 meter, or 10 centimeters, on a side. The adult is 2 meters tall and the child is 1 meter.

Example 1.13 Converting measures of volume

Express a volume of 364 mL in liters.

Solution

Using the unit-conversion method, first write the given quantity, and then multiply by the appropriate conversion factor:

$$364 \text{ mL} \left(\frac{1 \text{ L}}{1000 \text{ mL}} \right) = 0.364 \text{ L}$$

Problem 1.13 Express the quantity 3.97 L in milliliters.

1.9 DENSITY

The characterization of substances requires measurement of their physical properties—that is, properties that can be determined without changing the substance's nature. One of the most useful of these properties is **density**, a derived unit defined as mass per unit volume:

$$\text{Density} = \frac{\text{mass (g)}}{\text{volume (cm}^3)}$$

Density is a property of substances in any physical state—gaseous, liquid, or solid. It is used to evaluate the purity of solids and liquids (pure gold in the solid state, for example, has a density of 19.32 g/cm^3). Density is also used to estimate the amount of dissolved solids in solutions, as in the clinical analysis of urine samples to evaluate kidney function. The density of a solution rises as the amount of a dissolved solid increases. Failure of the kidneys to concentrate the urine can be detected by the fact that the urine density is abnormally low. Density also serves as a conversion factor for translating mass into volume or volume into mass.

It is important to note that mass does not change when the temperature changes, but volume does. The volumes of most liquids increase slightly as the temperature increases; as a result, they undergo a slight decrease in their densities. For this reason, any measurement of density must always be accompanied by the value of the temperature at which the measurement was made. An interesting case of the temperature dependence of density serving a biological need is discussed in Box 1.3.

>> **Temperature and its measurement are discussed in Section 1.10.**

Example 1.14   Calculating mass from density

Calculate the mass of 96.0 mL of a solution that has a density of 1.09 g/mL at 20°C.

Solution
Use the unit-conversion method, first writing the quantity to be converted and then multiplying it by the appropriate conversion factor—in this case, the density:

$$96.0 \text{ mL} \left(\frac{1.09 \text{ g}}{\text{mL}}\right) = 104.64 = 105 \text{ g}$$

Problem 1.14   Calculate the mass of 135.0 mL of a liquid that has a density of 0.8758 g/mL at 20°C.

To measure a substance's density, one must first determine the mass of a small volume of it. The mass of a solid is easily obtained on a balance. Its volume is then ascertained by filling a graduated cylinder about halfway with a liquid in which the solid will neither float nor dissolve. The solid is placed in the cylinder and the resulting rise in the liquid's level is equal to the volume of the solid.

The mass of a liquid can be obtained by using a special container whose volume and mass are already known. One then fills the container with a test liquid and determines the liquid's mass (the difference in mass between the filled and unfilled container). That mass divided by the known volume yields the test liquid's density. However, there is a simpler way to measure liquid density—that is, through the use of a hydrometer.

Objects placed in liquids either float on the surface, sink to the bottom, or remain at some intermediate position in which they have been placed and neither float nor sink. If an object floats, its density is less than that of the liquid. If it sinks, its density is greater than that of the liquid. If it neither floats nor sinks, its density is equal to that of the liquid. A **hydrometer** is a solid, vertical floating object that rises or falls to a level where its density is equal to the density of the liquid in which it is placed.

A hydrometer is a hollow, sealed glass vessel in the form of a narrow, graduated tube with a bulb at the lower end. The hollow tube floats in an upright,

1.3 CHEMISTRY AROUND US

Temperature, Density, and the Buoyancy of the Sperm Whale

The head of a sperm whale is huge, making up about one-third of the animal's total weight. Most of the head is composed of the whale's unique spermaceti organ, which plays a key role in the feeding behavior of the sperm whale. This organ contains about 4 tons of spermaceti oil, a mixture of triacylglycerols and waxes.

The sperm whale's diet consists almost exclusively of squid found in very deep waters, at depths of almost 1 mile or more. Squid are plentiful at these depths, and there are no competitors. The only problem for the sperm whale is staying at these depths to wait for the squid. It would take a tremendous amount of energy for a marine animal to stay at a given depth by swimming. The spermaceti organ is an energy-conserving solution to this problem.

The sperm whale's normal body temperature of about 37°C keeps the spermaceti oil in a liquid state as long as the whale is resting on the surface. As the whale dives to feeding depths, however, it pumps cold seawater through chambers in the spermaceti organ. The spermaceti oil cools and becomes partly solid. The density of the oil increases because its solid form is more dense than its liquid form. The increased density allows the whale to descend to the ocean depths without much swimming effort. The whale controls the depth to which it dives by controlling the temperature and hence the density of the spermaceti oil. After the whale feeds, an increase in the circulation of blood to the spermaceti organ warms the spermaceti oil, decreasing its density and thereby lifting the whale to the ocean surface.

vertical orientation because the bulb at the lower end is filled with weights. Graduations on the tube are calibrated so that the mark coinciding with the surface of the liquid indicates the liquid's density.

Hydrometers in common use in the clinical laboratory are graduated not in density units (such as grams per milliliter) but in values of **specific gravity,** a property defined as the ratio of the density of the test liquid to the density of a reference liquid:

$$\text{Specific gravity} = \frac{\text{density of test liquid}}{\text{density of reference liquid}}$$

Note that, because specific gravity is a ratio of densities, the units cancel, and specific gravity has no units. The standard reference liquid for measuring the specific gravity of aqueous solutions is pure water at 4°C. (Aqueous solutions are solutions containing water.) The density of water is at its maximum, 1.000 g/cm³, at that temperature (actually 3.98°C). The temperature dependence of water and its importance to living organisms is considered in Box 1.4 on the following page. Specific gravities of blood or urine (both are aqueous solutions) may be reported as 1.028, for example, which means that the sample's density was 1.028 times the density of pure water.

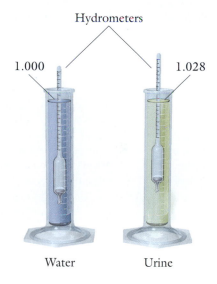

Hydrometers

1.000   1.028

Water   Urine

1.10 TEMPERATURE

If a hot object makes contact with a cold object, the hot object will cool and the cool object will warm. This effect is explained by proposing that heat always flows from a hot body to a cold body in much the same way that water flows downhill under the influence of gravity. Substances can either gain heat or lose it, depending on whether they are cooler or hotter than their environments. The measurement of heat itself will be described in Section 1.11 but, to measure heat, we must first have a way to measure how hot or cold an object is. In other words, we need a device capable of indicating the object's position on a scale of "hotness." This position on the scale is called a **temperature,** and the device is called a **thermometer.** It is important to keep in mind that the flow of heat and the concept of temperature are inextricably related but are not the same thing.

Many thermometers consist of a very small diameter glass tube partly filled with a fluid. The fluid expands when heated and contracts when cooled,

>> **The role of heat in chemistry is introduced in Chapters 6 and 8.**

1.4 CHEMISTRY AROUND US

Density and the "Fitness" of Water

In 1912, American physiologist Lawrence J. Henderson introduced a new idea into the study of ecology and ecosystems—the idea of the "fitness" of the environment to support life. He meant by this idea that evolution is a reciprocal process, requiring not only that living things adapt to their environments, but also that an environment must have certain unique characteristics that enable it to support life.

You are probably familiar with the fact that water is the principal component of living organisms and, furthermore, that life probably evolved in water. Water's density is one of the physical properties that explains this liquid's central role in living systems and in their environments. Specifically, water is the one notable exception to the rule that the density of ordinary solids is greater than the density of their corresponding liquids. If water behaved like other liquids, life on Earth would not exist in the form that we know it today.

In general, when a liquid cools, its density increases until the substance solidifies. If the solid and liquid forms of the substance are present simultaneously, the solid, which is denser, will rest at the bottom of its container with the liquid above it. As with other substances, the density of water also increases as the temperature decreases—but only until the temperature reaches 4°C. At that point, unlike the densities of other liquids, its density begins to decrease, and, at water's freezing point, the ice that forms floats on the remaining liquid water instead of sinking as another substance would. The density of water reaches its maximum at 4°C and is lower above and below that temperature. That fact is illustrated by the graph below, in which the density of water is plotted as a function of Celsius temperature. This unusual behavior will be explained in Chapter 6.

When natural bodies of water are gradually cooled from above-freezing temperatures to below-freezing ones, the surface water eventually reaches 4°C. Because the density is highest at that temperature, the surface water descends to the bottom, where it remains at 4°C (close to, but not quite, freezing). As the environmental temperature continues to drop, however, the density of the water that is still near the surface decreases, and that water rises and remains at the surface until it freezes. Thus, the ice floats on the surface of liquid water. The final result of this anomalous density curve is a stratification, or layering, of temperature zones, with the "heaviest" water at 4°C at the bottom, surmounted by "lighter" water at lower temperatures and, finally, ice at the surface. The period during which the water below the ice can persist in the liquid state depends on the depth of the water.

If ice sank in liquid water, Earth's first winters would have seen large quantities of solid water sinking to the bottom of rivers and lakes. The ice formed during those winters would not have melted, because of the time required to transfer sufficient heat through the water above it. In addition, the densities would have increased uniformly with depth, and so no mixing could have taken place.

Life is abundant under the antarctic ice. (G. L. Kooyman/ Animals Animals.)

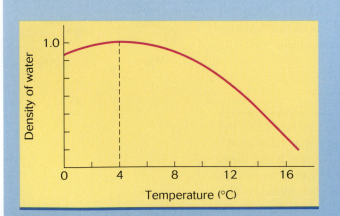

The density of water is plotted as a function of increasing temperature. Water melts at 0°C and, as it is warmed, its density increases to a maximum value of 1.0 g/cm³ at 4°C. Above that value, its density decreases as the temperature is raised.

This process would have been repeated yearly until eventually most or perhaps all of the planet's liquid water would have been transformed into ice. Because water has its maximum density at a temperature above its freezing point, most of Earth's water exists in the liquid form and allows the continued existence of life as we know it.

thereby rising or falling to different levels within the tube. The scale on the thermometer is created by determining the locations of two reference positions on the tube. The lower one shows the fluid's level at the temperature at which pure water freezes. The upper one shows the fluid's level at the temperature at which pure water boils. Then, a uniform scale of equal divisions is established between these two positions.

The freezing and boiling points of pure water have different numerical values in the three temperature scales commonly encountered in chemistry. On the **Fahrenheit scale** (on which the units are degrees), the freezing point is 32°F and the boiling point is 212°F (both are exact numbers). On the **Celsius scale** (also in degrees), the freezing point of water is 0°C and the boiling point is 100°C (both exact numbers). The SI temperature scale is the **Kelvin scale,** and its units are called kelvins, symbol K (no degree symbol is used with the symbol for kelvins). The size of the kelvin is identical with the size of the Celsius degree. However, there is a difference between the two scales in that the Kelvin scale recognizes a low temperature limit called absolute zero, the lowest theoretically attainable temperature, and gives it the value of 0 K. On the Kelvin scale, the freezing point of water is 273.15 K (Celsius scale, 0°C). For convenience, this value will be rounded to 273 so that Celsius degrees can be converted into kelvins by simply adding 273 to the Celsius value:

$$K = °C + 273$$

Conversion between Celsius and Fahrenheit scales is a bit more complicated. Figure 1.14 emphasizes that there are two problems in conversion:

1. The sizes of the degrees are different.
2. The numerical values of the reference points for the freezing and boiling points of water are not the same.

Let's consider these two problems one at a time.

A Celsius degree is larger than a Fahrenheit degree. There are 100 Celsius degrees between the freezing and the boiling points on that scale and 180 degrees between the freezing and the boiling points on the Fahrenheit scale. In other words, there are 180/100, or 9/5, Fahrenheit degrees per 1 Celsius degree. In this book, we will write the equations comparing the sizes of Fahrenheit and Celsius degrees in either of two ways:

$$\text{Fahrenheit degrees} = °C \times (9/5)(°F/°C)$$

or

$$\text{Celsius degrees} = °F \times (5/9)(°C/°F)$$

Because these equations are written in the form of unit-conversion calculations, they allow us to convert from one type of degree into another.

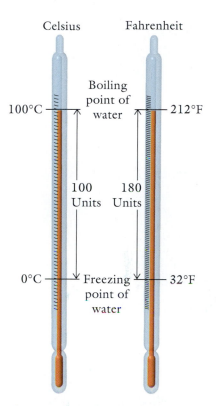

Figure 1.14 This drawing of two thermometers emphasizes the two pieces of information necessary for converting temperatures from one scale into another: (1) the relation between the sizes of degrees and (2) the difference in the numerical values of the freezing and boiling points in the two systems.

Example 1.15 **Converting number of Celsius degrees into the equivalent number of Fahrenheit degrees**

If a temperature is seen to rise 30 degrees on the Celsius scale, how many degrees will it have risen on the Fahrenheit scale?

Solution
We use the first of two conversion equations, °F = °C × (9/5)(°F/°C), and insert the given value:

$$°F = 30°C \times (9/5)(°F/°C) = 54°F$$

The temperature will have risen 54 Fahrenheit degrees.

Problem 1.15 If a temperature is seen to rise 65 degrees on the Fahrenheit scale, how many degrees will it have risen on the Celsius scale?

The preceding examples dealt only with the problem of equating numbers of Celsius degrees with numbers of Fahrenheit degrees or vice versa, not with the conversion of one thermometer reading into another. To interconvert temperature readings, we must take account of the fact that the numerical values of the reference points for the freezing and boiling points of water on the two scales are not the same. This concept can be best understood through the following examples.

Example 1.16 Converting Celsius temperature into Fahrenheit temperature

Convert a temperature reading of 50°C into the equivalent temperature on the Fahrenheit scale.

Solution

To do this conversion, let's consult the temperature-scale diagram in Figure 1.14. As the diagram indicates, 50°C is exactly halfway between the freezing and the boiling points of water on the Celsius scale. On the Fahrenheit scale, the equivalent halfway point is 90 Fahrenheit degrees higher than the Fahrenheit freezing point of 32°F. The reading on the Fahrenheit scale directly opposite the Celsius scale must then be 122°F, the result of adding 32°F to the 90 Fahrenheit degrees above the freezing point. Therefore, to convert a Celsius temperature into a Fahrenheit temperature, we first multiply °C by 9/5 and then add 32 to the result. In the form of an equation,

$$°F = °C \times (9/5)(°F/°C) + 32°F$$
$$= 50°C \times (9/5)(°F/°C) + 32°F = 122°F$$

Problem 1.16   What Fahrenheit temperature is equivalent to 100°C?

Now let's consider the conversion of a Fahrenheit reading on the thermometer into the equivalent Celsius reading.

Example 1.17 Converting Fahrenheit temperature into Celsius and Kelvin

A thermometer reads 122° on the Fahrenheit scale. What is its equivalent reading on both the Celsius and the Kelvin scales?

Solution

This Fahrenheit value was chosen so that we can simply reverse the steps that we took in Example 1.16. The value 122°F is exactly 90 Fahrenheit degrees above the freezing point of water. Therefore, to convert Fahrenheit temperature into Celsius temperature, first, subtract 32 and, second, multiply by 5/9. The result is, not surprisingly, 50°C. In the form of an equation,

$$°C = (°F - 32°F)(5/9)(°C/°F)$$
$$= (122°F - 32°F)(5/9)(°C/°F)$$
$$= 50°C$$

To calculate the equivalent Kelvin temperature, simply add 273 to the Celsius temperature:

$$K = °C + 273 = 50 + 273 = 323 \text{ K}$$

Problem 1.17   Convert 98.6°F into both Celsius and Kelvin temperatures.

1.11   HEAT AND CALORIMETRY

Heat is a form of energy. In the section on temperature, heat was described as flowing from a region of higher temperature to one of lower temperature. By

focusing on the loss or gain of heat, scientists have enlarged the list of properties that are useful for characterizing different substances.

Each substance has a different capacity to absorb heat. This capacity is
measured by noting the rise in temperature of a fixed mass of a substance that
has absorbed a known amount of heat. We know how to measure mass and
temperature, but how do we measure heat? A unit of heat is defined by its
effect (the rise in temperature) on a fixed mass of a reference substance.

A widely used non-SI unit of heat is called the **calorie,** abbreviated cal,
and the official SI unit is called the **joule,** abbreviated J. The relation between
them (a useful conversion factor) is 4.184 J = 1 cal. The temperature of 1.0 g
of water rises 1 Celsius degree when 1 cal of heat is absorbed. Compare this
effect with the effect of that same calorie on 1 g of aluminum: a temperature
rise of almost 5 Celsius degrees.

The characteristic response (increase in temperature) of a given mass of a
given substance exposed to a given amount of heat is expressed by a quantity
called the **specific heat** (symbol, C_p). If the mass is in grams, the temperature
on the Celsius scale, and the heat in joules, the units of the specific heat are

$$C_p = \frac{\text{joules}}{\text{grams} \times \Delta°C}$$

This equation reads as follows: the specific heat is equal to the heat absorbed
or lost per Celsius degree change in temperature per gram of substance. (The
change in temperature is the difference, Δ, between final and initial temperatures, or $\Delta t = t_{final} - t_{initial}$.) The higher a substance's specific heat, the more
slowly its temperature rises in response to heating. The specific heat is a unique
property of a substance, and Table 1.6 lists various materials along with their
specific heats. Liquid water has one of the highest values of specific heats of
liquids. Water's role in temperature control in mammals is discussed in Box 1.5
on the following page.

In summary, specific heat describes the capacity of 1 gram of a substance
to absorb heat. It should not be confused with a related term, heat capacity,
which varies according to how much of the substance is under study. **Heat
capacity** is the capacity of a given sample to absorb heat and therefore depends

TABLE 1.6	Specific Heats of Some Common Substances
Substance	Specific heat [J/(g × °C)]
METALLIC SOLIDS	
aluminum	0.89
copper	0.39
iron	0.46
platinum	0.13
NONMETALLIC SOLIDS	
coal	1.26
concrete	0.67
glass	0.68
rubber	2.01
LIQUIDS	
water	4.18
ethanol	2.51
gasoline	2.12
olive oil	1.97

1.5　CHEMISTRY AROUND US

Specific Heat and the "Fitness" of Water

In describing the "fitness" of water to support life, Lawrence J. Henderson also brought attention to the fact that water has the highest specific heat of any common liquid, a property that is central in the life of every cell. It enables all living organisms to keep their internal temperatures within certain bounds; below a certain temperature, chemical reactions take place too slowly to support life, and, above a certain temperature, the structures of proteins are destroyed. Species that derive their body temperatures from internal metabolic activity, such as mammals and birds, are called endotherms. They have adapted to far more stressful environmental conditions than ectotherms, species that absorb heat from their environment, such as reptiles. A constant, elevated temperature leads to a higher and therefore a more efficient rate of metabolism: endotherms are able to extract more energy per unit time from the nutrients in their diets than ectotherms can. An endotherm's internal temperature is produced by chemical reactions that generate heat. It is maintained at a consistent level—neither too high nor too low—through regulated avenues of heat loss.

Because of water's high specific heat, temperature increases due to chemical reactions that take place within water are kept to a minimum compared with the same reactions taking place in liquids with low specific heats. Let's compare the temperature rise of living tissue (which contains water) with the rise that might occur in another liquid, one having a typical specific heat of about 2.0 J/g × °C,

when we add the same number of joules of heat (say, 440 kJ, the heat generated by a 75-kg male in 1 h) to each. To simplify the exercise, we will assume that the tissue has the same specific heat as water, because water is tissue's chief component. We will calculate the temperature rises per hour by solving for Δt in the equation for specific heat:

$$\text{Heat (joules)} = C_p \times \text{mass of sample (grams)} \times \Delta t$$

$$\Delta t = \frac{\text{heat (J)}}{C_p \times \text{mass (g)}}$$

In living tissue,

$$\Delta t_{water} = \left(\frac{440{,}000 \text{ J}}{4.184 \, \dfrac{\text{J}}{\text{g} \times °\text{C}} \times 75 \text{ kg}} \right) = 1.4°\text{C/h}$$
$$= 2.5°\text{F/h}$$

In the other liquid,

$$\Delta t_{\text{"other"}} = \left(\frac{440{,}000 \text{ J}}{2.0 \, \dfrac{\text{J}}{\text{g} \times °\text{C}} \times 75 \text{ kg}} \right) = 2.9°\text{C/h}$$
$$= 5.2°\text{F/h}$$

Body-temperature-regulating mechanisms are able to handle a load of 1 to 3 Fahrenheit degrees per hour but would be overwhelmed at 5 to 6 Fahrenheit degrees per hour.

The problem of temperature control in living organisms, then, is minimized by the fact that water, with its peculiarly high specific heat, is the chief component of living cells. In Chapter 6, we will continue this consideration of temperature control in organisms when we examine another of water's biologically significant physical properties, the energy required for causing its evaporation.

---

not only on the substance's innate characteristics, but also on its mass. The greater the mass, the greater the amount of heat that the sample can absorb. The heat capacity of 100 g of water is 418.4 J/°C, and that of 200 g of water is 836.8 J/°C. Conversely, if we add the same amount of heat to two samples of the same substance, one of which has twice the mass of the other, the smaller mass will reach a higher temperature.

Because the amounts of heat encountered in chemical processes are in the range of thousands of joules or calories, the values are reported in kilojoules (kJ) or kilocalories (kcal). In discussions of nutrition, however, caloric values are instead measured and reported in Calories (capital C). A **Calorie** is equal to a kilocalorie, or 1000 "small" calories.

Example 1.18　Calculating specific heat

What is the specific heat of a substance if the addition of 6.00 joules of heat to 16.0 grams of that substance causes its temperature to rise from 20°C to 38°C?

Solution

Insert the given values into the formula for specific heat given previously:

$$C_p = \frac{6.00 \text{ J}}{16.0 \text{ g} \times (38°\text{C} - 20°\text{C})} = 0.021 \text{ J/(g} \times °\text{C)}$$

Problem 1.18 What is the specific heat of a substance if the addition of 334 J of heat to 52 g of that substance causes its temperature to rise from 16°C to 48°C?

Example 1.19   Using specific heat in calculations

How much heat must be added to 45.0 g of a substance that has a specific heat of 0.151 J/(g × °C) to cause its temperature to rise from 21.0°C to 47.0°C?

Solution

The relation between heat and temperature rise is obtained by solving the equation for specific heat:

$$\text{Heat} = C_p \times \text{mass of sample (g)} \times \Delta t$$

By substituting the appropriate values, we have

$$\text{Heat} = \left(0.151 \, \frac{J}{g \times °C}\right)(45.0 \text{ g})(47.0°C - 21.0°C) = 177 \text{ J}$$

Problem 1.19 How much heat must be added to 37 g of a substance that has a specific heat of 0.12 J/g × °C to cause its temperature to rise from 11°C to 22°C?

Heat is measured by the rise in temperature of a given mass of water. The equation for calculating the specific heat of a substance was rearranged in Example 1.19 into the form

$$\text{Heat} = C_p \times \text{mass of sample (g)} \times \Delta t$$

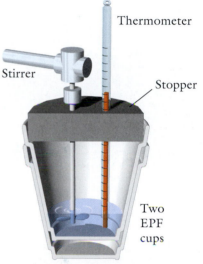

A simple calorimeter

When the heat produced by some physical or chemical process is absorbed into a given mass of water, the water's temperature rise will allow us to calculate the heat produced by the process. The same approach can be used if the process absorbs heat rather than producing it. In that case, the surrounding water's temperature will decrease. The laboratory device in which such an experiment is performed is called a **calorimeter.**

A calorimeter can be as simple as an expanded polystyrene foam (EPF) coffee cup outfitted with a thermometer. EPF is an excellent insulator and prevents loss of heat. If 50 mL of hydrochloric acid is added to 50 mL of potassium hydroxide in water in the calorimeter, the temperature of the water will rise and the heat of the reaction can be determined.

A much more complex calorimeter is necessary to measure the heat produced as a result of human basal metabolism. The **basal metabolic rate (BMR)** is the minimal metabolic activity of a human at rest and with an empty gastrointestinal tract. When quantities of heat are considered in the context of nutrition and metabolism, those quantities are commonly given in calories (1 cal = 4.184 J). Therefore in discussions about heat in the context of nutrition and metabolism, we will use specific heat in units of calories. The calorimeter for this measurement consists of a small room in which a patient reclines. The room is surrounded by a hollow jacket through which water flows at a fixed rate. The water temperature is measured at the inflow and at the outflow, which determines Δt—say, 2.0 Celsius degrees. The specific heat of water is known, and it is necessary to measure only the mass of water. If the water flow rate is 0.6 L per minute and the time for the experiment is 125 min, the volume of water whose temperature was raised by 2.0 Celsius degrees is 75 L. If we assume that the density of water is 1.0, then the mass of water is 75 kg. The heat produced in 125 min is

$$\text{Heat} = 1.0 \text{ cal/g} \cdot °C \times 75,000 \text{ g} \times 2.0° = 150 \text{ kcal}$$

$$\text{Basal metabolic rate} = 150 \text{ kcal}/125 \text{ min} = 1.2 \text{ kcal}/\text{min}$$

This patient is attached to a device that measures oxygen, carbon dioxide, and energy consumption to determine the basal metabolic rate. (St. Bartholomew's Hospital/Science Photo Library/Photo Researchers.)

Because this kind of calorimeter is quite elaborate and can be found at only a few research centers, basal metabolic rates are usually determined indirectly by measuring the oxygen uptake of a patient at rest. As the body consumes nutrients such as carbohydrates, fats, and proteins, it uses the same amount of oxygen as would be consumed by burning these nutrients in a test tube. An adult human male produces about 4.8 kcal of heat per liter of oxygen consumed. Therefore, measuring oxygen uptake can be equated with basal metabolic rate. An adult human male consumes about 250 mL of oxygen per minute at rest. At 4.8 kcal/L of oxygen, his basal metabolic rate is

$$\left(0.25\frac{\text{L O}_2}{\text{min}}\right)\left(4.8\,\frac{\text{kcal}}{\text{L O}_2}\right) = 1.2\,\frac{\text{kcal}}{\text{min}}$$

which corresponds quite well with the BMR obtained with the direct calorimetric method.

Example 1.20   Estimating minimal human caloric requirements

An equation for the estimation of minimal daily caloric needs (24-hour basal metabolic rate in kcal/day) for human males is

BMR = 66 + [6.22 × weight (lb)] + [12.7 × height (in.)] − [6.8 × age (yr)]

What is the daily minimal Calorie (kcal) requirement of a 25-year-old man, 70 inches tall and weighing 160 pounds?

Solution

$$\text{BMR} = 66 + (6.22 \times 160) + (12.7 \times 70) - (6.8 \times 25)$$
$$= 1.8 \times 10^3\,\text{kcal/day}$$

Problem 1.20   An equation for the estimation of minimal daily caloric needs (24-hour basal metabolic rate in kcal/day) for human females is

BMR = 655 + [4.36 × weight (lb)] + [4.32 × height (in.)] − [4.7 × age (yr)]

What is the daily minimal Calorie (kcal) requirement of a 22-year-old woman, 66 inches tall and weighing 132 pounds?

Summary

The Composition of Matter

• Substances are classified on the basis of their composition.

• The purity of a substance is established by verifying that one and only one substance is present.

• If a pure substance can be decomposed into simpler substances, it is a compound.

• If a substance cannot be decomposed into simpler substances, it is an element.

• Mixtures are composed of two or more pure substances in proportions that may vary.

• If the mixture appears uniform to the eye, it is homogeneous. If it is nonuniform, it is heterogeneous.

Measurement and the Metric System

• The modern metric system is called the SI system whose base units are the kilogram, meter, and second.

• Units of the older metric, or non-SI, system are the gram, centimeter, and second.

• Very large and very small quantities in the SI system are simplified by the use of prefixes denoting multiples of ten.

Uncertainty and Significant Figures

• The last digit in any measurement is an estimate, and therefore, is an unavoidable uncertainty or variability.

• The degree of uncertainty is expressed by employing the concept of significant figures.

Scientific Notation

• In scientific notation, a number is expressed as a coefficient (a number between 1 and 10) multiplied by an exponential factor (10 raised to a whole-number power).

• Numbers expressed in scientific notation are added or subtracted by converting all exponents into the same value.

The exponents are then added in multiplication and subtracted in division.

Calculations and Significant Figures

• The number of significant figures in a calculated quantity cannot be greater than the smallest number of significant figures in any of the measured quantities.

Use of the Unit-Conversion Method in Calculations

• This method requires you to identify the kind of units needed by the answer and the kind of units given in the problem.

• The two kinds of units are related with an equation that defines the conversion factor.

Fundamental Properties of Matter

• The mass of an object is the quantity of matter in that object.

• Volume is the amount of space that the object occupies. It is the product of three linear dimensions and is therefore a derived quantity.

• Density is defined as mass divided by volume and is determined by measuring the mass of a given volume of a substance.

• The density of a liquid or a solution can also be measured by its effect on the buoyancy of a float called a hydrometer.

Heat and Calorimetry

• The concept of temperature allows us to quantitatively specify how hot or cold an object is.

• There are three temperature scales: the Celsius, Kelvin, and Fahrenheit scales.

• Heat is determined in a calorimeter by measuring the change in temperature undergone by a given mass in response to the addition or loss of heat.

• Human basal metabolic rates can be determined directly by calorimetry or, more often, indirectly by oxygen consumption.

Key Words

basal metabolic rate (BMR), p. 29	element, p. 7	physical property, p. 5
calorie, p. 27	error, p. 11	scientific notation, p. 12
Celsius scale, p. 25	Fahrenheit scale, p. 25	significant figure, p. 11
chemical property, p. 5	heat, p. 26	temperature, p. 23
compound, p. 7	Kelvin scale, p. 25	unit conversion, p. 17
conversion factor, p. 17	mass, p. 20	volume, p. 21
density, p. 22	metric system, p. 9	
derived units, p. 9	mixture, p. 5	

Exercises

The Composition of Matter

1.1 Classify the following processes as either chemical or physical properties of oxygen: (a) oxygen reacts with carbon to form carbon dioxide; (b) oxygen boils at −183°C; (c) oxygen dissolves in water; (d) oxygen reacts with hydrogen to form water.

1.2 Classify the following processes as either chemical or physical properties of magnesium: (a) magnesium dissolves in hydrochloric acid, with the production of bubbles of hydrogen gas; (b) magnesium melts at 649°C; (c) magnesium is shiny and conducts electricity; (d) when heated in oxygen, magnesium changes from a shiny metal to a white solid that no longer conducts electricity.

1.3　Classify the following as pure substances or mixtures: (a) air; (b) mercury; (c) aluminum foil; (d) table salt.

1.4　Classify the following as pure substances or mixtures: (a) seawater; (b) table sugar; (c) smoke; (d) blood.

Measurement and the Metric System

1.5　What are the fundamental quantities and their units in the SI system that are encountered in chemistry?

1.6　Give examples of units that can be derived from the fundamental units of the SI system that we use in this book.

1.7　Write the following quantities in more convenient form by changing the units: (a) 630 m (meters); (b) 1440 ms (milliseconds); (c) 0.000065 kg (kilograms); (d) 1300 μg (micrograms).

1.8　Write the following quantities in more convenient form by changing the units: (a) 0.813 kilometers; (b) 0.367 seconds; (c) 7140 mg (milligrams); (d) 0.096 mg (milligrams).

Significant Figures

1.9　How many significant figures are in the following numbers? (a) 0.0945; (b) 83.22; (c) 106; (d) 0.000130.

1.10　Deduce the number of significant figures contained in the following quantities: (a) 16.0 cm; (b) 0.0063 m; (c) 100 km; (d) 2.9374 g; (e) 1.07 lb/in.^{2}.

1.11　How many significant figures are in the following measurements? (a) 25.9000 g; (b) 102 cm; (c) 0.002 m; (d) 2001 kg; (e) 0.0605 s.

1.12　How many significant figures are in the following measurements? (a) 21.2 m; (b) 0.023 kg; (c) 46.94 cm; (d) 453.59 g; (e) 1.6030 km.

Scientific Notation

1.13　Express the following numbers in scientific notation: (a) 0.00839; (b) 83,264; (c) 372; (d) 0.0000208.

1.14　Convert the following numbers into nonexponential form: (a) 2.3×10^{-3}; (b) 2.9×10^2; (c) 3.92×10^{-4}; (d) 1.73×10^4.

1.15　Write the following numbers in scientific notation: (a) 936,800; (b) 1638; (c) 0.0000568; (d) 0.00917.

1.16　Write the following quantities in scientific notation, along with correct metric abbreviations: (a) 275,000 kilograms; (b) 87,000 years; (c) 0.097 second.

Calculations Using Significant Figures

1.17　With careful concern for the correct number of significant figures in the answers, calculate (a) 3.21 cm × 15.091 cm; (b) 3.82 × 1.1 × 2.003.

1.18　Report the result of dividing (a) 13.87 by 1.23 and (b) 0.095 by 1.427, with strict attention to the correct number of significant figures in the answers.

1.19　What is the result of adding (a) 12.786 to 1.23 and (b) 3.961 to 24.6543?

1.20　What is the result of subtracting (a) 2.763 from 3.91 and (b) 54.832 from 89.2?

Unit-Conversion Method

1.21　How many (a) kilometers, (b) meters, (c) centimeters, and (d) millimeters are there in 0.800 miles calculated to the correct number of significant figures? The conversion factor is 1 mile = 1.6093 km.

1.22　How long would it take to drive 375 miles at 55 miles per hour? Calculate your answer in hours.

1.23　How many milliliters are in 3.75 gallons?

1.24　How many milligrams are in 2.95 ounces?

1.25　How many miles are run in a 10-K (10 kilometer) foot race? (Use 10-K as an exact number.)

1.26　World class times in the 100-meter foot race are about 9.10 s. How fast is that in miles per hour? (Use 100 meters as an exact number.)

Mass, Volume, Density, Temperature, and Heat

1.27　How many grams are in 1 oz?

1.28　How many milligrams are in 1 oz?

1.29　How many (a) kilograms, (b) grams, and (c) milligrams are there in 0.60 lb calculated to the correct number of significant figures? Use the conversion factor 453.59 g = 1 lb.

1.30　How many (a) kilometers, (b) meters, and (c) centimeters are there in 0.820 mi. (miles) calculated to the correct number of significant figures? Use the conversion factor 1 mile = 1.609 km.

1.31　Calculate the length in centimeters of the side of a cube whose volume is 1.0 L.

1.32　Calculate the SI equivalent of 1.0 L.

1.33　Calculate the mass of a rectangular solid, a three-dimensional object with six sides, all of which are rectangles, of density 2.2 g/cm^3 and dimensions 3.25 cm × 6.50 cm × 17.00 mm.

1.34　What is the volume of 250.0 grams of acetone whose density is 0.7899 g/cm^3?

1.35　Convert 68°F into °C.

1.36　Convert 45°F into °C.

1.37　Express 27°C in kelvins.

1.38　Express 125°C in °F.

1.39　Calculate the specific heat of 32 g of a substance that required 22.0 J to raise its temperature from 18°C to 28°C.

1.40　How many joules are required to raise the temperature of 0.750 kg of a substance whose specific heat is 0.220 J/(g × °C) from 13.0°C to 68.0°C?

1.41　A glass vessel calibrated to contain 19.84 mL of water at 4°C was found to weigh 31.962 g when empty and dry. Filled with a sodium chloride solution at the same temperature, it was found to weigh 54.381 g. Calculate the solution's density.

1.42　A sample of urine weighed 25.853 g at 4°C. An equal volume of water at the same temperature weighed 23.718 g. At 4°C, the density of water is exactly 1.000 g/cm^3. Calculate the density of the urine sample.

1.43　How many pounds of carbohydrate yielding 16.74 kJ/g must be consumed to obtain the 8.4×10^3 kJ of energy required by an average woman in a 24-h period?

1.44　The specific heat of the glass used in thermometer manufacture is 0.84 J/(g × °C). Calculate the heat in joules required to raise the temperature of a 100-gram thermometer 2.0 Celsius degrees.

1.45 A glass vessel that can contain 12.3 mL of water at 4°C was found to weigh 28.463 g when empty and dry. Filled with a sodium chloride solution at the same temperature, it was found to weigh 41.242 g. Calculate the solution's density.

1.46 A volume of glucose solution weighed 22.842 g at 4°C. An equal volume of water at the same temperature weighed 22.394 g. At 4°C, the density of water is exactly 1.000 g/cm³. Calculate the density of the glucose solution.

1.47 The density of gold is 19.32 g/cm³. Calculate the volume of a sample of gold that weighs 2.416 kg.

1.48 A urine sample has a density of 1.09 g/cm³. What volume will 15.0 g of the sample occupy?

Unclassified Exercises

1.49 Classify the following properties as either chemical or physical: (a) an electric current sent through water results in the production of bubbles of hydrogen gas; (b) lead melts at 327°C; (c) lead is a metal, has a silvery luster, and conducts electricity; (d) when lead is heated in oxygen, its appearance changes from a silvery luster to a white solid that no longer conducts electricity.

1.50 Write the following quantities in simpler metric forms: (a) 0.0045 L; (b) 2.87×10^{-8} s; (c) 0.0057 km; (d) 0.0000036 kg.

1.51 Write the following quantities in scientific notation, along with correct metric abbreviations:

(a) 9,620,000 kilograms; (b) 54,870 days; (c) 253 milliseconds; (d) 0.000274 kilometers.

1.52 Count the significant figures in the following numbers: (a) 0.00256; (b) 128.009; (c) 2.00730; (d) 201; (e) 0.09864.

1.53 How many ounces are in 45.8 mg?

1.54 What are the results (a) of multiplying 3.2×10^3 by 3.1×10^{-5}; (b) of dividing 9.47×10^{-3} by 2.32×10^{-2}; (c) of adding 5.6×10^{-3} and 2.3×10^{-2}; and (d) of subtracting 1.4×10^{-4} from 3.6×10^{-3}?

1.55 Perform the following calculations. Answers must have the correct number of significant figures.

(a) $(8.20 \times 10^2) + (3.75 \times 10^4)$
(b) $(5.21 \times 10^{-2}) + (2.74 \times 10^{-3})$
(c) $(1.01 \times 10^{-4}) + (7.23 \times 10^{-3})$

1.56 Perform the following calculations. Answers must have the correct number of significant figures.

(a) $(2.7 \text{ cm} + 1.08 \text{ cm}) \times 22.47 \text{ cm}$
(b) $(2.54 \text{ cm} - 0.541 \text{ cm}) \div 2.2 \text{ cm}$
(c) $(21.63 \text{ g} + 4.284 \text{ g}) \times 0.0372 \text{ g}$
(d) $(183.7 \text{ mL} - 38.57 \text{ mL}) \div 21.3 \text{ mL}$

1.57 What is an alternative SI equivalent for 1.0 L?

1.58 Calculate the specific heat of 12.7 g of an unknown metal whose temperature increased 25.0 Celsius degrees when 80.0 J of heat energy was absorbed by the metal.

Expand Your Knowledge

Note: The icons ▲ ▲ ▲ denote exercises based on material in boxes.

1.59 A 175-lb man was placed in a tank filled with 1000 gallons of water. The temperature of the water increased by 2.49 Celsius degrees in a 1.5-h period. Calculate the heat lost by the man during this experiment. Using that data, calculate the patient's basal metabolic rate—that is, the energy required in a 24-h period. (Hint: The heat lost is equal to the heat gained by the water.)

1.60 A prescription for nifedipine calls for 0.2 mg/kg of body weight, four times per day. The drug is packaged in capsules of 5 mg. How many capsules per dose should be given to a patient who weighs 75 kg?

1.61 You need 5.00 g of sodium chloride (NaCl) to add to water to prepare a mixture containing 5.00 g of NaCl per 100 mL of mixture. There is a solution of NaCl in water in the stockroom, but the label on the container specifies only the substances in the mixture and not its quantitative composition. How could you obtain the required amount of NaCl to prepare your mixture?

1.62 Two students were required to determine the area of a sheet of paper in units of square centimeters. One student had a centimeter ruler marked in centimeters, and the other an inch ruler marked in half inches. The results of their measurements were

	Length	Width	Area
Student A	28 cm	22 cm	6.2×10^2 cm²
Student B	11 in.	8.5 in.	6.0×10^2 cm²

Which student had the correct answer, and what mistake did the other student make?

1.63 On the basis of some previous experimental results, you design a new experiment and predict its outcome. The results of the new experiment do not agree with the results that you predicted. What is your next move?

1.64 What is the difference between a law and a theory?

1.65 You just bought a brand-name oven on the Internet. Something seemed wrong to you when the instructions called for roasting a chicken at 175 degrees. It was then that you realized that the Web site URL had a "uk" location. What was the problem and how did you fix it?

1.66 The Celsius and Fahrenheit temperature scales have the same numerical value at −40 degrees. Prove that mathematically.

1.67 Create a single conversion factor that allows you to calculate the following conversions. (a) 3.60 m to yards; (b) 26.0 inches to cm; (c) 2.00 light years to miles; (d) 64.0 ml to ounces of water. (Velocity of light = 3.0×10^8 m/s.)

1.68 You have been purchasing gasoline at your favorite station for 2 dollars a gallon. A friend calls and tells you that she has been purchasing gasoline at her station for 66 cents a liter. Should you switch?

1.69 Your dermatologist suggests a commercial shampoo to keep your scalp seborrhea under control. You find the recommended product at the pharmacy but also notice a generic product alongside it. The recommended product costs $6.50 for 185 mL, and the concentration of the active ingredient is 0.500%. The price of the generic product is $7.00 for 165 mL, and the concentration of the active ingredient is 0.700%. Which one should you buy?

1.70 Your son and daughter-in-law live in Germany, and you have just been informed of the birth of your first granddaughter. You're told that she weighed 3.53 kg and has a body length of 53.0 cm. Your friends plead with you to translate those values into more familiar units. What are the results of your efforts?

1.71 Ilean loves to have a 12-ounce can of sugar-sweetened soda every day with her lunch. That can of soda contains 162 Cal. Her metabolic energy requirements are 2000 Calories per day. What percent of her daily caloric requirement is filled with the soda that she has for lunch?

1.72 One pound of fat contains about 4200 calories of energy. If the soda that Ilean has each day for lunch (see Exercise 1.71) is in exact excess of her caloric daily requirement, how much weight will she gain in one year?

1.73 A metal whose density is 2.50 g/cm^3 is used to build a box whose dimensions are 1.00 m on a side and 2.50 mm in thickness. Calculate the density of the box, assuming that there is no air in it. Finally, calculate how much water (density 1.00 g/cm^3) you would have to add to the box so that the overall density of box plus water equals 1.00 g per cc.

1.74 In Box 1.3, the sperm whale was said to have the ability to vary its density by solidifying the oil in its spermaceti organ. Remember, the sperm whale cannot change its mass. Explain how the density of the sperm whale can change if it cannot change its mass.

1.75 The IRS informs you that, although your calculations were all correct, your method of calculation resulted in an additional cost to you of $3, which they expect to receive shortly. How could you have made the correct calculations and yet still have incorrect final results?

CHAPTER 2

ATOMIC STRUCTURE

(David J. Sams/Tony Stone.)

Chemistry in Your Future

When you were training as a sales representative for a pharmaceutical company, it surprised you to learn that there are many important medications whose mode of action is not at all well understood. A case in point is the drug lithium carbonate, which has long been used to control the psychiatric dysfunction known as bipolar disorder, or manic depression. Some scientists think its efficacy stems from lithium's similarity to the elements sodium and potassium, which are known to play major roles in the transmission of nerve impulses. Chapter 2 presents the fundamental concepts that explain the chemical relatedness of these elements.

For more information on this topic and others in this chapter, go to www.whfreeman.com/bleiodian2e

Learning Objectives

- Use Dalton's atomic theory to explain the constant composition of matter and the conservation of mass in chemical reactions.
- Define atomic mass.
- Describe the structure of the atom in reference to its principal subatomic particles.
- Identify main-group and transition elements, metals, nonmetals, and metalloids.
- Correlate the arrangement of the periodic table with the electron configurations of the valence shells of the elements.
- Describe the octet rule.

For thousands of years, people have speculated about the fundamental properties of matter. The basic question to be answered was, Why are there so many different substances in the world? Many explanations were proposed but the most enduring one—the one that has proved to be correct—was that all matter is composed of particles, or atoms. The atoms were assumed to combine in various ways to produce all the existing and future kinds of matter. This idea, first proposed by Greek philosophers in the fourth century B.C., remained more a philosophical argument than a scientific theory until the early nineteenth century. At that time, John Dalton, an English schoolmaster, proposed an atomic model that convincingly explained experimental observations on which chemists of the day had come to rely.

In this chapter, we examine the nature of atomic structure and correlate it with the chemical reactivity of the elements. By the chapter's end, we will know the answer to such questions as, Why does one atom of calcium react with one atom of oxygen but with two atoms of chlorine?

2.1   CHEMICAL BACKGROUND FOR THE EARLY ATOMIC THEORY

By the early 1800s, two experimental facts had been firmly established:

1. Mass is neither created nor destroyed in a chemical reaction. When substances react chemically to create new substances, the total mass of the resulting products is the same as the total mass of the substances that have reacted. This is called the **Law of Conservation of Mass.**

2. The elements present in a compound are present in fixed and exact proportion by mass, regardless of the compound's source or method of preparation. This is called the **Law of Constant Composition.**

Consider vitamin C, or ascorbic acid, which can be obtained from citrus fruits or synthesized in the chemical laboratory. No matter where vitamin C comes from, the relative amounts of carbon (symbol C), hydrogen (symbol H), and oxygen (symbol O) of which it is composed are the same. They are specified as the mass percent of each element, which taken together is called the **percent composition** of the compound.

The mass percentages of C, H, and O in vitamin C are

$$\text{Mass percent of carbon in vitamin C} = \frac{\text{mass of carbon}}{\text{mass of vitamin C}} \times 100\%$$

$$\text{Mass percent of hydrogen in vitamin C} = \frac{\text{mass of hydrogen}}{\text{mass of vitamin C}} \times 100\%$$

$$\text{Mass percent of oxygen in vitamin C} = \frac{\text{mass of oxygen}}{\text{mass of vitamin C}} \times 100\%$$

Multiplication of the mass ratios by 100% converts them into percentages.

Example 2.1   Calculating the mass percent of a compound

Analysis of 4.200 g of vitamin C (ascorbic acid) yields 1.720 g of carbon, 0.190 g of hydrogen, and 2.290 g of oxygen. Calculate the mass percentage of each element in the compound.

Solution

$$\text{Mass percent of carbon in vitamin C} = \frac{1.720 \text{ g}}{4.200 \text{ g}} \times 100\% = 40.95\%$$

$$\text{Mass percent of hydrogen in vitamin C} = \frac{0.190 \text{ g}}{4.200 \text{ g}} \times 100\% = 4.52\%$$

$$\text{Mass percent of oxygen in vitamin C} = \frac{2.290 \text{ g}}{4.200 \text{ g}} \times 100\% = 54.52\%$$

Problem 2.1   Analysis of 4.800 g of niacin (nicotinic acid), one of the B-complex vitamins, yields 2.810 g of carbon, 0.1954 g of hydrogen, 0.5462 g of nitrogen (symbol N), and 1.249 g of oxygen. Calculate the mass percentage of each element in the compound.

2.2   DALTON'S ATOMIC THEORY

In the years from 1803 to 1808, John Dalton showed how the atomic theory could be used to explain the quantitative aspects of chemistry considered in Section 2.1. Since that time, our knowledge of the nature of the atom has changed significantly; Dalton was not correct in every particular. Nevertheless, he succeeded in creating, on the basis of a small number of specific observations, the first workable general theory of the structure of matter.

Dalton's model consisted of the following proposals:

- All matter is composed of infinitesimally small particles called **atoms** (believed in Dalton's time to be indestructible but now known to be composed of even smaller parts).

- The atoms of any one element are chemically identical.

- Atoms of one element are distinguished from those of a different element by the fact that the atoms of the two elements have different masses.

- Compounds are combinations of atoms of different elements and possess properties different from those of their component elements.

- In chemical reactions, atoms are exchanged between starting compounds to form new compounds. Atoms can be neither created nor destroyed.

Later in this chapter, we will examine some of the ways in which these proposals have undergone modification, but the key ideas have survived to this day.

Dalton's idea of chemical reactions is illustrated in Figure 2.1. Atoms (called **elementary particles** by Dalton) that are combined in compounds (called **compound particles** by Dalton) rearrange to form new compounds. (You can see where the modern names **element** and **compound,** Section 1.1, come from.) This model shows that, in a chemical process, mass is conserved. The new compounds possess the mass of their constituent atoms. No mass has been gained or lost. The atoms have simply become rearranged.

Dalton's model explained why the elements in a compound are present in fixed and exact proportion by mass. If the atoms of each element have a definite and unique mass and if compounds consist of a fixed ratio of given atoms—say, one atom of sulfur to two atoms of hydrogen—then the ratio of their masses also must be constant.

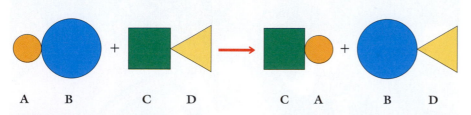

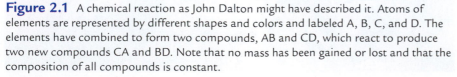

Figure 2.1   A chemical reaction as John Dalton might have described it. Atoms of elements are represented by different shapes and colors and labeled A, B, C, and D. The elements have combined to form two compounds, AB and CD, which react to produce two new compounds CA and BD. Note that no mass has been gained or lost and that the composition of all compounds is constant.

2.3 ATOMIC MASSES

In Dalton's definition, the principal difference between different elements is the different masses of their atoms. Because he could not weigh individual atoms to measure their mass, he proposed a relative mass scale instead. He identified the lightest element then known—hydrogen—and proposed that it be assigned a relative mass of 1.0. He then assigned masses to other elements by comparing them with hydrogen.

Dalton also examined compounds in which hydrogen combined with other elements in what Dalton believed were 1:1 ratios of the different atoms. By determining the percent compositions of these compounds, he could compare the masses of other elements with the mass of hydrogen in each compound and thus establish their relative atomic masses.

For example, the mass percentages of hydrogen and chlorine in the compound hydrogen chloride are 2.76% and 97.24%, respectively. If it is true that there is a 1:1 ratio of hydrogen and chlorine atoms in this compound, their relative masses in 100 grams of hydrogen chloride are

$$\frac{\text{Mass of a chlorine atom}}{\text{Mass of a hydrogen atom}} = \frac{97.24}{2.76} = \frac{35.2}{1}$$

or mass of a chlorine atom $= 35.2 \times$ mass of a hydrogen atom. Today the relative masses of atoms are determined by a technique called mass spectrometry. A mass spectrometer can be seen in Figure 2.2.

The relative masses of atoms are called the **atomic masses** of the elements, although chemists often call them **atomic weights,** because mass is determined by weighing. In this book, we will use the term atomic mass. Atomic masses were first assigned by assuming hydrogen's mass to be 1. Atomic masses are now defined relative to the mass of the most common isotope of the element carbon, whose mass is specified as exactly 12 **atomic mass units,** which now makes hydrogen's mass 1.008. (Isotopes are considered in Section 2.5.) The precise atomic masses of the elements are given in the table inside the back cover. The symbol for atomic mass unit before 1962 was **amu.** Since that time, a new symbol, **u,** has been slowly coming into use. We will continue to use the **amu** abbreviation because it is a more explicit reminder of its meaning.

It is important to note here that atoms are much too small to be handled individually, much less be seen. Chemists had to learn about them indirectly.

Figure 2.2 A mass spectrometer in action. (Ernest O. Lawrence/Berkeley National Laboratory.)

In the laboratory, mass is measured, not numbers of atoms. However, if the relative masses of atoms are known, it is possible to count the atoms. For example, suppose there are 19 g of fluorine and 1.0 g of hydrogen in a compound containing only hydrogen and fluorine. If the weighed masses of the two elements produce the same ratio as those elements' relative masses, there must be equal numbers of atoms of the two elements in the compound.

The relation between numbers of objects and their masses is easier to understand when familiar objects are used to illustrate it, as in Example 2.2.

Example 2.2   Relating mass to numbers of objects

The head chef at a large banquet requires 1000 oranges and 1000 cantaloupes for a dessert. Your job depends on being able to supply the correct numbers of each in the next 15 min, too short a time in which to count them. Luckily, you do have a balance large enough to accommodate such large numbers of items. The average mass per orange is 150 g and that per cantaloupe is 450 g. Calculate the total mass of each kind of fruit equivalent to the numbers required by the chef.

Solution

What you must do is weigh an amount of each fruit equivalent to the required numbers of items.

$$1000 \text{ oranges} \left(\frac{150 \text{ g orange}}{1 \text{ orange}} \right) = 150{,}000 \text{ g oranges} = 150 \text{ kg oranges}$$

$$1000 \text{ cantaloupes} \left(\frac{450 \text{ g cantaloupe}}{1 \text{ cantaloupe}} \right) = 450{,}000 \text{ g cantaloupes}$$
$$= 450 \text{ kg cantaloupes}$$

Slight differences in mass from orange to orange and cantaloupe to cantaloupe aside, weighing out 150 kg of oranges and 450 kg of cantaloupes should give you the required numbers.

Problem 2.2   Suppose, in one crate, there were 4500 g of oranges weighing 150 g each and, in another crate, 4500 g of cantaloupes weighing 450 g each. Calculate the numbers of oranges and cantaloupes in each crate.

The relation between numbers of atoms and their masses will be described in detail in Chapter 4.

2.4   THE STRUCTURE OF ATOMS

Since Dalton's time, scientists have learned that the atom is neither featureless nor indestructible. It is composed of subatomic particles, some that have electrical charges and some that do not. We will be concerned with three subatomic particles: protons, neutrons, and electrons.

Of these three types of particles, two are charged: the positively charged **proton** and the much lighter, negatively charged **electron.** The third is an electrically uncharged (neutral) particle, the **neutron,** that has about the same mass as the proton. These particles and their properties are listed in Table 2.1.

TABLE 2.1 | Properties of Subatomic Particles

Subatomic particle	Electrical charge	Mass (amu)	Location
proton	1+	1.00728	nucleus
neutron	0	1.00867	nucleus
electron	1−	0.0005486	outside nucleus

The protons and neutrons make up most of an atom's mass and are located together in a structure that lies at the center of the atom and is called the **nucleus.** The nucleus occupies only a small part, about $1 \times 10^{-13}\%$ of the atom's volume; therefore, the electrons occupy most of the volume of the atom.

Electrically charged particles repel one another if their charges are the same and attract one another if their charges are opposite. Even though electrons and protons have very different masses, the magnitudes of their respective charges are the same. That the atom is electrically neutral indicates that the oppositely charged particles are present in equal numbers: the number of protons in an atom is balanced by exactly the same number of electrons. The number of protons in the nucleus is called the **atomic number** and is unique for each element. Atomic numbers are listed on the inside back cover, along with the atomic masses.

Example 2.3   Calculating the charge of an atom

What are the sign and magnitude of the charge of an atom containing 12 protons, 11 neutrons, and 12 electrons?

Solution
Each proton has a single positive charge, neutrons have no charge, and each electron has a single negative charge. So, in this atom, the overall charge is the sum of all charges.

$$12 \text{ protons} \times [+1 \text{ (charge per proton)}] = +12$$
$$11 \text{ neutrons} \times [0 \text{ (charge per neutron)}] = 0$$
$$12 \text{ electrons} \times [-1 \text{ (charge per electron)}] = -12$$
$$\text{Charge of atom} = +12 + 0 + (-12) = 0$$

Problem 2.3   What are the sign and magnitude of the charge of an atom containing 9 protons, 10 neutrons, and 9 electrons?

Atoms of elements may acquire a charge by gaining or losing electrons in reactions with other compounds or elements. An atom bearing a net electrical charge is called an **ion.** A positively charged ion is called a **cation,** and a negatively charged ion is called an **anion.**

Example 2.4   Calculating the charge of an ion

What are the sign and magnitude of the charge of an ion containing 14 protons, 15 neutrons, and 12 electrons?

Solution

$$14 \times (1+) + 15 \times (0) + 12 \times (1-) = 2+$$

Problem 2.4   What are the sign and magnitude of the charge of an ion containing 13 protons, 14 neutrons, and 10 electrons?

Chemists assume that the electrons' contribution to the mass of an atom is negligible, because the mass of an electron is only about $1/2000$ the mass of either a proton or a neutron (see Table 2.1). Thus, for all practical purposes, all the mass of the atom is located in the nucleus. The following example demonstrates why chemists feel justified in considering the mass of an atom to be the sum of the masses of its protons and neutrons.

Example 2.5   **Adding the masses of subatomic particles**

Calculate and compare the masses of an atom of 6 protons, 6 neutrons, and 6 electrons with the mass of the same atom minus the mass contribution of the 6 electrons.

Solution

The mass of each kind of particle can be obtained from

$$\text{Mass of protons} = 6 \times 1.00728 \text{ amu} = 6.04368 \text{ amu}$$
$$\text{Mass of neutrons} = 6 \times 1.00867 \text{ amu} = 6.05202 \text{ amu}$$
$$\text{Mass of electrons} = 6 \times 0.0005486 \text{ amu} = 0.003292 \text{ amu}$$
$$\text{Total mass} = 12.09900 \text{ amu}$$

Then:

$$\text{Total mass} - \text{electron mass} = 12.09570 \text{ amu}$$
$$\text{Percent difference} = 0.027\%$$

Problem 2.5   What are the name and approximate mass of an element that has 22 protons and 26 neutrons?

2.5   ISOTOPES

Although all the atoms of a particular element have the same number of protons—hence the same atomic number—the number of neutrons in the atoms of that element can vary. The result is a family of atoms of the same element that have the same chemical properties but have slightly different masses. They are known collectively as the **isotopes** of the element. Isotopes are therefore atoms of the same element that all have the same atomic number but have different mass numbers. (This is one of the ways in which Dalton's theory required modification.) Isotopes are routinely used in medicine and biology to follow the movement of molecules through the body, as well as in therapeutic applications.

» **The medical use of isotopes is described in Chapter 10.**

An isotope is identified in chemical notation by its **mass number,** which is the sum of the protons and neutrons that it contains. We can distinguish one isotope from another in two ways. One approach is to write the mass number as a superscript in front of the symbol for the element and write the atomic, or proton, number as a subscript. An example is $^{17}_{8}\text{O}$. The other is to simply write the name of the element followed by its mass number. The atomic number is then implicitly stated in the element's name—for example, oxygen-17.

Example 2.6   **Identifying isotopes by symbolic notation**

Explain the difference between two isotopes of carbon, carbon-12 and carbon-14, and express that difference with a symbolic notation that uses superscripts and subscripts.

Solution

The two isotopes of carbon have mass numbers of 12 and 14, respectively, and the same atomic number—that is, the same number of protons, 6. The symbolic notation is

$$^{12}_{6}\text{C} \quad \text{and} \quad ^{14}_{6}\text{C}$$

Problem 2.6   What is the symbolic notation for the two isotopes of nitrogen that contain 7 and 8 neutrons, respectively?

The number of neutrons in an isotope can be calculated from the isotope's mass number and atomic number.

A PICTURE OF HEALTH

Percentage of Atoms of Different Elements in the Body

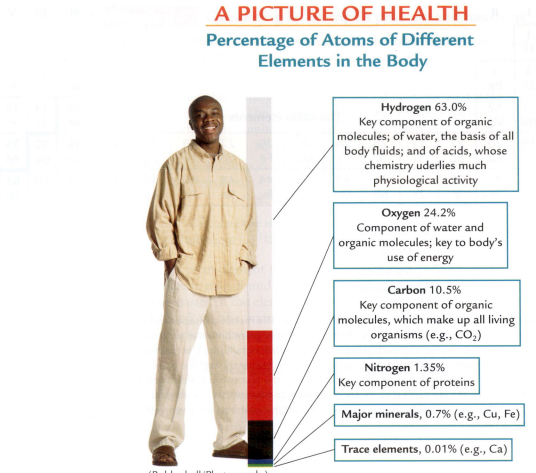

Hydrogen 63.0%
Key component of organic molecules; of water, the basis of all body fluids; and of acids, whose chemistry uderlies much physiological activity

Oxygen 24.2%
Component of water and organic molecules; key to body's use of energy

Carbon 10.5%
Key component of organic molecules, which make up all living organisms (e.g., CO_2)

Nitrogen 1.35%
Key component of proteins

Major minerals, 0.7% (e.g., Cu, Fe)

Trace elements, 0.01% (e.g., Ca)

(Rubberball/Photosearch.)

Figure 2.4 The violent reaction of the alkali metal potassium with water. The products of the reaction are hydrogen gas and potassium hydroxide, as well as lots of heat. The flames are a result of the hydrogen reacting with oxygen from the air. The sparks are pieces of the molten potassium driven into the air by the force of the explosion. (Alexander Boden.)

The elements in the first column react vigorously with water (Figure 2.4), producing hydrogen and chemical products called bases—for example, potassium hydroxide. Bases added to water form caustic, or alkaline, solutions. The production of caustic, or basic, solutions is the reason that these elements are called **alkali metals** (potassium hydroxide is an alkali metal hydroxide). Drain opener and oven cleaner (Figure 2.5) are extremely caustic products found in many kitchens.

Figure 2.5 Many home cleaning products consist of caustic, or basic, chemical compounds. (Richard Megna/ Fundamentals Photographs.)

Basic solutions react with chemicals called acids to produce salts. All salts produced by reaction of the alkali metal hydroxides with hydrochloric acid and isolated by removal of water are white solids and very soluble in water. Acids and bases will be considered briefly in Chapter 3 and more fully in Chapters 8 and 9. The acid that you are most familiar with is acetic acid, the acidic component of vinegar.

The elements—fluorine, chlorine, bromine, iodine, and astatine—in the vertical column, or group, that directly precedes that of the noble gases are called the **halogens.** Three members of this group can be seen in Figure 2.6. They are **nonmetals**—elements that cannot be cast into shapes and do not conduct electricity. When nonmetals react, they tend to gain electrons to form anions. All the halogens react with hydrogen to form compounds that are very soluble in water and whose solutions are acidic. These solutions can be neutralized with potassium hydroxide. If the water is then removed, the remaining product is a white potassium salt, which is quite soluble in water and, when dissolved in water, is an excellent conductor of electricity.

Other groups of elements that have similar chemical properties can be identified in the same way. For example, the group to the right of the alkali metals and consisting of beryllium, magnesium, calcium, strontium, and barium comprises the **alkaline earth metals.**

The first period, or row, has only two members: hydrogen and helium. Although hydrogen is not an alkali metal, it does have some of the properties of that family and is often placed in the first position in Group I. Helium is inert, and so it is placed in the first position in Group VIII. The second and third periods each contain eight elements, and, if there were no other kinds of elements, the periodic table would consist of eight columns. However, in the fourth period, a new feature emerges.

In period 4, a new set of ten elements appears. The properties of these elements are intermediate between Groups II and III of the preceding period 3. These intervening elements are called the **transition elements.** An important property of these elements is that they can form cations of more than a single electrical charge. For example, iron can form cations of charge 2+ and 3+, and copper can form cations of 1+ and 2+. Transition metals often form highly colored salts. The transition metals are found in biological systems where electrons are transferred in the course of metabolic processes. Iron is a transition element of great biological importance, the source of the deep red color of oxygenated blood. Molybdenum is a key factor in the biological process of atmospheric nitrogen fixation. Cobalt is the central feature of the structure of vitamin B_{12}.

Four series of transition elements of increasing complexity are in periods 4 through 7. For example, periods 6 and 7 include larger transition groups, each consisting of 14 elements, called **inner transition elements.** The elements numbered 57 through 70 are called the lanthanide inner transition elements, and the elements numbered 89 through 102 are called the actinide inner transition elements, after the first element in each series. The chemical properties of the elements in each inner transition series resemble one another so closely that each series is traditionally assigned a single position in the periodic table. Each series in its entirety is placed below the table, resulting in the table's compact arrangement.

In the version of the periodic table used in this chapter (Figure 2.3), the columns of transition elements are not numbered. The elements belonging to the first two and last five groups in the table, denoted by Roman numerals, are called the **main-group elements.** The noble gases are designated by the symbol VIII.

Figure 2.3 indicates the general locations of the metals, which occupy most of the left-hand side of the table, and the nonmetals, which occupy the extreme right-hand side of the table. An intermediate category of elements, called **metalloids** or **semimetals,** has properties between those of metals and

Figure 2.6 Members of the halogen family at room temperature. *Left to right:* Chlorine (Cl_2), a pale yellow gas; bromine (Br_2), a reddish liquid with a reddish vapor; and iodine (I_2), a dark purple solid with a purple vapor. (Chip Clark.)

TABLE 2.2	Essential Elements in the Human Body Listed by Their Relative Abundance

ELEMENTS COMPRISING 99.3% OF TOTAL ATOMS (MAJOR ELEMENTS)

hydrogen	carbon
oxygen	nitrogen

ELEMENTS COMPRISING 0.7% OF TOTAL ATOMS (MAJOR MINERALS)

calcium	sulfur
phosphorus	magnesium
potassium	chlorine
sodium	

ELEMENTS COMPRISING LESS THAN 0.01% OF TOTAL ATOMS (TRACE ELEMENTS)

iron	selenium
iodine	molybdenum
copper	fluorine
zinc	tin
manganese	silicon
cobalt	vanadium
chromium	

nonmetals and lies between the two larger classes. Table 2.2 is a list of elements essential in the nutrition of humans. The physiological consequences of nutritional deficiencies of those elements are listed in Table 2.3.

The reasons for this discussion of the periodic table are twofold. One reason is to point out that, once you are familiar with the chemistry of one member of a group, you will also have a good idea of the chemistry of the other members of the group. The other reason is to set the stage for considering the next great advance in chemical knowledge—the relation between the atomic structure of an element and the element's chemical reactivity.

TABLE 2.3	Elements Required for Human Nutrition and the Consequences of Their Deficiencies

Element	Result of nutritional deficiency
calcium	bone weakness, osteoporosis, muscle cramps
magnesium	calcium loss, bowel disorders
potassium	muscle weakness
sodium	muscle cramps
phosphorus	muscle and bone weakness
iron	anemia
copper	anemia
iodine	goiter
fluorine	tooth decay
zinc	poor growth rate
chromium	hyperglycemia
selenium	pernicious anemia
molybdenum	poor growth rate
tin	poor growth rate
nickel	poor growth rate
vanadium	poor growth rate

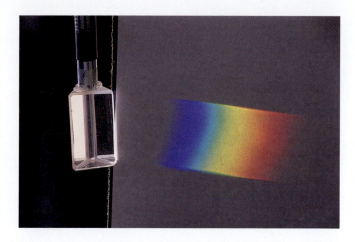

Figure 2.7 White light entering on the left side is resolved by a glass prism into a continuous spectrum, seen on the right side.

2.7 ELECTRON ORGANIZATION WITHIN THE ATOM

The picture of the atom, as we last left it, was of a tiny nucleus containing most of the atom's mass, surrounded by electrons equal in number to the atom's atomic number. Until the early twentieth century, the arrangement of those electrons was a mystery. Were they distributed in some sort of order or would we discover that they had no order at all? Today we know that an ordered arrangement of the electrons within atoms explains the structure of the periodic table and the chemical reactivity of the elements. Before we can profitably look into this arrangement, however, we must become familiar with some of the scientific developments that led to its discovery. The modern view of the atom's structure began with what scientists had observed about the colors emitted from elements made incandescent by flames or sparks.

Atomic Spectra

Light from the sun, a lit candle, or an incandescent bulb is called white light. It can be separated, or resolved, into its component colors, or frequencies, called its **visible spectrum,** by means of a prism, as shown in Figure 2.7.

The visible spectrum from incandescent sources such as the sun or a heated filament in a light bulb is a series of colors that continuously change and merge into one another, as we see in Figure 2.7. In contrast, when light is emitted by excited, or energized, atoms or individual elements that have been vaporized in a flame, it produces not a continuous spectrum but a series of sharply defined and separated lines of different color. Figure 2.8 shows the appearance of flames into which various compounds of Group I elements have been injected. The flame's colors are distinctive in regard to the identity of the

(a)   (b)   (c)

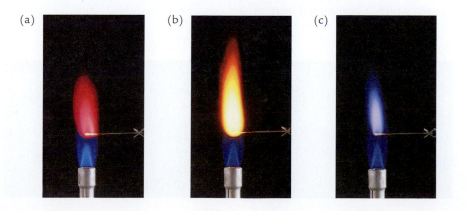

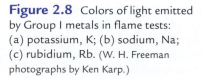

Figure 2.8 Colors of light emitted by Group I metals in flame tests: (a) potassium, K; (b) sodium, Na; (c) rubidium, Rb. (W. H. Freeman photographs by Ken Karp.)

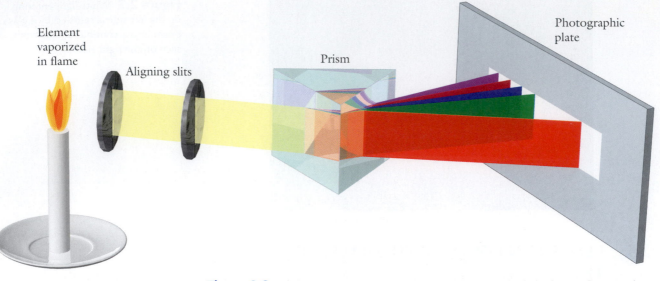

Figure 2.9 Schematic representation of a spectrometer. The light from a flame such as those in Figure 2.8 is shaped by a series of slits and passed through a prism that resolves the light into its constituent frequencies (colors). The resulting line spectrum is projected onto a photographic plate for recording.

elements. These emissions are then sent through an instrument called an atomic spectrometer (Figure 2.9) to be resolved into their component colors. The spectral lines produced by the spectrometer, called the element's **atomic emission spectrum,** are recorded on a photographic plate as in Figure 2.10. Every element has its own characteristic emission spectrum, and these spectra are used to identify and quantitatively analyze unknown materials.

Radiation is absorbed by an atom when that radiation corresponds to the same specific frequency of light emitted from the atom when it is excited in a flame. In this way, the emission of light of a specific frequency by excited atoms of an element is complemented by the absorption of radiation of that same frequency when the atoms of the element in an unexcited state are exposed to that frequency of light. The absorption of light of specific frequencies is called an **absorption spectrum.** Box 2.1 describes how atomic absorption spectra are obtained and what they are used for. Other types of absorption spectra are used to detect and characterize molecules and will be presented in succeeding chapters.

The meaning of these line spectra puzzled physicists for more than 100 years. Until early in the twentieth century, chemists had no idea that atomic spectra could be the key to an understanding of chemical reactivity. That key lies in the relation between the color (frequency) of light and its energy, which is considered next.

(a)   (b)

Figure 2.10 Atomic emission spectra of (a) hydrogen and (b) helium. (Department of Physics, Imperial College/Science Photo Library/ Photo Researchers.)

2.1 CHEMISTRY IN DEPTH

Absorption Spectra and Chemical Analysis

Absorption spectra are obtained by generating light of a specific frequency and passing it through a sample. Substances that absorb that frequency will remove energy from the incoming light. The specific frequency absorbed identifies the material, and the amount of energy that it removes from the incoming light depends on the amount of a particular substance in the sample. An absorption spectrum is therefore simultaneously a qualitative and a quantitative tool for chemical analysis.

The important components for obtaining absorption spectra are a device for producing light of a particular frequency and another device for determining the loss of energy in the incident light beam after it emerges from the sample. The first can be accomplished with the same kind of prism used to resolve light into its components (see Figure 2.7). By rotating a prism irradiated with white light, one can select particular frequencies and direct them through a sample. A photocell placed after the sample can measure the light energy before and after the sample is placed in the beam and thus determine the energy loss of the beam due to absorption by the sample. The photograph is of a modern atomic absorption spectrometer. It is used in the clinical laboratory to determine, in the blood, the presence and the amounts of ions such as sodium, potassium, and calcium.

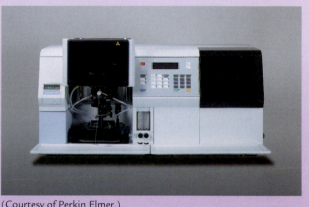

(Courtesy of Perkin Elmer.)

Electromagnetic Radiation and Energy

Radiation describes the transfer of energy from one point in space to another. For example, after an iron bar has been heated to incandescence, you do not need to touch it to feel its heat. We say that the heat "radiates" from the incandescent metal. Heat and visible light are only small parts of the full spectrum of **electromagnetic radiation,** which has its origin in the oscillation, or vibration, of charged particles. For this reason, electromagnetic radiation is quantitatively described by its frequency—that is, by how many times per second a vibration is completed.

Frequency is calculated in cycles per second, or cps. The SI unit of frequency is the hertz, abbreviated Hz, named after Heinrich Rudolf Hertz, a German physicist who contributed a great deal to this branch of knowledge. Table 2.4 shows the frequency ranges for the entire electromagnetic spectrum. Cosmic radiation, at the top of the list, has the highest frequency; radio waves, at the

TABLE 2.4 Frequency and Energy of Electromagnetic Radiation

Type of radiation	Frequency (Hz)	Energy relative to visible light
cosmic rays	10^{22}	100,000,000
γ-rays	10^{20}	1,000,000
X-rays	10^{18}	10,000
ultraviolet light	10^{16}	10
visible light	10^{15}	1
infrared light	10^{14}	0.1
microwave radiation (radar)	10^{10}	0.0001
FM radio	10^{8}	0.0000001
AM radio	10^{6}	0.000000001

bottom, have the lowest. The colors of visible light have frequencies ranging from 4×10^{14} to 8×10^{14} Hz, which is a very small part of the entire spectrum.

The energy of electromagnetic radiation is related to its frequency: the higher the frequency, the greater the energy. Ultraviolet (UV) radiation, produced by the sun, has an energy great enough to cause chemical damage to human skin (see Box 10.2). Therefore the use of UV-blocking preparations is strongly recommended by dermatologists. Medical practitioners keep close watch on their patients' exposure to diagnostic X-rays to prevent possible damage to deep tissue. Gamma radiation (γ-rays) is used to destroy cancers. The right-hand column in Table 2.4 compares the energies of the various radiation frequencies with that of visible light. Frequencies from the ultraviolet through the visible to the infrared are used to analyze organic compounds.

Atomic Energy States

The underlying reason for the line structure of atomic emission spectra was discovered by Niels Bohr, a Danish physicist. Several years before Bohr's discovery, physicists Max Planck and Albert Einstein demonstrated that light consists of small packages that they called **photons.** They further showed that the energy of these photons depends on the frequency of the light. Bohr knew that, for light energy to be emitted, energy must be absorbed. He proposed that the discrete lines of atomic spectra must represent unique energies of emitted light and, therefore, only certain energies can be absorbed. The absorption of energy raises the atom from a stable, low-energy state, or **ground state,** to a higher-energy **excited state.** Consequently, the lines of atomic spectra are the result of light emitted when an atom returns from an excited state to its ground state. In other words, the atom first absorbs and then emits a unique quantity of energy corresponding to the frequency of that spectral line.

Electromagnetic radiation energies that can be absorbed by very small systems such as atoms and electrons come in small individual packages called **quanta** (singular, quantum). The size of a quantum depends directly on the frequency of the radiation, which means that the energy of atoms can be increased only in discrete units, or small jumps. If the size of a quantum of energy striking an atom is equal to the energy difference between two of that atom's energy states, then the atom will absorb the energy and enter an excited state. If the quantum is larger or smaller than the energy difference between the two states, there will be no absorption and the atom will remain in its ground state. We might count our money down to the last penny, for example, but the Internal Revenue Service prefers not to recognize anything smaller than dollars.

Concept checklist

✓　The discrete lines of atomic emission spectra represent energy states of the atom. Remember that each color of light represents electromagnetic radiation of a specific energy.

✓　The existence of discrete lines of specific frequencies means that only certain energy states, and no others, are allowed.

The first model to incorporate the relation between line spectra and atomic energy levels pictured the atom as a miniature solar system, with electrons moving about the nucleus much as planets revolve about the sun. Because electrons in the atom are limited to certain permissible values of energy, it was argued that they must remain at fixed distances in paths about the nucleus and cannot occupy any position intermediate between these paths.

This "solar system" model was attractive because it used familiar images to describe the unseen. However, although the model worked well for hydrogen, it failed to predict the behavior of the other elements. The model treated the electron as a discrete particle whose speed and trajectory, like those of a rocket ship or bullet, could be known at any time. However, the electron is such a light and small particle that, when it is observed, its position is altered by the observation. The consequent uncertainty about position and motion was incorporated into a new model of the atom called quantum mechanics.

2.8   THE QUANTUM MECHANICAL ATOM

The modern theory that describes the properties of atoms and subatomic particles is called **quantum mechanics.** It is a complex mathematical theory that has had great success in predicting a wide variety of atomic properties. Because it is a mathematical theory, its results are very difficult to describe in reference to everyday experience. The most notable of these difficulties is that the ability to locate an electron with any precision had to be abandoned. The best that scientists can do is to estimate the probability of finding an electron in a given region of space.

The chief goal of this chapter is to show how the detailed structure of the atom explains the periodicity of the chemical properties of the elements. To do so, we will construct the periodic table by building elements from electrons and protons. The elements will be built by adding electrons one at a time to nuclei that increase in proton number in the same way. This cannot be done without a set of rules to guide us in the way in which the incoming electrons are to be organized within the atoms constructed in this way. Quantum mechanics has provided those rules and has shown that electrons are organized within the atom into shells, subshells, and orbitals, whose relation to one another depends on the energy state of the atom, which is defined by its **principal quantum number.**

Shell: Identified by a principal quantum number $(1, 2, 3 \ldots n)$ that specifies the energy level, or energy state, of the shell. The higher the quantum number, the greater the energy and the farther the shell electrons are from the nucleus.

Subshells: Locations within a shell, identified by the lowercase letters s, p, d, and f.

Orbital: The region of space within a subshell that has the highest probability of containing an electron.

The region of space in which an electron is most likely to be found is called an **atomic orbital.** Atomic orbitals are best visualized as clouds surrounding the nucleus. Their size and shape depend on the atom's energy state: the greater the energy, the larger and more complex the orbitals. The s and p types of atomic orbitals are illustrated in Figure 2.11. Remember that the

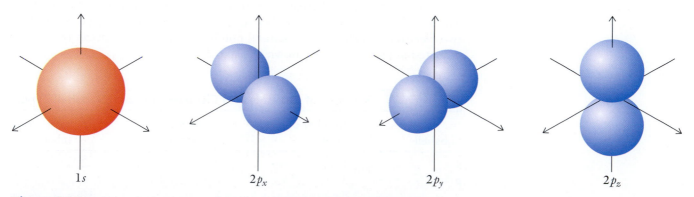

$1s$ $2p_x$ $2p_y$ $2p_z$

Figure 2.11 Atomic s and p orbitals. Their properties allow the representation of regions of space in which there is a greater than 90% probability of finding an electron. (See Section 2.9 for the significance of subscripts x, y, and z.)

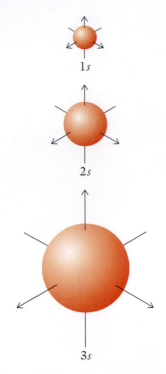

Figure 2.12 The 1*s*, 2*s*, and 3*s* atomic orbitals. The *s* orbitals all have the same shape but grow larger as their distance from the nucleus increases.

atom's electrons cannot be located with any precision, but the surface of an orbital encloses a space in which there is a high likelihood, or high probability, of finding a given electron. As the principal quantum number increases, electrons are farther from the nucleus and their orbitals occupy a larger volume. Figure 2.12 shows how *s* orbitals retain their overall shape but increase in volume. The same holds true for *p*, *d*, and *f* orbitals.

In Chapter 3, we will see that the formation of one kind of chemical bond is the connection (the bonding) of atoms by electrons contained in these orbitals. Because the orbitals are restricted to certain positions in space around the nucleus, the connections that they create between atoms give rise to the unique shapes and therefore the characteristic properties of molecules of different substances.

The relation between principal quantum number, number of subshells, and subshell names is tabulated in Table 2.5. You can see that, as the principal quantum number increases, the number of subshells increases. The number of orbital types within each subshell also increases. There is an *s* orbital in every shell; *s* and *p* orbitals are in the second shell; *s*, *p*, and *d* orbitals are in the third shell; and *s*, *p*, *d*, and *f* orbitals are in the fourth shell.

TABLE 2.5	Relation Between Principal Quantum Number and Number and Type of Orbital Within a Subshell	
Principal quantum number (shell)	Number of subshells	Subshell names (orbital types)
1	1	*s*
2	2	*s, p*
3	3	*s, p, d*
4	4	*s, p, d, f*

The numbers of orbitals within each subshell are listed in Table 2.6. The data in Tables 2.5 and 2.6 tell us that an atom in principal quantum state 1 has one *s* orbital available for electrons, an atom in principal quantum state 2 has one *s* and three *p* orbitals available for electrons, and an atom in principal quantum state 3 has one *s*, three *p*, and five *d* orbitals available for electrons.

The final feature of electron behavior is that electrons possess a property called **spin,** analogous to the rotation of a planet. A single electron within an orbital is called an **unpaired electron** or a **lone electron.** For an orbital to accommodate two electrons, their spins must be opposite. The electrons are then said to be **spin-paired.** The spin-pairing will be represented symbolically in this book by a pair of arrows pointing in opposite directions (Section 2.9). The significance of spin-pairing is that one orbital can contain no more than two electrons; on that basis, the total number of electrons that can be accommodated in each subshell is shown in Table 2.7.

TABLE 2.6	Relation Between Subshell Type and Number of Orbitals
Subshell type	Number of orbitals
s	1
p	3
d	5
f	7

TABLE 2.7	Number of Orbitals and Electrons in Each Subshell Type		
Subshell	Number of orbitals	Electrons per orbital	Total electrons per subshell
s	1	2	2
p	3	2	6
d	5	2	10
f	7	2	14

2.9   ATOMIC STRUCTURE AND PERIODICITY

To see how the modern quantum mechanical theory explains the arrangement of elements in the periodic table, we will imagine creating the elements starting with hydrogen by adding electrons one at a time around an atomic nucleus. To keep the resulting atom neutral, the atomic number of the nucleus must increase simultaneously every time that we add a new electron. This imaginary exercise has been given the German name *Aufbau,* which means "buildup." The result of the aufbau procedure is the complete description of the electron organization of the atom, also called the atom's **electron configuration.**

We will use two methods for representing electron configuration. Method 1 identifies electrons, first, by the principal quantum number (1, 2, and so forth); next, by the subshell type (*s, p,* and so forth); and, finally, by the number, written as a superscript, of electrons in each orbital ($1s^2$ and $2p^3$, for example). Method 2 uses boxes to represent orbitals at different energy levels and arrows in the boxes to represent electrons—for example, $\boxed{\uparrow}$.

Here is a list of rules for constructing an atom. We will begin to put them into practice in Example 2.10.

---

1. The larger the principal quantum number, the greater the number of subshells (see Table 2.5). An electron described as $2s$ is in the second shell and in an *s* subshell. A $3d$ electron is in the third shell and in a *d* subshell.
2. Each subshell has a unique number of orbitals (see Table 2.6).
3. The order in which electrons enter the available shells and subshells depends on the orbital energy levels. The orbitals fill according to the diagram in Figure 2.13 on the following page, with the lowest energy orbitals (closest to the nucleus) filling before those more distant. Boxes at each level correspond to the number of orbitals in the subshell.
4. There can be no more than two electrons per orbital, and the two electrons can occupy the same orbital only if they are spin-paired (their spins are opposite). This rule is known as the **Pauli exclusion principle.** Method 2 emphasizes the exclusion principle by showing that orbitals can contain no more than two electrons.
5. In a subshell with more than one orbital (for example, *p* or *d* orbitals), electrons entering that subshell will not spin-pair until every orbital in the subshell contains one electron. This rule is known as **Hund's rule.** All electrons have the same electrical charge and tend to repel one another. This repulsive force can be reduced if the electrons stay as far apart as possible (by occupying different orbitals whenever possible).

Aufbau rules

Example 2.9   Interpreting an atom's principal quantum number

If an atom is in an energy state of principal quantum number 3, how many and what kinds of subshells are allowed?

Solution
We see in Table 2.5 that the principal quantum number 3 is characterized by three subshells, the *s, p,* and *d* subshells.

Problem 2.9   How many and what kinds of subshells are allowed in principal quantum state 4?

Figure 2.13 emphasizes two important features of electron behavior. First, the allowed energies for electrons do not represent a gradually increasing continuum

Figure 2.13 The quantized energy levels of atomic subshells. Atomic orbitals are represented by boxes.

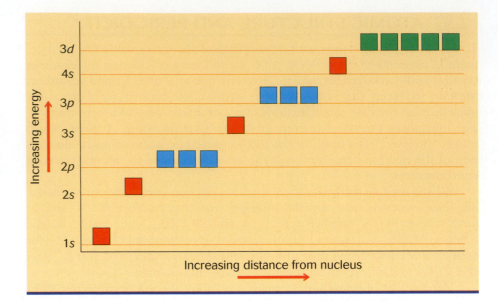

but a series of jumps from one level to another, with no intermediate energy levels allowed. This feature is what physicists mean when they say that energy is **quantized.** Second, beginning with principal quantum number 4, the s subshell is lower in energy than the d subshell of the preceding period. Therefore the $4s$ level fills before the $3d$ level.

Example 2.10 **Diagramming electron configurations by using method 1**

Diagram the electron configuration of elements 1, 2, 3, 4, 5, and 6. Remember, the atomic number and the number of electrons are identical.

Solution

Use Table 2.7 on page 52 to determine the principal quantum number ("shell"), available subshells, and number of electrons per subshell, and construct a table as shown here for each of these elements. An s subshell can contain a maximum of two electrons. Because we fill the orbitals one electron at a time, the first two elements will exhaust the first s subshell, $1s$. In the third and fourth elements, electrons will fill the second s subshell, $2s$; electrons in the fifth and sixth elements must go into the next available (empty) subshell, $2p$, which can accommodate a maximum of six electrons in its three orbitals.

Element	Shell	Subshell	Number of electrons	Electron configuration
hydrogen	1	s	1	$1s^1$
helium	1	s	2	$1s^2$
lithium	1	s	2	
	2	s	1	$1s^2 2s^1$
beryllium	1	s	2	
	2	s	2	$1s^2 2s^2$
boron	1	s	2	
	2	s	2	
	2	p	1	$1s^2 2s^2 2p^1$
carbon	1	s	2	
	2	s	2	
	2	p	2	$1s^2 2s^2 2p^2$

Hydrogen's electron configuration is $1s^1$; that is, there is one electron (indicated by the superscript) in the s orbital of the first shell (indicated by 1). Helium's configuration is $1s^2$; helium's two electrons fill and complete the s orbital of the first shell. Lithium, atomic number 3, with three electrons, follows helium. Its third electron must enter the second shell, where the first available empty subshell is the $2s$ orbital. Thus, the configuration of lithium is $1s^2 2s^1$. With the fourth element, beryllium, the $2s$ subshell is complete. Beryllium's electron configuration is $1s^2 2s^2$. Boron, atomic number 5, follows beryllium. Its fifth electron must enter the next available empty orbital. The electron configuration of boron is $1s^2 2s^2 2p^1$. Carbon's electron configuration is $1s^2 2s^2 2p^2$.

Problem 2.10   Diagram the electron configuration of element 16 by using method 1.

The electron configuration of carbon in Example 2.10 must be looked at more closely. The three p orbitals lie perpendicularly to one another, and so chemists label them according to the Cartesian axes on which they lie: x, y, and z (see Figure 2.11). This designation is added as a subscript. Thus, the three p orbitals are known as p_x, p_y, and p_z. According to rule 5, these orbitals are occupied singly until each contains one electron, after which an electron of opposite spin may enter each one. Therefore, the complete electron configuration of carbon is $1s^2 2s^2 2p_x^1 2p_y^1$.

Example 2.11 | **Diagramming electron configuration by using method 2**

Use method 2 to diagram the electronic configuration of the first four elements.

Solution
Represent each orbital by a box at an appropriate energy level (see Figure 2.13), and indicate the presence of electrons by arrows pointing up or down. Remember, rule 4 states that an orbital can contain no more than two spin-paired electrons. The atomic number is equal to the number of protons, and the number of protons is equal to the number of electrons.

Element	Atomic number	Orbital 1s	2s
hydrogen	1	↑	
helium	2	↑↓	
lithium	3	↑↓	↑
beryllium	4	↑↓	↑↓

Problem 2.11   Using method 2, diagram the electronic configuration of element 5. (Hint: Because the $1s$ and $2s$ orbitals of element 5 are filled, the fifth electron must go into a new orbital.)

Example 2.12 | **Diagramming electron configurations by using method 2**

Diagram the electronic configuration of elements 6 and 8 by using method 2.

Solution
Elements 6 and 8, carbon and oxygen, possess six and eight electrons, respectively. Therefore, because the $1s$ and $2s$ subshells hold a maximum of

four electrons, both those subshells will be filled, and electrons 5 through 8 must enter the p subshell. In diagramming these structures, we must pay close attention to rule 5 (Hund's rule).

Element	Atomic number	Orbital				
		1s	2s	$2p_x$	$2p_y$	$2p_z$
carbon	6	↑↓	↑↓	↑	↑	☐
oxygen	8	↑↓	↑↓	↑↓	↑↓	↑

Problem 2.12 Using method 2, diagram the electronic configuration of element 7.

TABLE 2.8 **Electron Configurations of the First Eleven Elements of the Periodic Table**

Element	Atomic number	Symbol	Electron configuration
hydrogen	1	H	$1s^1$
helium	2	He	$1s^2$
lithium	3	Li	$1s^2 2s^1$
beryllium	4	Be	$1s^2 2s^2$
boron	5	B	$1s^2 2s^2 2p^1$
carbon	6	C	$1s^2 2s^2 2p_x^1 2p_y^1$
nitrogen	7	N	$1s^2 2s^2 2p_x^1 2p_y^1 2p_z^1$
oxygen	8	O	$1s^2 2s^2 2p_x^2 2p_y^1 2p_z^1$
fluorine	9	F	$1s^2 2s^2 2p_x^2 2p_y^2 2p_z^1$
neon	10	Ne	$1s^2 2s^2 2p^6$
sodium	11	Na	$1s^2 2s^2 2p^6 3s^1$

Table 2.8 is a list of the first 11 elements in the periodic table along with their electron configurations represented in the notation of method 1. You can see that, by the time we reach elements with atomic numbers of 20 or so, this notation can take up quite a bit of space. A more compact method for writing electron configurations uses the fact that every period in the table ends with a noble gas, an element whose outer shell contains the maximum number of electrons that the shell can hold. We can use the bracketed name of a noble-gas element to represent its full electron configuration, thus shortening the notation for all the elements in the period below. This notation is demonstrated in the following example.

Example 2.13 **Diagramming electron configurations by using the compact method**

Show the compact method of writing the electronic configuration of element 19, potassium.

Solution
The electron configuration of element 19, potassium, is $[1s^2 2s^2 2p^6 3s^2 3p^6]4s^1$. Brackets have been placed around the part of the electron configuration that matches the configuration of argon, (symbol Ar), the preceding noble gas.

This configuration is called the argon **core** of the potassium atom. The argon core of the potassium atom is denoted as [Ar], and the potassium electron configuration is written in compact notation as $[Ar]4s^1$.

Problem 2.13   Write the electron configuration of element 16, sulfur, by using the compact method.

In compact notation, lithium is $[He]2s^1$, sodium is $[Ne]3s^1$, and calcium is $[Ar]4s^2$. (You should be able to locate these elements in Figure 2.3.)

At the beginning of this discussion, we selected three families of elements to illustrate the periodicity of chemical properties: the noble gases, the halogens, and the alkali metals. Table 2.9 presents the electron configurations of the elements in these families for periods 2 through 6. The electrons of the outermost shell are printed in color.

TABLE 2.9   **Electron Configurations of Some of the Alkali Metals, the Halogens, and the Noble Gases**

Alkali metals (Group I)		Halogens (Group VII)		Noble gases (Group VIII)	
Li	$[He]\,2s^1$	F	$[He]\,2s^22p^5$	Ne	$[He]\,2s^22p^6$
Na	$[Ne]\,3s^1$	Cl	$[Ne]\,3s^23p^5$	Ar	$[Ne]\,3s^23p^6$
K	$[Ar]\,4s^1$	Br	$[Ar]\,3d^{10}4s^24p^5$	Kr	$[Ar]\,3d^{10}4s^24p^6$
Rb	$[Kr]\,5s^1$	I	$[Kr]\,4d^{10}5s^25p^5$	Xe	$[Kr]\,4d^{10}5s^25p^6$
Cs	$[Xe]\,6s^1$	At	$[Xe]\,4f^{14}5d^{10}6s^26p^5$	Rn	$[Xe]\,4f^{14}5d^{10}6s^26p^6$

Notice that (1) each noble gas has eight electrons in its outer shell, (2) each alkali metal has a single electron in its outermost shell, and (3) each halogen has seven electrons in its outermost shell. In addition to the fact that the elements in a family (group) have the same number of outer-shell electrons, note that the number of outer-shell electrons is the same as the main-group elements' group number in Figure 2.3. This important relation is further emphasized in Table 2.10.

In the first three periods, we easily see that the number of electrons in the outer electronic shell (electrons outside the inner noble-gas shell) is identical with the element's group number in Figure 2.3. Things become more complex in period 4, however, when the inner $3d$ subshell begins to fill to form the transition elements after the outer $4s$ subshell has filled.

The numbers of outer-shell electrons in elements of the transition series do not correlate simply with the main-group elements' group numbers; so, in Figure 2.3, group numbers are not assigned to these elements. We will examine the properties of specific members of the transition series—iron or copper, for example—in Chapters 3 and 8.

TABLE 2.10   **Electron Configurations of the Main-Group Elements in the Second and Third Periods**

	Group							
	I	II	III	IV	V	VI	VII	VIII
period 2	Li $1s^22s^1$	Be $1s^22s^2$	B $1s^22s^22p^1$	C $1s^22s^22p^2$	N $1s^22s^22p^3$	O $1s^22s^22p^4$	F $1s^22s^22p^5$	Ne $1s^22s^22p^6$
period 3	Na $[Ne]3s^1$	Mg $[Ne]3s^2$	Al $[Ne]3s^23p^1$	Si $[Ne]3s^23p^2$	P $[Ne]3s^23p^3$	S $[Ne]3s^23p^4$	Cl $[Ne]3s^23p^5$	Ar $[Ne]3s^23p^6$

2.10 ATOMIC STRUCTURE, PERIODICITY, AND CHEMICAL REACTIVITY

When an alkali metal atom, such as sodium, reacts with a halogen atom, such as chlorine, the sodium atom loses its single electron and the chlorine atom gains a single electron. Method 1 can be used to show the changes in the electron configurations of these elements as a result of their reaction:

$$Na \quad + \quad Cl \quad \longrightarrow \quad Na^+ \quad + \quad Cl^-$$
$$1s^22s^22p^63s^1 + 1s^22s^22p^63s^23p^5 \longrightarrow 1s^22s^22p^6 + 1s^22s^22p^63s^23p^6$$

The electron configuration of the resulting positively charged sodium atom (sodium ion) is identical with that of the noble gas preceding sodium in the periodic table—neon. The electron configuration of the resulting negatively charged chlorine atom (chloride ion) is that of the noble gas immediately following chloride in the table—argon. By reacting with one another, both atoms have achieved a noble-gas electron configuration. In general, chemical reactions of the elements achieve the same results: they lead to a noble-gas electron configuration in the atoms' outer electron shells.

Because the noble gases are inert and chemically stable, chemists have concluded that a filled outermost electron shell is the most stable configuration that an atom can have. Because a noble gas has eight electrons in its outer shell, this conclusion is summarized as the **octet rule.** Helium, a noble gas with only two electrons in its outer shell, is an exception to this rule. The outermost electron shell (in a given period) is called the **valence shell.** The electrons in the outer shell are called the **valence electrons.**

Chemists use a symbolic representation first proposed by American chemist Gilbert N. Lewis and called **Lewis symbols.** The nucleus and inner-shell electrons are represented by the element's symbol, and dots around it represent the valence-shell electrons. The Lewis symbols for elements in periods (rows) 1, 2, and 3 of the periodic table are

>> The names of anions, such as lithium ions and fluoride ions, will be discussed in Chapter 3.

H·							:He
Li·	·Be·	:B·	·C·	·N·	·O·	:F·	:Ne:
Na·	·Mg·	:Al·	·Si·	·P·	·S·	:Cl·	:Ar:

With the exception of helium, the number of valence-shell electrons is equal to the group number of the main-group element.

Example 2.14 Portraying a chemical reaction with Lewis symbols

Use Lewis symbols to illustrate the reaction of lithium and fluorine—that is, the formation of lithium and fluoride ions.

Solution

Lithium's Lewis symbol suggests that lithium atoms have a tendency to lose one electron, resulting in the formation of a cation with a single charge. The formation of a lithium ion can be represented as

$$Li· - e \longrightarrow Li^+$$

in which "e" stands for electron. Fluoride's Lewis symbol suggests that fluoride atoms are driven to gain an electron, resulting in the formation of an anion with a single charge:

$$:F· + e \longrightarrow :F:^-$$

An atom's loss of an electron cannot occur without another atom's gain of that electron. Therefore, we will show that the two processes take place at the same time:

$$Li \cdot \overset{\cdot\cdot}{\underset{\cdot\cdot}{F}} : \longrightarrow Li^+ + : \overset{\cdot\cdot}{\underset{\cdot\cdot}{F}} :^-$$

Problem 2.14 Use Lewis symbols to illustrate the formation of sodium and chloride ions.

Concept checklist

✓ All members of the same group of the main-group elements have the same outer-shell electron configuration. Periodicity in chemical properties is the result of these identical outer-shell electron configurations.

✓ The chemical properties of elements within the same group are almost identical, an observation indicating that it is the outer-shell electron configuration that chiefly dictates the chemical reactivity of the elements.

✓ The octet rule states that, with the exception of helium, elements react to attain an outer-shell electron configuration of eight valence electrons or, in other words, the electron configuration of the nearest noble gas.

✓ Lewis symbols serve to highlight the electrons of the valence shells of the elements and to predict the loss or gain of electrons in chemical reactions.

In the next chapter, we will further explore the tendency of atoms to attain the noble-gas outer-shell electron configuration and its role as a driving force behind chemical reactions between the elements.

Summary

Dalton's Atomic Theory

• All matter consists of particles called atoms, which are indestructible.

• The atoms of any one element are chemically identical.

• Atoms of different elements are distinguished from one another because they have different masses.

• Compounds consist of combinations of atoms of different elements. In chemical reactions, atoms trade partners to form new compounds.

• The theory explained (1) the conservation of mass in chemical reactions and (2) the constant composition of matter.

Atomic Masses

• The relative mass of an element is called its atomic mass.

• Atomic masses are calculated in relation to the mass of the most common form of the element carbon, exactly 12 atomic mass units (amu).

The Structure of Atoms

• Atoms contain three kinds of particles: protons, electrons, and neutrons.

• The protons and neutrons are located together in the nucleus.

• In an electrically neutral atom, the number of protons is balanced by exactly the same number of electrons.

• An atom bearing a net electrical charge is called an ion. A cation is a positively charged ion, and an anion is a negatively charged ion.

Isotopes

• Naturally occurring elements consist of isotopes.

• Isotopes of a given element have the same atomic number but have different numbers of neutrons.

• An isotope's mass number is the sum of the numbers of protons and neutrons in its nucleus.

The Periodic Table

• The properties of the elements repeat periodically if the elements are arranged in order of increasing atomic number.

• In the periodic table, elements of similar chemical properties are aligned in vertical columns called groups.

• Horizontal arrangements, called periods, end on the right with a noble gas.

• Each of the second and third periods contains eight elements. But the fourth period includes a new group of ten transition elements.

• Elements are designated as either main-group or transition elements.

• The metals occupy most of the left-hand side of the periodic table, and the nonmetals occupy the extreme right-hand side of the table.

- Elements called metalloids, or semimetals, have properties between those of metals and nonmetals and lie between the two larger classes.

Electron Organization Within the Atom

- Electrons in atoms are confined to a series of shells around the nucleus.
- The farther away the electrons are from the nucleus, the greater their energy.
- Electron energy states are characterized by an integer called the principal quantum number.
- Electrons are contained in shells that consist of subshells, and electrons in these subshells reside in orbitals.
- Only two electrons are allowed in any orbital, and, to do that, they must have opposite spins—they must be spin-paired.

- The atom's orbitals define the probability of finding an electron in a given region of space around the nucleus.
- With an increase in the number of shells, the number of subshells and the complexity of their shapes increase.

Atomic Structure, Periodicity, and Chemical Reactivity

- The symbolic notation that indicates the distribution of electrons in an atom is called its electron configuration.
- An alternative method of representation, Lewis symbols, emphasizes the number of electrons in an atom's valence shell.
- The valence electrons dictate the chemical reactivity of each element.
- The octet rule summarizes the tendency of atoms to attain noble-gas outer-shell electron configurations through chemical reactions.

Key Words

anion, p. 40	electron, p. 39	nucleus, p. 40
atom, p. 37	emission spectrum, p. 48	octet rule, p. 58
atomic mass, p. 38	isotope, p. 41	periodic table, p. 42
atomic number, p. 40	Lewis symbol, p. 58	proton, p. 39
Aufbau, p. 53	mass number, p. 41	quantum mechanics, p. 51
cation, p. 40	neutron, p. 39	valence electron, p. 58

Exercises

Atomic Theory

2.1 Explain the concept of conservation of mass.

2.2 Explain the concept of the constant composition of chemical compounds.

2.3 Analysis of 3.800 g of glucose yields 1.520 g of carbon, 0.2535 g of hydrogen, and 2.027 g of oxygen. Calculate the mass percent of each element in the compound.

2.4 Analysis of 5.150 g of ethanol (ethyl alcohol) yields 2.685 g of carbon, 0.6759 g of hydrogen, and 1.789 g of oxygen. Calculate the mass percent of each element in the compound.

2.5 If carbon is assigned a mass of 12.0000 amu and oxygen weighs four-thirds as much as carbon, what is the atomic mass of oxygen?

2.6 If sulfur weighs twice as much as oxygen, what is sulfur's atomic mass?

2.7 If nitrogen weighs 0.875 the mass of oxygen, what is nitrogen's atomic mass?

2.8 If beryllium is 0.75 the mass of carbon, what is beryllium's atomic mass?

Atomic Structure

2.9 Describe the mass and electrical charges of the three principal subatomic particles making up the atom.

2.10 If an electrically neutral atom can be shown to possess 16 electrons, what must its atomic number be? Identify the element.

2.11 Explain the fact that the mass of a newly discovered element is about twice its atomic number.

2.12 Complete the following table by filling in the blanks.

Protons	Neutrons	Electrons	Mass (amu)	Element
19	20	___	___	___
34	___	___	79	___
___	20	___	40	___
___	___	11	23	___

2.13 What would be an important feature of an atom consisting of 23 protons, 28 neutrons, and 20 electrons?

2.14 What would be an important feature of an atom consisting of 9 protons, 10 neutrons, and 10 electrons?

Isotopes

2.15 Calculate the atomic mass of strontium, which has four naturally occurring isotopes:

Natural abundance (%)	Atomic mass (amu)
0.56	83.91
9.86	85.91
7.02	86.91
82.56	87.91

2.16 Element X consists of two isotopes, with masses 25 amu and 26 amu, and its mass is reported as 25.6 in the table of atomic masses. What are the percent abundances of its two isotopes?

2.17 Write the symbolic isotopic notation, with superscripts and subscripts, for the following pairs of isotopes: oxygen 16 and 17; magnesium 24 and 25; silicon 28 and 29.

2.18 Write the symbolic isotopic notation, with superscripts and subscripts, for the following pairs of isotopes: sulfur 32 and 33; argon 36 and 40; chromium 52 and 54.

Periodic Table

2.19 What are the names of the elements having the following symbols? (a) Li; (b) Ca; (c) Na; (d) P; (e) Cl.

2.20 What are the names of the elements having the following symbols? (a) He; (b) Mg; (c) K; (d) N; (e) F.

2.21 Identify the period and group of the following elements: Li, Na, K, Rb, Cs.

2.22 Identify the period and group of the following elements: Si, Ge, As, Sb.

2.23 Can the elements in Exercise 2.21 be described as metals or nonmetals? Explain your answer.

2.24 Can the elements in Exercise 2.22 be described as metals or nonmetals? Explain your answer.

2.25 For which elements (main-group or transition) do the numbers of electrons in the outer, or valence, shell correspond to the group number?

2.26 Which group of the periodic table is known as the alkali metals?

2.27 Which group of the periodic table is known as the halogens?

2.28 Which group of the periodic table is known as the alkaline earth metals?

Atomic Structure and Periodicity

2.29 Can one observe the emission spectrum of an atom in the ground state? Explain your answer.

2.30 True or false: Atomic spectra indicate the changes in energy states of electrons. Explain your answer.

2.31 On the basis of the quantum mechanical model of the atom, state the main features of the organization of an atom's electrons.

2.32 State the number and names of the subshells of the first three electron shells—principal quantum numbers 1, 2, and 3.

2.33 What is the definition of an orbital?

2.34 What is the maximum number of electrons that can occupy any orbital?

2.35 How many orbitals are there in an s subshell?

2.36 What is the maximum number of electrons that a p subshell can accommodate?

2.37 Is there any spatial orientation of an s subshell?

2.38 Is there any spatial orientation of orbitals in p subshells? If so, describe it.

2.39 Which elements are described by the following electron configurations?

(a) $1s^2 2s^2 2p^6 3s^2$
(b) $1s^2 2s^2 2p^6 3s^2 3p^4$
(c) $1s^2 2s^2 2p^6 3s^2 3p^6 4s^1$

2.40 Which elements are described by the following electron configurations? (a) [He]$2s^2 2p^1$; (b) [He]$2s^2 2p^5$; (c) [Ar]$4s^2$; (d) [Kr]$5s^2$.

2.41 What positively charged ion (cation) is represented by the following electron configuration? $1s^2 2s^2 2p^6$

2.42 What negatively charged ion (anion) is represented by the following electron configuration? $1s^2 2s^2 2p^6$

2.43 Which group of the periodic table is known as the alkali metals, and what electron configuration do they have in common?

2.44 Which group of the periodic table is known as the halogens, and what electronic configuration do they have in common?

2.45 Which group of the periodic table is known as the alkaline earth metals, and what electronic configuration do they have in common?

2.46 Which group of the periodic table is known as the noble gases, and what electronic configuration do they have in common?

Unclassified Exercises

2.47 True or false: The number of protons in an atom is called the atomic number.

2.48 What is the name of the light emitted by the atoms of an element that have been injected into a flame?

2.49 How does the energy of an electron change as its distance from the nucleus increases?

2.50 What is the maximum number of electrons that can occupy an electron shell with principal quantum number $n = 3$?

2.51 What is the principal characteristic of the elements in the same group of the periodic table?

2.52 Is sulfur a metal or a nonmetal?

2.53 Is calcium a metal or a nonmetal?

2.54 What is the chief property of the elements at the right-hand end of every period of the periodic table?

2.55 What is the symbolic notation of the two isotopes of oxygen whose masses are 16 and 18?

2.56 What is the octet rule?

2.57 The element neptunium possesses 93 electrons and weighs 237 amu. Calculate the contribution (in %) of its electrons to its total mass.

2.58 What are the similarities and differences between the two principal isotopes of carbon, atomic number 6, whose atomic masses are 12 and 13, respectively?

2.59 If the frequency of blue light is greater than that of red light, which has the greater energy?

2.60 Will the emission spectrum of an atom contain all frequencies of light? Explain your answer.

2.61 What is the name of the rule stating that, when the outermost s and p subshells of elements in any period are filled (that is, have a total of eight electrons), the elements are in the greatest state of stability?

Expand Your Knowledge

2.62 In Example 2.1, the mass percentages of carbon, hydrogen, and oxygen in vitamin C were determined. Calculate the mass in grams of each of these elements in an 8.200-g sample of pure vitamin C.

2.63 Titanium has five naturally occurring isotopes of atomic masses 45.95, 46.95, 47.95, 48.95, and 49.95, respectively. What do you need to know to account for the fact that titanium's mass in the table of atomic masses is listed as 47.88?

2.64 If necessary, correct the following statement: Most of the elements listed in the periodic table have equal numbers of proton, electrons, and neutrons.

2.65 Criticize the following statement: All of Dalton's atomic theory is considered valid today.

2.66 Can you think of a method to detect sodium, potassium, or calcium in body fluids? Explain your answer.

2.67 The presence of an element in materials of unknown composition can be detected by its emission spectrum. How is this detection possible?

2.68 How many unpaired electrons are in each of the following atoms or ions in their ground states? O; O^-; O^+; Fe.

2.69 Define each of the following terms: (a) photon of light; (b) quantum number; (c) ground state; (d) excited state.

2.70 Which of the following electron configurations correspond to an excited state? Identify the atoms and write the ground-state electron configurations where appropriate:

    (a) $1s^2\ 2p^2\ 3s^1$
    (b) $1s^2\ 2s^2\ 2p^6$
    (c) $1s^2\ 2s^2\ 2p^4\ 3s^1$
    (d) $[Ar]4s^2\ 3d^5\ 4p^1$

2.71 What electron interactions are responsible for Hund's rule?

2.72 What is the most important relation between elements in the same group in the periodic table?

2.73 Write the ground-state electronic configurations for the following neutral atoms: (a) Si; (b) Ni; (c) Fe; (d) As.

2.74 Give an example of (a) an atom with a half-filled subshell; (b) an atom with a completed outer shell; (c) an atom with its outer electrons occupying a half-filled subshell; and (d) an atom with a filled subshell.

2.75 Use Hund's rule to write ground-state electronic configurations for the following ions: (a) O^+; (b) C^-; (c) F^+; (d) O^{2-}.

2.76 Determine the number of unpaired electrons in the ground state of the following species: (a) F^+; (b) Sn^{2+}; (c) Bi^{3+}; (d) Ar^+.

CHAPTER 3

MOLECULES AND CHEMICAL BONDS

(Bill Aron/PhotoEdit.)

Chemistry in Your Future

One of your tasks as a nutritionist is to instruct people placed on special diets by their physicians. For example, people with chronically high blood pressure are often put on a low-sodium diet. Most of these clients understand that this diet means limiting their use of table salt (sodium chloride), but many are not aware that it also means reducing their use of foods preserved with sodium nitrite, flavored with monosodium glutamate, and so forth. The ingestion of these substances contributes to the body's total concentration of sodium ions—but how? The answer to this question and similar ones can be found in Chapter 3.

For more information on this topic and others in this chapter, go to www.whfreeman.com/bleiodian2e

Learning Objectives

- Describe the difference between ionic and covalent bonds.
- Describe the relation between the octet rule and the formation of ions.
- Predict the formulas of binary ionic compounds.
- Construct the names of ionic compounds.
- Construct the names of covalent compounds.
- Create Lewis structures.
- Describe the difference between polar and nonpolar covalent bonds.
- Use VSEPR theory to predict molecular shape.

How many different substances are there? The number of substances known is estimated to be 1×10^7, but the precise number can never be known, because new substances are created daily by chemists throughout the world. Most of the synthetic fibers of your clothes, the preservatives in your food, the antibiotics prescribed by your doctor, and the human insulin now produced by bacteria did not exist 70 years ago. Biological structures such as membranes, chromosomes, and muscle can now be described in chemical detail. Medicine has acquired a more comprehensive scientific basis because of the many chemical laboratory tests now available for the diagnosis of pathological conditions. All these remarkable accomplishments have resulted from advances in the understanding of the nature of chemical bonds.

To explain bonding, we begin with the octet rule presented in Chapter 2. All elements, with the exception of the noble gases, are unstable and react chemically to achieve the outer-electron configuration of the noble gases. The result of that chemical process is the formation of compounds in which the atoms are connected to one another by chemical bonds.

There are two major types of chemical bonds, ionic and covalent. An **ionic bond** is a strong electrostatic attraction between a positive ion and a negative ion. It results when electrons are transferred from the valence shell of one atom into the valence shell of another. A **covalent bond** is a bond formed when electrons are shared between atoms. In general, ionic bonds are formed between metals and nonmetals, and covalent bonds are formed between nonmetals. As we proceed in our consideration of chemical bonding, however, you will find that the differences between ionic and covalent bonds are more of degree than of kind.

3.1   IONIC VERSUS COVALENT BONDS

The kind of bond—ionic or covalent—that will form as a result of a chemical reaction depends on the relative abilities of the reacting atoms to attract electrons. When one atom has a much greater ability to attract electrons than the other atom does, electron transfer will occur, and the resulting bond will be ionic. Such is the case when metals react with nonmetals. When two atoms are similar in their ability to attract electrons, electron transfer is no longer possible. In those cases, octet formation is accomplished through electron sharing, and the resulting bond is covalent. This kind of bonding generally takes place in reactions between the nonmetals.

The ease or difficulty of removing electrons from the outer, or valence, shell of an element depends on how strongly the positively charged nucleus interacts with the outer electrons. It takes work to overcome the attraction of the positively charged nucleus for the negatively charged electron, and the energy for that process is called the **ionization energy.** If the ionization energy of metals were prohibitively high, so that electrons could not be removed from a metal's valence shell, then iron would never rust and silver would never tarnish.

Ionization energy is one of the chemical properties that shows periodicity when elements are arranged by increasing atomic number. In period 1, for example (Figure 3.1), the ionization energy increases across the period from Li through Be, B, C, N, O, F, and Ne. In general, the ionization energies increase across any given period, proceeding from Group I to Group VIII. The reason for this trend is that, within a period, the nuclei grow in positive charge, and, at the same time, the distance between each nucleus and the electrons going into that outer shell remains nearly the same. As a result, the intensity of electrical attraction between nuclei and outer electrons increases with atomic number across each period, and it becomes more and more difficult to remove electrons from the valence shell. (See Box 3.1 on page 66.)

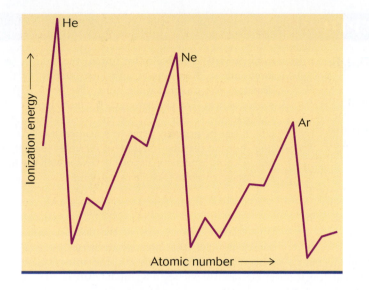

Figure 3.1 The ionization energies of the elements through the first four periods to scandium (atomic number 21). Ionization energy is the work required to remove an electron from an element's valence shell. It is greatest for elements at the end of a period (the noble gases) and smallest for elements at the beginning of a period (the Group I metals).

On the other hand, ionization energy decreases from the top to the bottom of any group of the periodic table. Within Group I, for example, the energy needed to form cations is greatest for lithium and decreases in the order lithium > sodium > potassium > rubidium > cesium. It is easier to form a cesium ion than a lithium ion, because the number of electron shells increases down each group. Thus the outer electrons become increasingly insulated from the influence of the positively charged nucleus (see Box 3.1).

Concept checklist

✓ An ionic bond will form if one of the bonding partners has a much greater electron-attracting power than the other.

✓ A covalent bond will form if the electron-attracting powers of the bonding partners are similar in strength.

3.2 IONIC BONDS

The chemical reactivity of the elements depends, then, on their different tendencies to gain, lose, or share the electrons in their valence shells. After reaction, the valence shells have the configuration characteristic of a noble gas—they have acquired a complete octet.

Lewis Symbols and Formulas of Ionic Compounds

Lewis symbols were used to illustrate these changes in Example 2.14, where we observed that octet formation in the reaction between a metal, lithium, and a nonmetal, fluorine, created electrically charged atoms called ions, specifically lithium ions and fluoride ions. Let's extend that approach to metals in Groups II and III of the periodic table. We will use Lewis symbols to analyze the reactions between the metals lithium, magnesium, and aluminum and the nonmetal fluorine.

The group numbers of these elements tell us the number of electrons in the valence shell. Lithium, in Group I, has one electron in its valence shell; magnesium, in Group II, has two electrons in its valence shell; and aluminum, in Group III, has three electrons in its valence shell. Fluorine, a nonmetal, is in Group VII and has seven electrons in its valence shell. The convention used to describe ionic charge is to write the number first and the sign of the charge second.

3.1 CHEMISTRY IN DEPTH

On the Intensity of Electrical Fields

We have seen that ionization energy increases from left to right across the periodic table and decreases from top to bottom. The key to understanding this pattern is the variation, from element to element, in the attraction that the nucleus exercises over the valence electrons. Many of the properties of molecules depend on the intensity of the electrical field surrounding their component atoms, and so it is worthwhile to consider this phenomenon in more detail. To simplify the concept, we will use the intensity of light as a stand-in, or model, for the intensity of the field surrounding an electrical charge.

Imagine a library containing a circular reading table with a seating capacity of eight. The reading lamp in the center of the table is fitted with a bulb whose wattage can be varied and depends (magically) on the number of readers at the table. Each time a reader sits at the table, the lamp's intensity miraculously increases by 25 watts. When the first reader sits at the table, the bulb lights with an intensity of 25 watts. With the seating of the next reader, the lamp's intensity automatically increases by an additional 25 watts. Both readers experience the increase in intensity of light. Six additional readers join the others at the table, and, with each new reader, 25 additional watts are added to the reading lamp. Each time a new reader sits down, the already seated readers experience an increase in the lamp's intensity. The table is filled when eight readers are seated, each of whom experiences the maximum intensity of the lamp, now at 200 watts.

Unfortunately, the library has provided no other reading light. Therefore any new readers must sit in a circle outside the first. The light's intensity continues to increase by 25 watts for each new reader, but the new readers must sit at a greater distance from the source than the first eight readers.

The intensity of light from a source such as an electric light bulb decreases rapidly with distance. We have all experienced what happens as we move away from a light source or

move toward it. As we move toward the light, it becomes more intense. As we move away, the intensity of the light decreases. We experience the same effect with respect to a source of heat. The closer we get, the more intensely we feel the heat; the farther away, the less we feel the heat. These effects can be precisely expressed mathematically by a relation showing that the intensity of radiation such as light or heat (electromagnetic radiation) decreases with the square of the distance from the source of radiation:

$$\text{Intensity} \propto \frac{1}{(\text{distance})^2}$$

This relation tells us that, if the distance from the light source doubles, the intensity decreases by a factor of 4. The same relation holds for the intensity of electrical attraction. We will see it again in Chapter 10.

Because of their distance from the source, the intensity of light is less for those in the second circle than for those in the first, even though the intensity of the light source has increased. In fact, the light intensity experienced by the readers in the second circle is even lower than distance would predict because, in addition to the distance factor, the readers in the first circle obscure some of the light simply by getting in the way.

This decrease in intensity is precisely what electrons experience as they are added to a growing nucleus in the aufbau process. Instead of light intensity, however, it is the positive charge of the nucleus that reaches out to them, with diminishing effect. The electron that enters the fluorine atom to form a fluoride ion is the eighth "reader" sitting down at the table, and it experiences close to the maximum electrical attraction that any nucleus in the periodic table could exert over an additional electron. For this reason, fluorine's ionization energy is the greatest of all the elements (except the noble gases). At the other extreme is cesium, whose outer electron is very weakly held by its nucleus despite the fact that the nuclear charge is 55+. There are four "circles of readers" intervening between cesium's outer electron and its "reading lamp."

Lewis symbols (Section 2.10) depict the reaction between lithium and fluorine, with arrows indicating the transfer of electrons, as

$$\text{Li} \cdot \quad \cdot \ddot{\text{F}} : \longrightarrow \text{Li}^+ + : \ddot{\text{F}} :^-$$

The reaction between magnesium and fluorine is

$$\text{Mg} \quad \begin{matrix} \cdot \ddot{\text{F}} : \\ \cdot \ddot{\text{F}} : \end{matrix} \longrightarrow \text{Mg}^{2+} + \begin{matrix} : \ddot{\text{F}} :^- \\ : \ddot{\text{F}} :^- \end{matrix}$$

The reaction between aluminum and fluorine is

$$\text{Al} \begin{matrix} \cdot \ddot{\text{F}} : \\ \cdot \ddot{\text{F}} : \\ \cdot \ddot{\text{F}} : \end{matrix} \longrightarrow \text{Al}^{3+} + \begin{matrix} : \ddot{\text{F}} :^- \\ : \ddot{\text{F}} :^- \\ : \ddot{\text{F}} :^- \end{matrix}$$

In the reaction between lithium and fluorine, lithium needed to lose only one electron to achieve the helium electron configuration. (Helium's electron configuration is a duet rather than the octets of succeeding periods, but helium is a noble gas nonetheless.) Remember, the electron loss can occur only if the electron can be accepted by another atom. For that reason, the process is more properly described as **electron transfer** rather than electron loss. Fluorine required only one electron for octet formation in its valence shell. The electron transfer therefore required 1:1 ratio of reacting atoms, 1 Li to 1 F.

In the reaction between magnesium and fluorine, magnesium needed to lose two electrons to achieve the neon electron configuration. Fluorine again required only one electron for octet formation in its valence shell. The transfer of two electrons from magnesium therefore required acceptance of those two electrons by two fluorine atoms. This transfer resulted in a 1:2 ratio of reacting atoms, 1 Mg to 2 F.

In the reaction between aluminum and fluorine, aluminum needed to lose three electrons to achieve the neon electron configuration. Fluorine again required only one electron for octet formation in its valence shell. The transfer of three electrons from aluminum therefore required acceptance of those three electrons by three fluorine atoms. This transfer resulted in a 1:3 ratio of reacting atoms, 1 Al to 3 F.

Although no oxygen takes part, chemists consider the loss of electrons to be an oxidation and the gain in electrons a reduction.

A chemical compound is identified by its **formula,** which is a statement about the combining ratio of its constituent elements. In the formula, the symbols of all the elements in the compound are combined with subscripts that tell us the relative numbers of atoms of each element. When the formula of any ionic compound contains only one cation or anion, its subscript is left out, and the 1 is understood. Thus, when lithium reacts with fluorine, the 1:1 ratio of atoms in the resulting product is expressed as the formula of the newly formed compound, LiF. The 1:2 ratio of ions in the product of the magnesium–fluorine reaction is described in the formula MgF_2. Similarly, the 1:3 ratio of reacting aluminum and fluorine atoms leads to a product of formula AlF_3. Two-element compounds such as these are called **binary compounds.** The first column in the following table lists the relative numbers of atoms of each element as a ratio, and the second column lists the corresponding formulas of each of the three compounds.

Ratio of atoms in each compound	Formulas of each compound
1 Li:1 F	LiF
1 Mg:2 F	MgF_2
1 Al:3 F	AlF_3

When we look at the reaction of Group III metals with Group VII, Group VI, and Group V nonmetals, a similar pattern emerges. The first reaction is a repeat of the aluminum–fluorine reaction:

Look next at the reaction between aluminum and oxygen (Group VI):

Finally, examine the reaction between aluminum and nitrogen (Group V):

$$\text{Al} \cdot \overset{\cdot\cdot}{\underset{\cdot\cdot}{\text{N}}} : \longrightarrow \text{Al}^{3+} + : \overset{\cdot\cdot}{\underset{\cdot\cdot}{\text{N}}} :^{3-}$$

Aluminum must lose three electrons to achieve the neon electron configuration. It therefore requires partners that can accept those three electrons. We have already seen how the transfer takes place in the aluminum–fluorine reaction. In the aluminum–oxygen reaction, oxygen can accept only two electrons to complete its octet, and so the electron exchange requires that three oxygen atoms react with two aluminum atoms. Nitrogen (Group V), however, requires three electrons to complete its octet, and aluminum needs to lose three electrons to achieve noble-gas configuration; so, when these two elements react, a 1:1 exchange takes place. The first column in the following table lists the relative numbers of atoms of each element as a ratio, and the second column lists the corresponding formulas of each of the three compounds.

Ratio of atoms in each compound	Formulas of each compound
1 Al:3 F	AlF_3
2 Al:3 O	Al_2O_3
1 Al:1 N	AlN

Formulas of Ionic Compounds

To predict the formula of a binary ionic compound, one first needs to know the ionic charges of all component elements.

Example 3.1 Predicting the electrical charges of ions

What are the electrical charges of ions formed from elements of Groups I, II, III, V, VI, and VII?

Solution

Remember that the elements in Groups I, II, and III are metals. They lose electrons from their valence shells to form cations, whose positive charge is equal to their group number. In contrast, elements in Groups V, VI, and VII gain electrons to form anions, whose charge is equal to the group number minus 8:

Group number	I	II	III	V	VI	VII
Ion charge	1+	2+	3+	3−	2−	1−

Problem 3.1 Predict the charge of ions formed from the following elements: (a) potassium; (b) calcium; (c) indium; (d) phosphorus; (e) sulfur; (f) bromine.

The next step in determining the formulas of binary ionic compounds is to satisfy the requirement that the net electrical charge of an ionic compound must be zero. In other words, the magnitude of the total positive charge must be equal to the magnitude of the total negative charge. Therefore, once we know the charges of the individual ions, we use that information to calculate how many of each ion are needed to achieve a net, or overall, charge of zero. We call it the **net-charge approach** to determine the formulas of ionic compounds. Dot diagrams allowed us to find the answers by visual examination, but there are less-cumbersome ways to do it as well. Take another look at the aluminum–oxygen reaction.

In Example 3.1, we determined that aluminum forms ions of 3+, and oxygen of 2−. To find a combination of aluminum and oxygen ions in which the net charge is zero, we can multiply those charge numbers by factors that will make the total positive and total negative charges equal. If we multiply the charge on the aluminum ion by 2, and the charge on the oxygen ion by 3, the positive and negative charges are the same, and the net charge is zero.

$$Al^{3+} \times 2 = 6+$$
$$O^{2-} \times 3 = 6-$$

We take the factors that were used to multiply each elemental ion and insert them into the formula as subscripts of those same elements. The formula for the ionic compound formed by aluminum and oxygen is therefore Al_2O_3.

You can take the guesswork out of finding the right multipliers by simply using the number of the cation charge as the anion subscript and using the number of the anion charge as the cation subscript. This shortcut has been called the **cross-over approach.** For the aluminum–oxygen compound,

$$Al^{3+} \diagdown O^{2-} \longrightarrow Al_2O_3$$

When the ions have the same charge coefficient (for example, Ca^{2+} and O^{2-}), a literal use of the cross-over approach would yield the formula Ca_2O_2. Therefore, when the charge values are the same, assume a 1:1 ratio of ions, and the formula for calcium oxide is CaO.

Example 3.2 Predicting the formulas of ionic compounds by using the net-charge approach

Use the net-charge approach to predict the formulas for the ionic compounds formed from (a) Li and Cl; (b) Al and I; (c) Mg and N.

Solution
Use the elements' group numbers to discover their ionic charges. Multiply these ionic charges by factors that result in a net charge of zero for each compound.

(a) $Li^+ \times 1 = 1+$
    $Cl^- \times 1 = 1-$
    Formula: LiCl

(b) $Al^{3+} \times 1 = 3+$
    $I^- \times 3 = 3-$
    Formula: AlI_3

(c) $Mg^{2+} \times 3 = 6+$
    $N^{3-} \times 2 = 6-$
    Formula: Mg_3N_2

Problem 3.2 Use the net-charge approach to predict the formulas for the ionic compounds formed from (a) K and Br; (b) Ga and F; (c) Ca and P.

Example 3.3 Predicting the formulas of ionic compounds by using the cross-over approach

Use the cross-over approach to predict the formulas of the ionic compounds formed from (a) Na and I; (b) Ca and N; (c) Al and S.

Solution
Use the elements' group numbers to discover their ionic charges. Use the number of the charge on one element as the other element's subscript.

(a) $Na^+ \diagdown I^- \longrightarrow Na_1I_1 \longrightarrow NaI$

(b) $Ca^{2+} \diagdown N^{3-} \longrightarrow Ca_3N_2$

(c) $Al^{3+} \diagdown S^{2-} \longrightarrow Al_2S_3$

Problem 3.3 Use the cross-over approach to predict the formulas of the ionic compounds formed from (a) K and Cl; (b) Mg and P; (c) Ga and O.

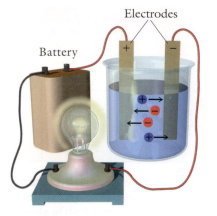

Electrodes

Battery

Figure 3.3 Diagram of an electrical conductivity cell. It consists of a vessel containing pure water (to which experimental substances can be added); a pair of electrodes (one negative, one positive) attached to an electric power source; and an electric light bulb that lights up when an electric current is able to flow through the medium between the electrodes.

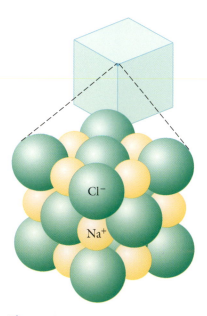

Cl⁻

Na⁺

Figure 3.4 A drawing of the crystal structure of sodium chloride. Because each sodium ion is surrounded by six chloride ions and each chloride ion is surrounded by six sodium ions, it is not possible to identify a unique structural unit.

Solution

Find the polyatomic ion from each compound in Table 3.2 and note its formula and ionic charge:

Acetate	$C_2H_3O_2$	$1-$
Ammonium	NH_4	$1+$
Phosphate	PO_4	$3-$
Hydrogen phosphate	HPO_4	$2-$
Dihydrogen phosphate	H_2PO_4	$1-$

Remember that sodium and potassium, both in Group I, form cations of charge $1+$; and calcium, in Group II, forms a cation of charge $2+$. Using Figure 3.2 or the methods illustrated in Examples 3.1 and 3.2, we therefore determine the formulas of these compounds to be (a) $NaC_2H_3O_2$; (b) $(NH_4)_3PO_4$; (c) K_2HPO_4; (d) $Ca(H_2PO_4)_2$.

Problem 3.5 Write the names of the following compounds: (a) NH_4HSO_3; (b) $CaCO_3$; (c) $Mg(CN)_2$; (d) $KHCO_3$; (e) $(NH_4)_2SO_4$.

3.5 DOES THE FORMULA OF AN IONIC COMPOUND DESCRIBE ITS STRUCTURE?

Figure 3.3 represents an experimental arrangement called a conduction cell. It consists of a vessel containing a medium that will not conduct electricity—for example, pure water; a pair of electrodes, one negative and the other positive, attached to an electrical power source; and a device that can reveal the flow of an electric current through the cell. This device can be as simple as an electric light bulb that will light up whenever an electric current flows through the medium between the electrodes.

Ionic solids do not conduct electricity; neither does pure water. However, when ionic solids, such as sodium chloride, are dissolved in water or melted, the solutions and the melted solids do conduct electricity. For this reason, ionic compounds are called **electrolytes.** Water—and compounds such as glucose, which when dissolved in water do not conduct electricity—are called **nonelectrolytes.**

In their solid form, the ions of an ionic compound are locked into place and all their charges are neutralized. Only when the solid melts or is dissolved in water, thereby freeing the individual ions, can the ions carry an electric current by moving to an electrode of opposite charge. These phenomena reveal that ionic bonds exist only in the solid state. Once the solid is melted or dissolved in water, the constituent ions become virtually independent of one another.

That independence explains why we can speak of sodium or potassium cations in blood or tissue without referring to a corresponding anion. The sodium and potassium ions, not their compounds, are of primary importance in their physiological function. Table 3.3 lists several of the cations found in blood. Their cationic charges are balanced by the presence of chloride ions, bicarbonate ions, and proteins, which normally have a negative ionic charge in blood and tissue.

Ionic solids consist of three-dimensional arrays of positive ions surrounded by negative ions and negative ions surrounded by positive ions. This arrangement makes it impossible to identify a discrete structural unit that can be represented by the formula of the solid (Figure 3.4). There is no such structural unit as a molecule for an ionic compound.

TABLE 3.3	Cations Found in Blood and Tissue	
Element	Cation	Biological function
sodium	Na^+	extracellular water, electrolyte, acid–base balance
potassium	K^+	intracellular water, electrolyte, acid–base balance
calcium	Ca^{2+}	bones, teeth, muscle activity
magnesium	Mg^{2+}	essential factor for the function of many enzymes
zinc	Zn^{2+}	essential factor for the function of many enzymes
iron	Fe^{2+}, Fe^{3+}	essential factor for hemoglobin, oxidative enzymes
copper	Cu^{2+}	synthesis of heme proteins, oxidative enzymes

When we consider the nature of covalent bonds, you will see that this distinction becomes very important. The formulas of compounds such as water (H_2O), carbon dioxide (CO_2), and oxygen (O_2), whose chemical bonds are covalent, represent not only the combining ratios of component elements, but also the actual, ultimate structural units that we can call molecules.

Concept checklist

✓ The formula of an ionic compound represents only the combining ratios of the component ions, not a unique structural unit.

✓ In contrast with ionic solids, the chemical identities of molecules, such as water, are retained whether they are in the solid, liquid, or gaseous state.

A PICTURE OF HEALTH
Examples of Molecules and Ions in the Body

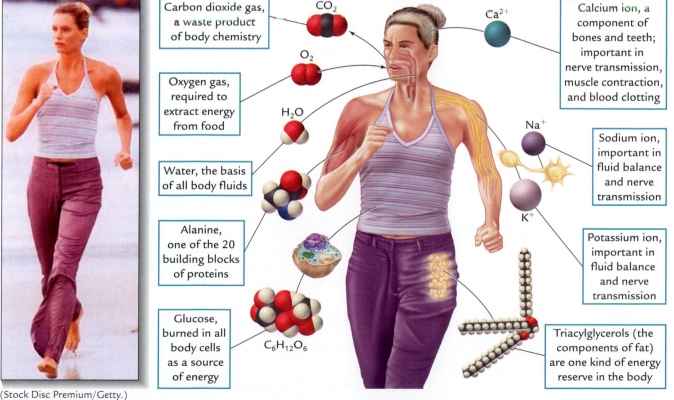

Carbon dioxide gas, a waste product of body chemistry — CO_2

Oxygen gas, required to extract energy from food — O_2

Water, the basis of all body fluids — H_2O

Alanine, one of the 20 building blocks of proteins

Glucose, burned in all body cells as a source of energy — $C_6H_{12}O_6$

Calcium ion, a component of bones and teeth; important in nerve transmission, muscle contraction, and blood clotting — Ca^{2+}

Sodium ion, important in fluid balance and nerve transmission — Na^+

Potassium ion, important in fluid balance and nerve transmission — K^+

Triacylglycerols (the components of fat) are one kind of energy reserve in the body

(Stock Disc Premium/Getty.)

3.6 COVALENT COMPOUNDS AND THEIR NOMENCLATURE

One of the simplest covalent bonds unites two hydrogen atoms to form the hydrogen molecule, H_2. Hydrogen atoms possess only one electron, and they can achieve the noble-gas configuration of helium by gaining one electron more. The helium configuration calls for a $1s$ orbital containing two electrons. Hund's rule tells us that two electrons can share an orbital only if the electrons have opposite spins. Therefore, a single covalent bond uniting two atoms consists of two spin-paired electrons shared by both atoms.

Before proceeding to a detailed discussion of covalent bonds, let's first consider some aspects of the system used for naming covalent compounds. As with ionic compounds, we will find that the names themselves provide considerable information on which to base a first acquaintance.

Most of the compounds that we shall refer to in this chapter—such as ammonia, carbon dioxide, and water—have common names with which you are already familiar, names adopted before the development of systematic methods of nomenclature, and those are the names that we shall be using for them. However, should an unfamiliar substance be referred to, a systematic method is available to us.

The systematic method for naming simple, binary molecular compounds is based on the fact that some pairs of elements can form more than one covalently bonded compound. Two such pairs, for example, are (1) nitrogen and oxygen and (2) carbon and oxygen. The system adopted for naming their compounds uses Greek prefixes (see the table in the margin) and is illustrated in Table 3.4.

GREEK PREFIXES IN CHEMICAL NOMENCLATURE	
1	mono-
2	di-
3	tri-
4	tetra-
5	penta-
6	hexa-
7	hepta-
8	octa-
9	nona-
10	deca-

TABLE 3.4 Systematic Names of Covalent Compounds

Nitrogen and oxygen compound	Systematic name	Carbon and oxygen compound	Systematic name
N_2O	dinitrogen monoxide	CO	carbon monoxide
NO	nitrogen monoxide	C_2O_3	dicarbon trioxide
NO_2	nitrogen dioxide	CO_2	carbon dioxide
N_2O_3	dinitrogen trioxide		
N_2O_4	dinitrogen tetroxide		
N_2O_5	dinitrogen pentoxide		

You can take your first clue concerning a covalent compound's name from the order in which its elements appear in the chemical formula: the first element in the formula is named first; the second element is named second, but the name is altered to accommodate the suffix **-ide** (as in the naming of simple ionic compounds). The Greek prefixes are added to the elements' names to indicate the numbers of different kinds of atoms in each molecule of the compound. Note, however, that NO is called nitrogen monoxide, not mononitrogen monoxide, and CO is called carbon monoxide, not monocarbon monoxide. The prefix **mono-** is never used for naming the first element.

Example 3.6 Naming covalent compounds

Name the following compounds: (a) CCl_4; (b) PCl_3; (c) PCl_5; (d) SO_2.

Solution

(a) Carbon tetrachloride; (b) phosphorus trichloride; (c) phosphorus pentachloride; (d) sulfur dioxide.

Problem 3.6   Name the following compounds: (a) NI_3; (b) P_2O_5; (c) S_2Cl_2; (d) SO_3.

3.7   REPRESENTATION OF COVALENT BONDS

Chemists use a number of different notation methods to represent covalent bonds and the ways in which they are organized within molecules. One such method is electron-dot notation. It was used in Chapter 2 to represent the structure of the valence shells of the elements, but it can be effectively adapted for representing covalent bonds as well.

When electron-dot notation is used to demonstrate the structure of molecules, the resultant diagrams are called **Lewis structures.** These diagrams are named after their originator, Gilbert N. Lewis, who used them to predict how atoms are connected to one another within a given molecule. The key concept in building Lewis structures lies in satisfying the octet rule, the rule stating that an element achieves stability by filling its outer shell with eight electrons. However, recall that hydrogen forms a duet to achieve stability.

When an element reacts to form covalent bonds, the number of bonds formed to complete its octet is called its **combining power.** For example, carbon, in Group IV, requires four additional electrons to fill its octet. It can do so by forming four covalent bonds with hydrogen to form methane, CH_4, or with chlorine to form carbon tetrachloride, CCl_4. Nitrogen and other elements in Group V require three additional electrons and can form three covalent bonds. The Group VI elements, such as oxygen and sulfur, require two additional electrons and can form two bonds to complete their octets. Group IV elements are called tetravalent; Group V elements, trivalent; Group VI elements, divalent; and Group VII elements, monovalent (the Greek prefixes again). Table 3.5 contains nonmetals from Groups IV through VII and indicates their bonding requirements for octet formation.

TABLE 3.5	Bonding Requirements for Octet Formation of Some Nonmetals*			
	Group			
IV	V	VI	VII
$-C-$	$-N-$	$-O-$	F—
	$-P-$	$-S-$	Cl—
			Br—
			I—

*The lines indicate the number of bonds needed for octet formation.

Example 3.7   Predicting formulas of simple molecular compounds

For each of the following pairs of elements, predict the formula of the product of the reaction between the two elements. The order of the elements in each formula is that given for each pair of elements in the example. (a) Carbon and bromine; (b) nitrogen and iodine; (c) carbon and sulfur.

Solution

Refer to Table 3.5 for the bonding requirements (combining power) of nonmetals.

(a) Carbon is tetravalent and bromine is monovalent; therefore, carbon will react with four bromine atoms, and the compound formed is CBr_4.

(b) Nitrogen is trivalent and iodine is monovalent; therefore, nitrogen will react with three iodine atoms, and the compound formed is NI_3.

(c) Carbon is tetravalent and sulfur is divalent; therefore, carbon will react with two sulfur atoms, and the compound formed is CS_2.

Problem 3.7   For each of the following pairs of elements, predict the formula for the product of the reaction between the two elements. The order of the elements in each formula is that given for each pair of elements in the problem. (a) Carbon and hydrogen; (b) phosphorus and chlorine; (c) carbon and oxygen.

Let's try out the Lewis method of representing molecular structure by using it to examine how two fluorine atoms react to form a diatomic molecule. If each fluorine atom shares one electron with the other, then, in effect, they will both have eight. In this way, each fluorine atom can achieve the stable electron configuration of neon. Note that the result of their bonding is the sharing of a pair of electrons between them:

$$:\ddot{F}\cdot + \cdot\ddot{F}: \longrightarrow :\ddot{F}(:)\ddot{F}:$$

On the left side of the arrow, the reacting atoms are represented with numbers of valence electrons equal to their group number. On the right side of the arrow, we see the fluorine molecule, with each atom surrounded by a completed octet. (Overlapping circles have been drawn around the bonded fluorine atoms to indicate that the shared pair of electrons is part of each atom's octet.)

The Lewis structure (the molecule drawn with Lewis symbols) is simplified by representing the bonding pair of electrons as a line between the atoms and the other pairs as dots surrounding the atoms: $:\ddot{F}—\ddot{F}:$. The pairs of electrons not shared in the covalent bond are called **nonbonded electrons** or **lone pairs.** Lewis structures can be further simplified by leaving out all nonbonded electrons, in which case they are called **structural formulas,** F—F.

It is important to recognize that hydrogen is one of several exceptions to the octet rule. Hydrogen requires only two electrons, or a **duet,** to achieve the nearest noble-gas configuration, helium. For this reason, hydrogen will always form a covalent bond to only one other atom and will therefore never be found between two other atoms in Lewis structures. In the formation of the hydrogen molecule, each hydrogen atom achieves the helium configuration by sharing the other's single electron.

Other exceptions to the octet rule are encountered with beryllium and boron, which form compounds, such as BeH_2 and BF_3, that contain four and six electrons, respectively, in the central atoms' completed valence shells. However, the octet rule applies to all the other elements in the first and second periods of the periodic table, the principal focus of our interest.

Drawing Lewis Structures

The derivation of a Lewis structure requires (1) knowledge of a compound's molecular formula and (2) a set of simple rules (illustrated in Example 3.8). The **molecular formula** (Chapter 4) tells us the number of atoms for each element in a molecule of the compound. It does not tell us, as the Lewis structure does, how they are connected to one another.

Example 3.8 Drawing Lewis structures: I

Construct the Lewis structure for ammonia, NH_3.

Solution

Step 1. Place the symbols for the bonded atoms into an arrangement that will allow you to begin distributing electrons. In ammonia (as well as in many other compounds), there is only one atom of one type (here, N) and several of another (here, three H atoms); so as a first approximation choose N as the central atom to which the rest are bonded.

$$\begin{array}{ccc} & H & \\ H & N & H \end{array}$$

Step 2. Determine the total number of valence electrons in the molecule. To do so, add up the valence electrons contributed by each one of the atoms in the molecule.

Element	Valence electrons	Atoms per molecule	Number of electrons
nitrogen	5	1	5
hydrogen	1	3	3
		Total valence electrons available =	8

Step 3. Represent shared pairs of electrons by drawing a line between the bonded atoms. For ammonia,

Each line (bond) represents two electrons, accounting for six of the eight available valence electrons.

Step 4. Position the remaining valence electrons (two in this case) as lone pairs to satisfy the octet rule for each atom in the compound (remember that hydrogen's "octet" consists of only two electrons).

$$\begin{array}{c} H \\ | \\ H-\underset{..}{N}-H \end{array}$$

Problem 3.8 Construct the Lewis structure for phosphorus trichloride, PCl_3.

Example 3.9 Drawing Lewis structures: II

Construct the Lewis structure for water, H_2O.

Solution

Step 1. Position the bonded atoms in a likely arrangement. In water, there is only one atom of oxygen and two of hydrogen, and so a good guess is that oxygen is the central atom to which the others are bonded:

$$\begin{array}{ccc} H & O & H \end{array}$$

Step 2. Determine the total number of valence electrons in the molecule by adding up the valence electrons contributed by each of the atoms in the molecule.

Element	Valence electrons	Atoms per molecule	Number of electrons
oxygen	6	1	6
hydrogen	1	2	2
		Total valence electrons available =	8

Step 3. Represent electron-pair bonds by drawing a line between bonded atoms. For water:

$$H\!\!-\!\!O\!\!-\!\!H$$

This structure accounts for four of the eight available valence electrons.

Step 4. Position the remaining four valence electrons as lone pairs around the atoms to satisfy the octet rule for each atom:

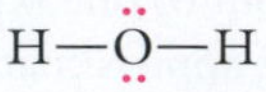

$$H\!\!-\!\!\overset{..}{\underset{..}{O}}\!\!-\!\!H$$

Problem 3.9   Construct the Lewis structure for sulfur dichloride, SCl_2.

The next step in learning to use Lewis structures is to diagram a compound in which there is no unique central atom.

Example 3.10   **Writing Lewis structures for more-complex compounds: I**

Write the Lewis structure for ethane, C_2H_6. Ethane is one of a large family of compounds called **alkanes,** also called hydrocarbons because they contain only the elements hydrogen and carbon. Remember that hydrogen cannot be inserted between the two carbon atoms, because it is not capable of participating in more than one bond at a time.

Solution

Step 1. Place the carbon atoms so that they are joined together, with the hydrogen atoms in terminal positions:

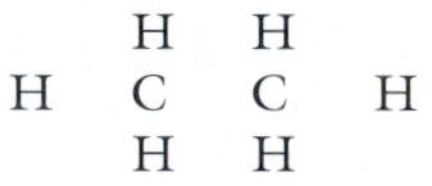

$$
\begin{array}{ccc}
 & H & H \\
H & C & C & H \\
 & H & H
\end{array}
$$

Step 2. Calculate the total number of valence electrons.

Element	Valence electrons	Atoms per molecule	Number of electrons
carbon	4	2	8
hydrogen	1	6	6
		Total valence electrons available =	14

Step 3. All these electrons are accommodated by drawing the seven bonds required to connect all the atoms of the molecule.

Step 4. Because there are no lone pairs, the final structure is

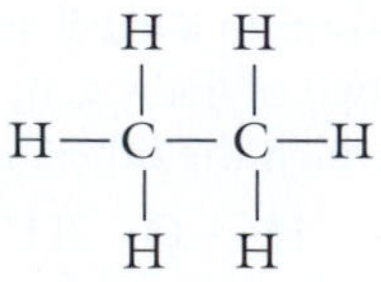

$$
\begin{array}{ccc}
 & H & H \\
 & | & | \\
H\!\!-\!\!&C\!\!-\!\!&C\!\!-\!\!H \\
 & | & | \\
 & H & H
\end{array}
$$

Problem 3.10   Write the Lewis structure for C_3H_8.

Lewis Structures Containing Multiple Bonds

Some interesting cases arise when there appear to be too few electrons to satisfy the octet rule. This apparent deficiency occurs when we try to write the Lewis structures for compounds such as ethylene (C_2H_4), acetylene (C_2H_2), nitrogen (N_2), carbon dioxide (CO_2), and hydrogen cyanide (HCN).

Example 3.11   **Writing Lewis structures for compounds with multiple bonds: I**

Write the Lewis structure for ethylene, C_2H_4.

Solution

Step 1. As we guessed with ethane, we can assume that the carbon atoms are joined and the hydrogen atoms occupy terminal positions:

$$\begin{array}{ccc} & \text{H} & \text{H} \\ \text{H} & \text{C} & \text{C} & \text{H} \end{array}$$

Step 2. Calculate the total number of valence electrons.

Element	Valence electrons	Atoms per molecule	Number of electrons
carbon	4	2	8
hydrogen	1	4	4
		Total valence electrons available =	12

Step 3. We can insert 10 of the 12 electrons by drawing five bonds to connect all the atoms, with the result that 2 electrons are left over:

$$\begin{array}{ccc} & \text{H} & \text{H} \\ & | & | \\ \text{H}—\text{C}—\text{C}—\text{H} \end{array}$$

Step 4. The hydrogen atom duets are satisfied. However, if we are to satisfy the octet rule, two pairs of electrons are needed for each carbon atom. That would seem to mean that 4 more electrons are required, but only 2 electrons are available. This apparent dilemma can be solved by adding the 2 remaining electrons as an additional bonding pair between the carbon atoms, to form a **double bond.** The shared double bond provides an octet around each carbon atom:

$$\begin{array}{ccc} & \text{H} & \text{H} \\ & | & | \\ \text{H}—\text{C}=\text{C}—\text{H} \end{array}$$

Problem 3.11   Write the Lewis structure for carbon dioxide, CO_2.

Example 3.12   **Writing Lewis structures for compounds with multiple bonds: II**

Write the Lewis structure for nitrogen, N_2.

Solution

Step 1. The gaseous element nitrogen is a diatomic molecule whose structure consists of two nitrogen atoms joined by a covalent bond.

$$\text{N}—\text{N}$$

Step 2. Calculate the total number of valence electrons.

Element	Valence electrons	Atoms per molecule	Number of electrons
nitrogen	5	2	10
		Total valence electrons available =	10

Step 3. When we attempt to place one bond between the atoms and add the remaining electrons as lone pairs on each nitrogen atom,

$$:\ddot{\text{N}}—\ddot{\text{N}}:$$

we find that we are short two pairs of electrons for the necessary octets.

Step 4. However, we can move two of the lone pairs to a position between the nitrogen atoms to form two more bonds; then we add the remaining four electrons as lone pairs on each of the atoms to obtain

$$:N\equiv N:$$

Three pairs of electrons are shared to complete the octets, and the bond between the nitrogen atoms is called a **triple bond.**

Problem 3.12   Write the Lewis structure for the rocket fuel hydrazine, N_2H_4.

When you begin your study of organic chemistry, you will find that many important organic molecules contain single, double, or triple bonds. Your vision depends on the properties of a double bond. Methods for the characterization of multiple bonds are described in Box 3.2.

3.8   LEWIS STRUCTURES OF POLYATOMIC IONS

Lewis structures are also used to depict polyatomic ions, in much the same way as we used them in the preceding worked examples. However, in calculating the total number of valence electrons, the charge on the ion must be taken into account. If it is a negative ion, electrons must be added; if it is a positive ion, electrons must be subtracted. For example, the ammonium ion possesses a charge of $1+$, and the carbonate ion has a charge of $2-$, and so the total number of valence electrons is calculated:

Ion	Total number of valence electrons
NH_4^+	$N\,(1 \times 5) + H\,(4 \times 1) - 1 = 8$
CO_3^{2-}	$C\,(1 \times 4) + O\,(3 \times 6) + 2 = 24$

With reference to preceding examples for details, the Lewis structures of the ammonium ion and the carbonate ion are:

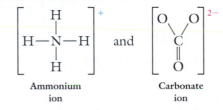

Ammonium   Carbonate
ion   ion

Polyatomic ions do not have their origin in the direct transfer of electrons from element to element. It is therefore important to recognize that ions can be formed in other types of chemical reactions.

Most of the polyatomic ions of interest to us arise from reactions taking place in aqueous solutions—mixtures consisting mostly of water, with other substances dissolved in it. All the polyatomic ions listed in Table 3.2 are created by reactions in water, when a dissolved substance reacts either with the water itself or with some other dissolved substance. For example, two substances, HCl (hydrogen chloride) and NH_3 (ammonia), react with water in the following ways:

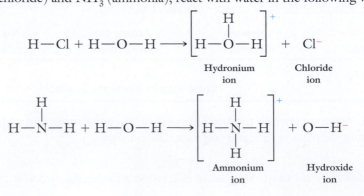

3.2 CHEMISTRY IN DEPTH

Molecular Absorption Spectra and Molecular Structure

Other aspects of molecular structure can be clarified by using radiation of lower energy than X-rays. In Section 2.7, we learned that the absorption of particular frequencies of electromagnetic energy causes changes in atomic energy states. Molecules as well as atoms possess unique energy states, and these energy states generally correspond to particular molecular structural features. Just as elements can be detected and quantified by atomic absorption spectra, molecules and parts of molecules can be characterized by molecular absorption spectra.

In your study of organic chemistry, you will learn about molecules that possess a number of multiple bonds. The electrons in those types of molecules possess energy states corresponding to frequencies in the ultraviolet and visible spectrum (see Table 2.4). Other molecular energy states correspond to the motions of atoms attached to other atoms—that is, to vibrations within molecules. These energy states are excited by infrared frequencies. A technique called nuclear magnetic resonance depends on the response of nuclear energy states to radiofrequencies. Nuclear magnetic resonance is of primary importance in the elucidation of complex molecular structures and has become an important tool in medical diagnosis, where it goes by the name of MRI (Chapter 10).

The application of these techniques requires the ability to generate the frequencies of interest and the ability to detect changes in the energy of the beam emerging from a sample. The absorption of ultraviolet, visible, infrared, and radiofrequencies is in daily use in all modern laboratories engaged in the characterization and chemical analysis of molecular structure. Some of these techniques are useful in the analysis of organic compounds.

In the first reaction, the product H_3O^+ ion, called the hydronium ion, is the result of the transfer of a hydrogen ion (H^+, not an electron) to water. This reaction always takes place when compounds called acids react with water. In the second reaction, a hydrogen ion (not an electron) is transferred from water to ammonia with the resultant formation of the hydroxide ion, which, in water, forms a basic solution. In each case, the newly formed covalent bond is created from a lone pair contributed by the central atom and is called a **coordinate covalent bond.** Once formed, such a bond cannot be differentiated from any of the other covalent bonds. Note carefully that the bonds uniting all atoms in these polyatomic ions are covalent.

We will explore the details of the chemistry of acids and bases in Chapter 9.

3.9 POLAR AND NONPOLAR COVALENT BONDS

When an electron pair is shared by two atoms of the same element, as in molecules such as H_2 and F_2, the electrons are attracted equally to both atomic nuclei and are found midway between them. Other examples are O_2, N_2, and Cl_2. This type of covalent bond is called a **nonpolar covalent bond.** When molecules consist of different elements, however, such as the molecules NO and HCl, one atom participating in the bond is likely to exert a greater attraction for the electrons than the other atom. In these cases, the electrons tend to reside closer to the nucleus with the greater ability to attract them. A bond of this type is called a **polar covalent bond.**

Actually, polar bonds are distributed along a continuum. At one extreme of this continuum is the ionic bond, in which both electrons reside totally on one of the ions of a cation–anion pair. At the other extreme is the pure covalent bond, in which the electron pair spends most of its time halfway between the bonded atoms. Most covalent chemical bonds are polar, however, and lie somewhere between these two extremes.

The ability of an atom within a molecule to draw electrons toward itself is called its **electronegativity.** Figure 3.5 on the next page provides an abbreviated list of electronegativities of various elements. Electronegativity is used to determine the order of elements in the name of a binary compound. That is why the name of the binary compounds formed between oxygen and nitrogen and

Figure 3.5 An abbreviated table of electronegativities.

H 2.2								He
Li 0.98	Be 1.6		B 2.0	C 2.6	N 3.0	O 3.4	F 4.0	Ne
Na 0.93	Mg 1.2		Al 1.6	Si 1.9	P 2.2	S 2.6	Cl 3.2	Ar
K 0.88	Ca 1.0		Ga 1.8	Ge 2.0	As 2.2	Se 2.6	Br 2.8	Kr
Rb 0.82	Sr 0.95		In 1.8	Sn 2.0	Sb 1.9	Te 2.1	I 2.7	Xe
Cs 0.79	Ba 0.89		Tl 1.8	Pb 1.9	Bi 1.9	Po 2.0	At 2.2	Rn

between carbon and oxygen are called NO and CO. The less-electronegative element comes first in the compound's name. Notice that the electronegativities of fluorine, oxygen, and nitrogen are among the largest of all the elements. This fact will become important when we consider weak interactions between molecules in Chapter 6. The electronegativity values of the main-group elements in Figure 3.5 are plotted three-dimensionally in Figure 3.6. This plot gives a clearer picture of the increase in electronegativity in the procession from cesium in the lower left to fluorine in the upper right of the periodic table.

A useful (though not infallible) rule of thumb is that, if the difference between the electronegativities of two elements is greater than 2.0, the bond between their atoms will probably be ionic; if the difference is less than 1.5, the bond will probably be covalent. This rule of thumb is demonstrated schematically in Figure 3.7. Using the values listed in Figure 3.5, we find the difference in electronegativities of lithium and fluorine to be 3.0 and may conclude that the LiF bond is probably ionic. The electronegativity difference

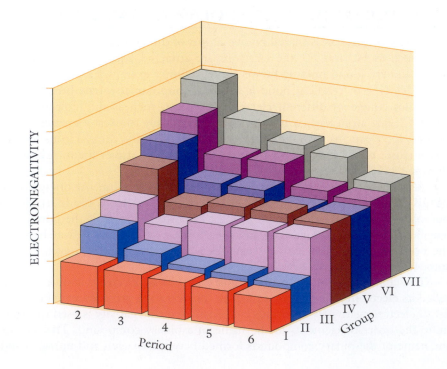

Figure 3.6 A three-dimensional plot of the electronegativity values of the main-group elements in Figure 3.5. Note the general increase in values from the minimum at cesium (lower right) to the maximum at fluorine (upper left).

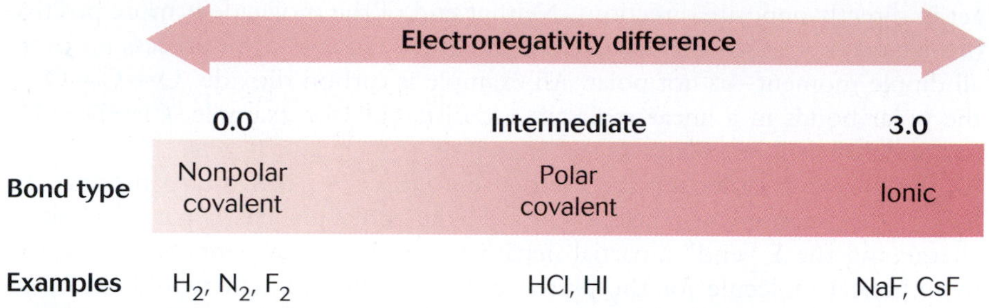

Electronegativity difference				
	0.0		Intermediate	3.0
Bond type	Nonpolar covalent		Polar covalent	Ionic
Examples	H_2, N_2, F_2		HCl, HI	NaF, CsF

Figure 3.7 A continuum showing the effect of electronegativity difference on the type of bond formed between elements. At one extreme is the nonpolar covalent bond, which can form only between identical atoms (electronegativities are therefore identical). At the other extreme is the ionic bond, which forms when the difference between the elements' electronegativities exceeds about 2.0.

between boron and carbon is 0.5; therefore a bond between the atoms of those two elements will be covalent. When the difference in electronegativities is in the range of 2.0 to 1.5, the bond will be ionic if the compound is a metal–nonmetal combination and will be covalent if it is a nonmetal–nonmetal compound.

When the two bonded atoms of a binary molecule (a two-atom molecule) have different electronegativities, the electrical asymmetry that is produced causes one end of the molecule to possess a negative charge and the other end a positive charge. The electrical charges are not full charges; rather they are partial charges and are indicated by lowercase Greek symbols $\delta+$ or $\delta-$. The covalent bond of such a molecule is electrically polarized, and, for convenience, chemists call that kind of bond a polar covalent bond. The H—F molecule is a good example. When we want to show the existence of partial charges on its atoms, we write the formula $\overset{\delta+}{H}—\overset{\delta-}{F}$. Because the molecule possesses only two atoms, the polar covalent bond of HF makes it a **polar molecule** as well. It is important to note that, although the HF molecule is electrically asymmetrical, it is still electrically neutral overall, because the collective positive charges of the combined nuclei are exactly balanced by the collective negative charges of the combined electrons.

A polar molecule is attracted to a neighboring polar molecule, positive end to negative end. The electrical force causing the attraction is measured by a quantity called the **dipole moment.** The magnitude of the dipole moment of a molecule depends on the distance separating the partial charges within the molecule and on the difference in electronegativities of the combined atoms. The greater the distance of separation and the greater the difference in electronegativities, the larger the dipole moment. It is customarily represented by an arrow pointing along the bond from the positive end toward the negative end, $+\longrightarrow$; from the less-electronegative element to the more-electronegative element.

A molecule may contain two or more polar bonds and not be polar. Its overall shape may be such that the dipole moments cancel each other out. As an example, imagine a generalized molecule consisting of three atoms, of formula XY_2. In this molecule, X is the central atom with covalent bonds to each Y atom. Because the joined atoms X and Y are different, they must have different electronegativities. Therefore each individual bond in the molecule must be polar and must contribute to the overall polarity of the molecule. There are two possibilities for the molecular shape: linear and bent, as shown in the margin.

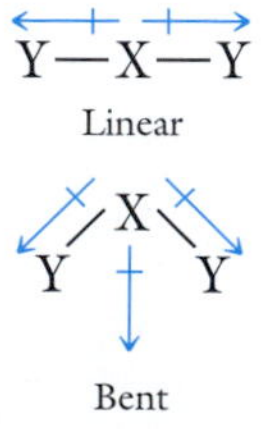

In the linear molecule, the two internal dipoles cancel each other's effect because the bonds are identical, and therefore they are equal in magnitude and

$$\overset{\longleftarrow + \quad + \longrightarrow}{Y—X—Y}$$
Linear

Bent

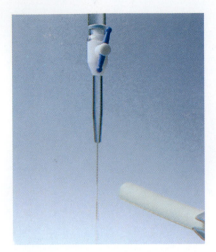

(a)

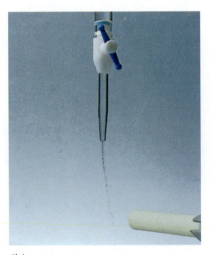

(b)

Figure 3.8 (a) The buret contains hexane, a nonpolar substance.
(b) The buret contains water, a polar substance. When the substances are allowed to trickle past an electrically charged rod, only the polar substance responds to the electrical field.
(W. H. Freeman photos by Ken Karp.)

act in directly opposite directions. Neither end of the molecule is more positive or negative than the other. Therefore the linear molecule will possess no overall dipole moment—is not polar. An example is carbon dioxide, $O{=}C{=}O$. If the polar bonds in a linear molecule are different (for example, $Cl{-}Be{-}F$), the molecule will be polar.

In the bent molecule, the internal dipoles act in concert to produce an overall dipole moment: the X "end" of the molecule has a partial positive charge and the Y "end" a partial negative one. An experiment that tested the hypothetical molecule for the presence of a dipole moment would therefore reveal whether the molecule was bent or straight. The absence of a dipole moment would mean that the molecule was linear, like CO_2. The presence of a dipole moment in a molecule with two identical bonds would mean the molecule was bent, like H_2O.

The consequences of molecular polarity can be seen primarily in physical properties. Molecules with large dipole moments tend to interact strongly with one another, as well as with other molecules possessing dipole moments. As a result, substances composed of polar molecules have higher melting and boiling points than do those composed of nonpolar molecules. At room temperature, methane, CH_4, which has no dipole moment, is a gas, but water, H_2O, which has a large dipole moment, is a liquid at the same temperature. Moreover, water, because of its polarity, is an excellent solvent for polar substances. Many of the physical and chemical properties of organic compounds depend on their overall molecular polarity, which is the result of the polarity of covalent chemical bonds within their molecules. Figure 3.8 shows how the polarity of a covalent bond can can be manifested in a substance's properties. A more thorough discussion of the effects of structure on physical properties will be found in Chapter 6.

3.10 THREE-DIMENSIONAL MOLECULAR STRUCTURES

Lewis structures provide us with two kinds of information about molecular structure:

- How atoms within a molecule are connected to one another
- Whether the connecting bonds are single, double, or triple

However, Lewis structures cannot describe the actual spatial arrangements of atoms within molecules. The important point to remember is that, although Lewis structures allow us to decide which atoms are connected and tell us about the nature of the bonds holding them together (single, double, or triple; polar or nonpolar), Lewis structures cannot reveal the three-dimensional arrangements of the atoms. It is the three-dimensional, or spatial, arrangement of atoms in a molecule that provides a molecule's unique characteristics. Those spatial arrangements are of central importance in all physiological functions, such as enzymatic activity, membrane transport, nerve transmission, and antigen–antibody interactions.

VSEPR Theory

A simple and very useful theory enables us to predict the shapes of both simple and complex molecules. The theory states that the shape of a molecule is affected by each of the valence-electron pairs surrounding a central atom, not just the pairs participating in bonds but also the nonbonded electron pairs.

The central idea of the theory is that, because electrons tend to repel one another, the mutual repulsion of all the electron pairs, bonded and nonbonded, forces the atoms within the molecule to take positions in space that minimize those mutual repulsions. In other words, the electrons strive to get as far away from one another as possible. The name of the method is the **valence-shell electron-pair repulsion theory,** or the **VSEPR theory.**

Because the most stable arrangement for the pairs of electrons in a molecule is to be as far away from one another as possible, they must occupy points in space that define symmetrical geometrical figures called regular polyhedra, as shown in Table 3.6. A linear molecule such as $BeCl_2$ (beryllium chloride) takes its linear shape so that the two pairs of electrons can be as far apart as possible. Boron trifluoride, BF_3, takes the shape of a planar, or flat, **equilateral triangle** for the same reason. The **tetrahedron** is a three-dimensional structure and is the result of the symmetrical arrangement of four pairs of electrons around a central atom. Each of the four pairs is as far from the others as possible (while still remaining attached to the central atom).

TABLE 3.6 **Arrangements of Electron Pairs Around a Central Atom Resulting in Minimum Repulsion Energy**

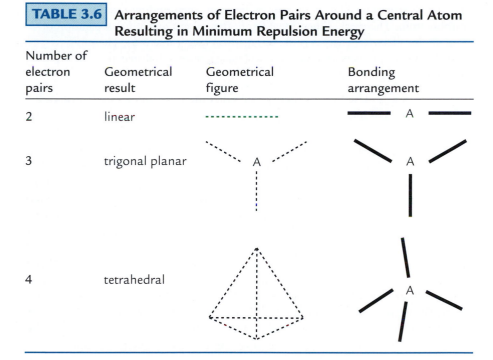

Number of electron pairs	Geometrical result	Geometrical figure	Bonding arrangement
2	linear		
3	trigonal planar		
4	tetrahedral		

To illustrate how these shapes emerge from considerations of the repulsion of the electron pairs of covalent bonds, let's apply VSEPR theory to predict the three-dimensional structure of methane, CH_4.

Example 3.13 **Using VSEPR theory to predict molecular structure: I**

Use VSEPR theory to predict the three-dimensional structure of methane, CH_4.

Solution

The most direct way to go about this task is to first determine the Lewis structure for methane.

Step 1. Most likely the carbon atom is central, with the hydrogen atoms terminal:

$$\begin{array}{ccc} & H & \\ H & C & H \\ & H & \end{array}$$

Step 2. Calculate the total number of valence electrons.

Element	Valence electrons	Atoms per molecule	Number of electrons
carbon	4	1	4
hydrogen	1	4	4
		Total valence electrons available =	8

Step 3. These electrons are all accommodated by drawing four single bonds between all atoms of the molecule.

Step 4. Because there are no lone pairs, the final structure is

The Lewis structure shows four pairs of electrons surrounding the central carbon atom. The three-dimensional figure that allows maximum separation of four similar electrical charges is the tetrahedron (see Table 3.6). Therefore, VSEPR theory predicts that methane will have a tetrahedral structure, as illustrated in Figure 3.9.

Problem 3.13   Use VSEPR theory to predict the three-dimensional structure of carbon tetrachloride, CCl_4.

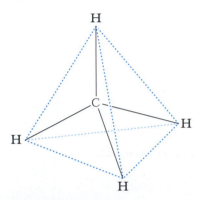

Figure 3.9 Methane has a tetrahedral molecular shape. The dotted lines indicate the outline of a regular tetrahedron, and the solid lines represent the C–H bonds.

A **bond angle** is the angle between two bonds that share a common atom. Figure 3.9 shows the carbon atom of CH_4 in its central location, with single bonds (separated by bond angles of 109.5°) connecting it to each of four hydrogen atoms. The dotted lines indicate the outline of the hypothetical solid figure enclosing the carbon atom, and the solid lines represent the carbon–hydrogen single bonds. Although each C–H bond is very slightly polar, the four identical bonds are arranged symmetrically; therefore there can be no overall molecular polarity. Experiments have confirmed that the structure of methane is tetrahedral and that the molecule has no dipole moment. X-ray diffraction is used to determine the structure of molecules. In this technique, X-rays are passed through crystals of a compound, and the pattern of emerging X-rays is recorded as in Figure 3.10. Mathematical treatment of the data locates the positions of all atoms within the structure—hence the molecule's shape.

Figure 3.10 The X-ray diffraction pattern produced by passing X-rays through a crystal of NaCl. The pattern of spots provides information regarding the arrangement of the ions within the crystal. (Hakôn Hope.)

Nonbonding Electrons

In contrast with the structure of methane, there are situations in which the arrangement of four electron pairs is not symmetrical. This occurs in molecules such as ammonia and water that possess nonbonded or lone pairs of electrons as part of the octet around the central atom. VSEPR theory explains this symmetry-altering effect of nonbonding electron pairs by pointing out that electron pairs in bonds are localized because they are fixed between two positively charged nuclei. Nonbonded pairs are not subjected to a similar localizing influence, because they are not fixed between two positively charged nuclei. As a result, they occupy a greater volume of space than a bonded pair does. In doing so, they push the bonded pairs closer to one another than the tetrahedral geometry would predict. Be careful not to confuse the shapes of molecules with bond angles. Even though the bond angles of ammonia and of water are all very close to the tetrahedral angle of methane, the molecular shapes of these compounds differ significantly. You can see that the shape of a molecule is primarily a function of the number of bonded atoms in the molecule, not of the bond angles. As an example, let's see what VSEPR theory predicts for the structure of ammonia.

Example 3.14 — Using VSEPR theory to predict molecular structure: II

Use VSEPR theory to predict the structure of the ammonia molecule.

Solution
The Lewis structure for ammonia, NH_3, was worked out in Example 3.8. As in methane, four pairs of electrons surround the central atom. VSEPR theory therefore predicts that the structure will be tetrahedral; but, in this case, one of the pairs is nonbonded, which alters the perfect symmetry exhibited by methane. Experiments have shown that the bonds between the central nitrogen atom and the hydrogens are separated by angles of 107°, rather than the 109.5° angles of a perfect tetrahedron.

We consider the shape of a molecule to be defined by the positions of the bonded atoms. Therefore, the ammonia molecule forms a pyramid (outlined by the blue dotted lines) with three hydrogen atoms forming the base. This shape is called **trigonal pyramidal** because the pyramid has a triangle for its base. The asymmetrical arrangement of bonds and one lone pair of electrons causes this molecule to have a strong dipole moment.

Problem 3.14   Use VSEPR theory to predict the structure of nitrogen triiodide, NI_3.

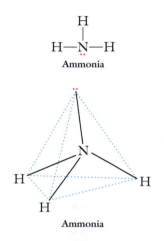

Ammonia

Ammonia

Example 3.15 — Using VSEPR theory to predict molecular structure: III

Use VSEPR theory to predict the three-dimensional structure of water, H_2O.

Solution
The Lewis structure for water, with its two lone pairs on the oxygen atom, was worked out in Example 3.9. From VSEPR theory, we know that the total of four electron pairs, two bonded and two nonbonded, surrounding the oxygen atom must be accommodated by a roughly tetrahedral configuration (outlined by blue dotted lines). However, because two of the pairs are nonbonding, they will occupy even more space around the central atom than did the one lone pair in ammonia. The result is that the two hydrogen atoms of water are squeezed even closer together than the hydrogen atoms in ammonia, with the two lone electron pairs directed at the other two corners of a distorted tetrahedron. The bond angle between the oxygen atom and the two hydrogen atoms has been determined experimentally to be 104.5°. The water molecule has a bent appearance because the two lone pairs cannot be "seen." This molecule has an even greater overall polarity than that of ammonia.

Problem 3.15   Use VSEPR theory to predict the three-dimensional structure of oxygen difluoride, OF_2.

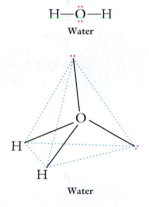

Water

Water

For a molecule with no central atom, such as ethane (see Example 3.10), we can assume that each of the carbon atoms possesses tetrahedral symmetry; so the shape of the molecule must be defined by two tetrahedrons joined at one apex.

VSEPR theory treats double and triple bonds as though they were single bonds. The Lewis structure for acetylene, C_2H_2, for example, is $H\!-\!C\!\equiv\!C\!-\!H$ and the molecule possesses a triple bond. Because VSEPR theory treats a triple bond as though it were a single bond, we proceed as though there were only two bonding pairs of electrons to be accommodated around each carbon atom. The geometrical arrangement that allows maximal separation of two bonding pairs is a straight line separating the bonding pairs by 180°. Acetylene must therefore be a linear molecule with all atoms in a straight line and an appearance similar to that depicted in its Lewis structure. Experiments have shown that acetylene has no dipole moment, thus verifying the VSEPR prediction.

The steps for determining the shapes of molecules by VSEPR theory are:

- Determine the Lewis structure.
- Determine the number of electron pairs (bonded and nonbonded) surrounding a central atom.
- Use VSEPR theory to decide what symmetrical three-dimensional shape will accommodate all bonded and nonbonded electron pairs.
- Classify the shape of the molecule by considering the arrangement of the joined atoms only, ignoring the nonbonded electron pairs. That is why, although the CH_4, NH_3, and H_2O molecules each have four pairs of electrons around the central atom, the shape of methane is tetrahedral, that of ammonia is trigonal pyramidal, and that of water is bent.

Three-Dimensional Representations of Molecules

VSEPR theory allows us to deduce the overall three-dimensional shapes of molecules, but we must use a different approach to illustrate three-dimensional internal molecular structure on a two-dimensional page. Although there are variations associated with each approach, there are two fundamental ways to illustrate three-dimensional molecular structure. One is called the ball-and-stick model, and the other is the space-filling model. Both are illustrated in Figure 3.11.

In the ball-and-stick model of methanol, the atoms are the "balls" and are differentiated by color; the bonds are the "sticks" and are shown about equal in length. The bond angles are correct, but the atom size and bond length do not represent the actual atom size to bond length ratio. Actual bond lengths are much shorter than those represented by the ball-and-stick models. In the space-filling models, the atoms and bond lengths are drawn to precise relative sizes and are disposed in space at the correct angles with respect to one another. The surfaces of the space-filling "atoms" in the models represent, in relative terms, the closest that the atoms can approach one another. However, using space-filling models for instruction presents difficulties because the bonds cannot be seen (the atoms hide them).

When it seems important to illustrate a particular molecular structure in three dimensions, one of these two models will be used. For example, the space-filling model of cholesterol shown in Figure 3.12 helps us to comprehend the molecular shape of an important biochemical. However, most of the molecular structures that you will encounter in this book will be two-dimensional Lewis structures minus the nonbonded electrons. We have chosen this style because the Lewis structures are easy to draw and well within the capability of those of us who are not artists.

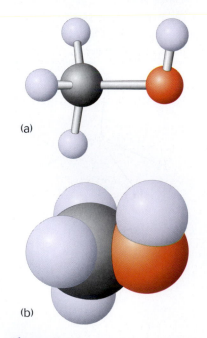

(a)

(b)

Figure 3.11 (a) Ball-and-stick and (b) space-filling models of methanol, CH_3OH. C = black, H = light blue, O = red.

www.whfreeman.com/bleiodian2e

LOOKING AHEAD

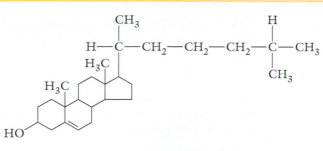

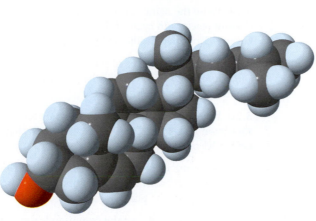

Figure 3.12 A space-filling molecular model of cholesterol, along with its planar molecular representation.

Summary

Ionic Bonds

• Electron transfer between a metal and a nonmetal completes the valence-shell octet for each atom and results in the formation of ions.

• The strong attraction bonding oppositely charged ions is called an ionic bond.

• The compounds thus formed are called ionic compounds.

Naming Binary Ionic Compounds

• A binary ionic compound results from the reaction between a metal and a nonmetal.

• In binary compounds, the metal (cation) takes the name of the element.

• The nonmetal's (anion's) name begins with that of the element but takes the ending -ide.

• The systematized method of nomenclature of compounds with ions of variable valence is called the Stock system.

• Some ions consist of combinations of atoms and are called polyatomic ions.

Covalent Compounds

• When elements acquire an octet of valence electrons by sharing electrons between them, a covalent bond is formed.

• Electron pairs shared by atoms of the same elements, as in H_2 and F_2, form nonpolar covalent bonds.

• Electron pairs shared by atoms of different elements form polar covalent bonds.

• The degree to which an atom within a molecule can draw electrons toward itself is defined as its electronegativity.

Three-Dimensional Molecular Structures

• Lewis formulas reveal only the connections between atoms in a molecule, not its shape.

• Molecular shape depends on the total number of electron pairs surrounding a central atom and their mutual repulsion.

• Valence-shell electron-pair repulsion (VSEPR) theory defines how the mutual repulsion of all the electron pairs causes them to assume the shapes of symmetrical geometrical figures.

Key Words

binary compound, p. 67
bond angles, p. 86
covalent bond, p. 64
dipole moment, p. 83
double bond, p. 79

electronegativity, p. 81
-ide, p. 70
ionic bond, p. 64
Lewis structure, p. 75
lone pair of electrons, p. 76

molecular formula, p. 76
Stock system, p. 70
valence-shell electron-pair repulsion (VSEPR) theory, p. 84

Exercises

3.1   What is the basis for the formation of chemical bonds?

3.2   What is the octet rule?

3.3   What are the ways in which elements can satisfy the octet rule?

3.4   What are the results of the two ways of satisfying the octet rule?

Nomenclature

3.5   Write the systematic name of each of the following compounds: (a) K_2SO_4; (b) $Mn(OH)_2$; (c) $Fe(NO_3)_2$; (d) KH_2PO_4; (e) $Ca(C_2H_3O_2)_2$; (f) Na_2CO_3.

3.6   Write the formulas of the following compounds: (a) ferric chloride; (b) stannic oxide; (c) cupric nitrate; (d) mercurous chloride; (e) plumbous chloride.

3.7   In the blank space to the right of the ion given, write the name of the ion.

Ion	Name
(a) NH_4^+	_____
(b) NO_3^-	_____
(c) SO_4^{2-}	_____
(d) PO_4^{3-}	_____
(e) $C_2H_3O_2^-$	_____

3.8   In the left-hand column, write the formula of the ion named in the right-hand column.

Ion	Name
(a) _____	Carbonate
(b) _____	Hydrogen carbonate
(c) _____	Hydroxide
(d) _____	Hydrogen phosphate
(e) _____	Dihydrogen phosphate

3.9   Write the formulas of the compounds formed in the reaction between aluminum and (a) chlorine; (b) sulfur; (c) nitrogen.

3.10   Write the names of the compounds formed in the reaction between aluminum and (a) chlorine; (b) sulfur; (c) nitrogen.

3.11   What are the formulas of the compounds formed from the following elements? (a) Lithium and oxygen; (b) calcium and bromine; (c) aluminum and oxygen; (d) sodium and sulfur.

3.12   What are the names of the compounds formed from the following elements? (a) Lithium and oxygen; (b) calcium and bromine; (c) aluminum and oxygen; (d) sodium and sulfur.

Ionic Bonds

3.13   What are the combining ratios of elements in Group I with elements in Group VII of the periodic table?

3.14   What are the combining ratios of elements in Group I with elements of Group VI of the periodic table?

3.15   What are the combining ratios of elements in Group II with elements in Group VII of the periodic table?

3.16   What are the combining ratios of elements in Group III with elements in Group VI of the periodic table?

3.17   Write the formulas of the compounds formed when calcium reacts with each of the first three elements of Group VI of the periodic table.

3.18   Write the names of the compounds formed when calcium reacts with each of the first three elements of Group VI of the periodic table.

3.19   Write the formulas of the compounds formed when aluminum reacts with each of the first three elements of Group VI of the periodic table.

3.20   Write the names of the compounds formed when aluminum reacts with each of the first three elements of Group VI of the periodic table.

3.21   Write the formulas of the compounds formed when aluminum reacts with each of the first four elements of Group VII of the periodic table.

3.22   Write the names of the compounds formed when aluminum reacts with each of the first four elements of Group VII of the periodic table.

Covalent Bonds and Lewis Structures

3.23   Write the Lewis structure of carbon tetrachloride, CCl_4.

3.24   Write the Lewis structure of chloroform, $CHCl_3$.

3.25   What is the Lewis structure of OF_2?

3.26   What is the Lewis structure of NCl_3?

3.27   Write the Lewis structure of carbon dioxide, CO_2.

3.28   Alkanes are compounds that contain only carbon and hydrogen. In most alkanes, the carbon atoms are connected to one another to form a chain. Write the Lewis structure for butane, C_4H_{10}. Does this Lewis structure have another name?

Three-Dimensional Molecular Structure

3.29   Use VSEPR theory to predict the angular orientation of fluorine atoms around the central boron atom in the polyatomic ion BF_4^-.

3.30   Use VSEPR theory to predict the three-dimensional structure of OF_2.

3.31   For each of the following molecules or ions, (1) describe the molecular shape and (2) predict whether the molecule or ion has a dipole moment. (a) CCl_4; (b) BF_4^-; (c) CO_2.

3.32   For each of the following molecules, (1) describe the molecular shape and (2) predict whether the molecule has a dipole moment. (a) $CHCl_3$; (b) NCl_3; (c) H_2O.

Unclassified Exercises

3.33   Describe a covalent chemical bond.

3.34   Describe an ionic bond.

3.35 Write the systematic name of each of the following compounds: (a) $AgNO_3$; (b) $HgCl_2$; (c) Na_2CO_3; (d) $Ca_3(PO_4)_2$.

3.36 Write the formulas of the compounds formed in the reaction between magnesium and (a) chlorine; (b) sulfur; (c) nitrogen.

3.37 Write the names of the compounds formed in the reaction between magnesium and (a) chlorine; (b) sulfur; (c) nitrogen.

3.38 Write the formulas of the compounds formed when barium reacts with the first three elements of Group VI of the periodic table.

3.39 Write the names of the compounds formed when barium reacts with the first three elements of Group VI of the periodic table.

3.40 Write the formulas of the compounds formed when potassium reacts with the first three elements of Group V of the periodic table.

3.41 Write the names of the compounds formed when potassium reacts with the first three elements of Group V of the periodic table.

3.42 Write the Lewis structure of carbon monoxide.

3.43 Use VSEPR theory to predict the angular orientation of the atoms in $TeBr_2$.

3.44 Write the formulas of the compounds formed by the cations and the anions in the following table.

	chloride	hydroxide	sulfate
Sodium	(a)	(b)	(c)
Calcium	(d)	(e)	(f)
Iron(II)	(g)	(h)	(i)

3.45 Indicate what kinds of bonds will be formed by the reaction of the following pairs of elements: (a) Si and O; (b) Be and C; (c) C and N; (d) Ca and Br; (e) B and N.

3.46 An organic compound containing on OH group is called an alcohol. Write the Lewis structure for the alcohol, CH_3OH (methanol). (Hint: The OH group is connected to a carbon atom and is not a hydroxide ion.)

3.47 Write the formulas of the compounds formed when lithium reacts with the first four elements of Group VII of the periodic table.

3.48 Write the names of the compounds formed when lithium reacts with the first four elements of Group VII of the periodic table.

3.49 Are there any exceptions to the octet rule?

3.50 Write the formulas of the compounds formed in the reaction between magnesium and (a) iodine; (b) selenium; (c) arsenic.

Expand Your Knowledge

3.51 Your instructor has given you a pure compound and has asked you to determine whether it is a molecular or ionic substance. Suggest an experiment to answer that question.

3.52 Is it possible for a substance to contain both ionic and covalent bonds? Explain your answer.

3.53 Write the Lewis structure of hydrogen peroxide, H_2O_2.

3.54 The Ag^+ ion reacts with cyanide ion, CN^-, to form a polyatomic ion $Ag(CN)_2^-$. Predict the three-dimensional structure of that ion.

3.55 Write the Lewis structure of nitrogen tetroxide, N_2O_4.

3.56 Predict the order of increasing electronegativity in each of the following groups of elements: (a) C, N, O; (b) Mg, Ca, Sr; (c) C, Ge, Pb; (d) Se, Te, Po.

3.57 Predict which bond in each of the following groups will be the most polar: (a) C–F, Si–F, Ge–F; (b) C–H, Si–H, Sn–H; (c) Al–Br, Ga–Br; (d) As–Br, Sb–Br.

3.58 Predict the formulas of the ionic compounds formed from the following pairs of elements. Name each compound. (a) Li and N; (b) Ga and O; (c) Rb and Cl; (d) Ba and S.

3.59 Write Lewis structures for each of the following molecules and ions. In each case, the first atom listed is the central atom. (a) PCl_3; (b) SO_4^{2-}; (c) PO_4^{3-}; (d) ClO_4^-

3.60 Predict the molecular structure of each molecule or ion in Exercise 3.59.

3.61 Use electron configurations to describe the formation of the following ionic compounds from the atoms: (a) GaF_3; (b) $AgCl$; (c) Li_3N.

3.62 Use Lewis electron-dot formulas for all species to describe the following equations: (a) $3Li + N \longrightarrow Li_3N$; (b) $Na + H \longrightarrow NaH$; (c) $Al + 3I \longrightarrow AlI_3$.

3.63 Write the chemical formula for (a) thallium chloride; (b) zinc nitride; (c) cadmium sulfide; (d) gallium oxide.

3.64 Write the chemical formula for (a) magnesium telluride; (b) aluminum selenide; (c) germanium bromide; (d) rubidium nitride.

3.65 Write the Lewis formulas for (a) SCl_2; (b) $GeCl_4$; (c) $AsBr_3$; (d) PH_3.

3.66 Write the Lewis formulas for (a) methane, CH_4; (b) methylamine, CH_3NH_2; (c) fluoromethane, CH_3F.

3.67 Write the Lewis formulas for (a) acetylene, C_2H_2; (b) phosgene, $COCl_2$.

3.68 Polyvinyl chloride is manufactured from an important industrial chemical named vinyl chloride whose chemical formula is C_2H_3Cl. Write vinyl chloride's Lewis formula.

3.69 Predict the shapes of the following molecules: (a) TeF_2; (b) $SnBr_2$; (c) OF_2.

3.70 Write the Lewis formulas for the following molecules, and predict their shapes. (a) CCl_4; (b) NF_3.

3.71 Predict the geometries of the following molecules: (a) SOF_2; (b) OsO_4; (c) SnF_2.

CHAPTER 4

CHEMICAL CALCULATIONS

Chemistry in Your Future

One of your customers at the pharmacy is a strict vegetarian as well as lactose intolerant. He wants to take vitamin B_{12} supplements to prevent coming down with pernicious anemia from the lack of animal products in his diet. He reads the label on the bottle of tablets that you hand him and remarks that 125 micrograms per tablet seems a tiny amount. (He says that his vitamin C tablets contain vitamin C in milligram amounts.) You assure him that a minute amount of vitamin B_{12} is enough for the body to function healthily. To reassure him further, you look up the compound's molecular formula, $C_{63}H_{88}CoN_{14}O_{14}P$, punch some buttons on your calculator, and tell him that 125 micrograms provides 5.6×10^{16} molecules of the vitamin, a very large number. Chapter 4 will show you how to do this calculation.

(Donald C. Johnson/The Stock Market.)

For more information on this topic and others in this chapter, go to www.whfreeman.com/bleiodian2e

Learning Objectives

- Calculate the formula mass of a compound.
- Define the mole.
- Use the mole as a unit-conversion factor for converting mass into moles and moles into mass.
- State the magnitude of Avogadro's number and what it implies about the size of atoms and molecules and the numbers of them in a weighable sample.
- Calculate the empirical and molecular formulas of a compound.
- Write a chemical reaction as an equation.
- Balance a chemical equation.
- Use a balanced chemical equation to predict the masses of compounds produced and used up in chemical processes

In this chapter, we will use our knowledge of atoms and molecules to answer some practical questions about the conservation of mass in chemical processes. For example, the technician who measures a patient's oxygen uptake to determine his or her basal metabolic rate can compare the amount of oxygen being used with the amount of carbon dioxide produced and discover the composition of the patient's diet. When a chemist weighs out 83 grams of sodium carbonate in the laboratory, she can calculate the number of atoms in the sample. She knows that, if 111 g of calcium chloride is added to the 83 g of sodium carbonate, 100 g of calcium carbonate will form. Both the economics of the manufacture of drugs and the control of pollution depend on a detailed understanding of the quantitative aspects of chemical reactions. These connections are the kind that we will explore in the following pages.

The ability to predict the quantitative outcome of chemical processes is based on one of the most fundamental ideas in chemistry: the concept of the mole. To define the mole, we must first explain how a compound's formula is used to calculate its relative mass. We will then use the mole concept to calculate the number of formula units, given the mass of a compound, and to determine empirical and molecular formulas. Lastly, we will explain how to balance chemical equations and to use the mole concept to predict the quantitative outcomes of reactions.

4.1 CHEMICAL FORMULAS AND FORMULA MASSES

The **formula** of a chemical compound states how many atoms of each element are in a fundamental unit of the compound. The fundamental unit is called a **formula unit.** Examples of formula units that you have seen before are $NaCl$, $CaBr_2$, and Al_2O_3.

We have seen that relative masses have been assigned to the different elements (Section 2.3). A relative mass may also be assigned to a compound by simply adding up the relative masses of all the atoms in the formula unit. This sum is called the **formula mass.** To determine the formula mass of water, H_2O, we add the relative masses (taken from the inside back cover of this book) of its two hydrogen atoms and one oxygen atom: 1.008 amu + 1.008 amu + 16.00 amu = 18.02 amu.

In Section 3.5, we made an important distinction between ionic compounds, such as $NaCl$, and compounds that exist as molecules, such as H_2O. The formula unit represents the compound's composition only. Accordingly, the mass units cannot be in grams but must be in atomic mass units, amu. For molecular substances such as H_2O and CO_2, the **formula mass** is identical to the quantity called **molecular mass.** Molecular mass is the mass in amu of a molecule of a molecular substance. The definition of formula mass using the idea of a formula unit permits us to ignore those distinctions when we are concerned primarily with mass relations. All we need to know from a formula unit is what elements are present in the compound and how many atoms of each element are present.

A compound's formula mass can be calculated only if its formula is known. The best way to illustrate the procedure is to calculate the formula masses of a few representative compounds. All masses in this chapter will be expressed to four significant figures, unless otherwise noted.

Example 4.1 Calculating the formula mass of a compound

Calculate the formula masses of the following compounds: $NaCl$; $CaCl_2$; Al_2O_3; $C_6H_{12}O_6$ (glucose, a sugar).

Solution

Formula mass is defined as the sum of atomic masses of a compound's constituent atoms. Therefore,

$$\text{Formula mass of NaCl} = \text{atomic mass of Na} + \text{atomic mass of Cl}$$
$$= 22.99 \text{ amu} + 35.45 \text{ amu} = 58.44 \text{ amu}$$

$$\text{Formula mass of CaCl}_2 = \text{atomic mass of Ca} + 2 \times \text{atomic mass of Cl}$$
$$= 40.08 \text{ amu} + (2 \times 35.45 \text{ amu}) = 111.0 \text{ amu}$$

$$\text{Formula mass of Al}_2\text{O}_3 = 2 \times \text{atomic mass of Al} + 3 \times \text{atomic mass of O}$$
$$= (2 \times 26.98 \text{ amu}) + (3 \times 16.00 \text{ amu}) = 102.0 \text{ amu}$$

$$\text{Formula mass of C}_6\text{H}_{12}\text{O}_6 = 6 \times \text{atomic mass of C}$$
$$+ 12 \times \text{atomic mass of H}$$
$$+ 6 \times \text{atomic mass of O}$$
$$= (6 \times 12.01 \text{ amu}) + (12 \times 1.008 \text{ amu})$$
$$+ (6 \times 16.00 \text{ amu})$$
$$= 180.2 \text{ amu}$$

Problem 4.1 Calculate the formula masses of the following compounds: (a) $Zn(HPO_4)_2$; (b) Fe_3O_4; (c) H_2O; (d) H_2SO_4; (e) $C_3H_6O_2N$ (the amino acid alanine, a component of proteins).

In Chapter 2, we saw that samples of different elements must contain equal numbers of atoms when the mass of each sample, in grams, is identical with each element's relative mass in atomic mass units. This means that there are equal numbers of atoms in 55.85 g of iron, 22.99 g of sodium, 16.00 g of oxygen, and so forth. The definition of formula mass allows us to make the same statement about the formula mass of a compound. That is, the mass of any compound in grams that is equal to its formula mass in atomic mass units contains the same number of formula units as the mass of any other compound in grams that is equal to its formula mass in atomic mass units. Therefore, the number of formula units in 58.44 g of NaCl is equal to the number of formula units in 180.2 g of glucose. Figure 4.1 shows the masses of a number of compounds in which there are equal numbers of formula units.

Example 4.2　Converting formula units into mass

Calculate the masses, in grams, of NaCl, $CaCl_2$, Al_2O_3, and $C_6H_{12}O_6$ that will contain equal numbers of formula units of each compound.

Solution

The formula masses of these compounds were calculated in Example 4.1, and those masses in grams will all contain the same numbers of formula units.

Figure 4.1 Masses of some molecular compounds. The mass of each substance is equal to its formula mass in grams. The substances from left to right are water, H_2O (18.02 g); ethanol, C_2H_6O (46.07 g); glucose, $C_6H_{12}O_6$, (180.2 g); and sucrose, $C_{12}H_{22}O_{11}$ (342.3 g). (W. H. Freeman photo by Ken Karp.)

The masses are 58.44 g of NaCl, 111.0 g of $CaCl_2$, 102.0 g of Al_2O_3, and 180.2 g of $C_6H_{12}O_6$.

Problem 4.2 Calculate the masses, in grams, of (a) Na_2SO_4, (b) KBr, (c) MgO, and (d) C_6H_{14} that will contain equal numbers of formula units of each compound.

4.2 THE MOLE

As noted in Chapter 2, it is impossible to weigh out one atom, but we can weigh out the mass of an element in grams that is equal to the relative mass of an atom, and we can do the same with a compound. The number of atoms contained in the atomic mass (in grams) of an element and the number of formula units contained in the formula mass (in grams) of a compound are fundamental chemical quantities called the **mole.** When numerical amounts are specified, the mole is abbreviated **mol.** The mole allows us to know the number of formula units in a sample of a given mass.

A mole of any element or compound contains the same number of atoms or formula units as a mole of any other element or compound. The mass in grams of 1 mol of a compound is called its **molar mass.** Remember, molar mass is given in grams, but formula mass and molecular mass are given in amu. The molar mass of sodium chloride is 58.44 g. It has the same number of formula units as 1 mol of glucose, which weighs 180.2 g. Table 4.1 illustrates the relations between formula mass (atomic mass units), molar mass (grams per mole), particles per mole, and number of moles for hydrogen atoms, hydrogen molecules, water, and calcium chloride.

The definition of molar mass,

1 formula unit of a compound in grams = 1 mole of that compound,

allows us to write the following conversion factors:

$$\frac{1 \text{ formula unit of any compound in grams}}{1 \text{ mol of that compound}} = 1$$

or

$$\frac{1 \text{ mol of any compound}}{1 \text{ formula unit of that compound in grams}} = 1$$

By using these conversion factors, we can calculate the number of formula units in any sample of a compound if we know both the formula mass and the mass of the sample in grams.

TABLE 4.1 **Relations Between Molar Quantities**

Substance	Formula	Formula mass (amu)	Molar mass (g/mol)	Particles per mole	Moles
atomic hydrogen	H	1.008	1.008	6.022×10^{23} hydrogen atoms	1 mol H atoms
molecular hydrogen	H_2	2.016	2.016	6.022×10^{23} hydrogen molecules 12.044×10^{23} hydrogen atoms	1 mol H_2 molecules 2 mol H atoms
water	H_2O	18.02	18.02	6.022×10^{23} water molecules 6.022×10^{23} oxygen atoms 12.044×10^{23} hydrogen atoms	1 mol H_2O molecules 1 mol O atoms 2 mol H atoms
calcium chloride	$CaCl_2$	111.0	111.0	6.022×10^{23} $CaCl_2$ formula units 6.022×10^{23} Ca^{2+} ions 12.044×10^{23} Cl^- ions	1 mol $CaCl_2$ formula units 1 mol Ca^{2+} ions 2 mol Cl^- ions

Concept check

✓ The mole permits the conversion of mass into number of formula units and the conversion of number of formula units into mass.

Example 4.3   Converting mass into moles

Calculate the number of moles in 270.0 g of glucose.

Solution

The conversion factor required for this calculation uses the fact that 1 formula mass of a compound is equal to 1 mol of that substance. The formula mass of glucose was calculated in Example 4.1 and is 180.2 amu, and so we know that, by definition, 1 mol of glucose weighs 180.2 g. To convert grams of glucose into moles of glucose, let's first look at the conversion from the point of view of units only:

$$\text{g glucose} \left(\frac{\text{mol glucose}}{\text{g glucose}} \right) = \text{mol glucose}$$

Next, we insert the relevant data and calculate the answer:

$$\text{mol glucose} = 270.0 \text{ g glucose} \left(\frac{1 \text{ mol glucose}}{180.2 \text{ g glucose}} \right) = 1.498 \text{ mol glucose}$$

It is interesting to note that, although it is not possible to weigh out 1.498 molecules of glucose (because any fraction of a molecule of glucose is no longer glucose), 1.498 mol of glucose can be weighed out with no difficulty. Remember, a mole is not a molecule.

Problem 4.3   Calculate the number of moles of $CaCl_2$ in 138.8 g of $CaCl_2$.

Example 4.4   Converting moles into mass

How many grams are in 0.3300 mol of Al_2O_3?

Solution

To calculate the mass of a given number of moles of Al_2O_3 (or any compound or element), we derive a conversion factor from the fact that 1 formula mass of Al_2O_3 contains 1 mol of Al_2O_3 and weighs 102.0 g.

$$\text{g } Al_2O_3 = 0.3300 \text{ mol} \left(\frac{102.0 \text{ g } Al_2O_3}{1 \text{ mol } Al_2O_3} \right) = 33.66 \text{ g } Al_2O_3$$

Problem 4.4   How many moles are in 0.3600 g of glucose?

>> The mole concept will be important in Chapter 7, when we learn to describe the concentration of solutions.

4.3   AVOGADRO'S NUMBER

The actual number of fundamental units in a mole has been determined by experiment and is called **Avogadro's number.** (Amedeo Avogadro was the first to clarify the difference between atoms and molecules.) It is a very large number, 6.022×10^{23}. It is defined as the number of carbon atoms in exactly 12 g of the pure isotope of carbon of mass 12 amu. One mole of anything—pencils, books, students, molecules, atoms—consists of 6.022×10^{23} of that thing.

Example 4.5   Calculating the number of moles of elements in a compound

How many moles of ions are there in 1 mol of each of the following ionic compounds? (a) NaCl; (b) $CaCl_2$; (c) $AlCl_3$; (d) Mg_3N_2.

Solution

Moles of each kind of ion	Number of moles of all ions
(a) 1 mol of Na^+ ions and 1 mol of Cl^- ions	2 mol
(b) 1 mol of Ca^{2+} ions and 2 mol of Cl^- ions	3 mol
(c) 1 mol of Al^{3+} ions and 3 mol of Cl^- ions	4 mol
(d) 3 mol of Mg^{2+} ions and 2 mol of N^{3-} ions	5 mol

Problem 4.5 How many moles of atoms of each kind are there in 1 mol of each of the following substances? (a) $ZnHPO_4$; (b) H_2SO_4; (c) Al_2O_3; (d) Ca_3P_2.

The magnitude of Avogadro's number is difficult to grasp, but perhaps the following example will help. Imagine a flat area the size of the United States. One mole of dimes spread out uniformly on this area would reach a height of about 10 miles. Mt. Everest, in comparison, is only about 6 miles high, and the highest altitude of a transcontinental airplane trip is about 8 miles. Because a mole of carbon occupies a volume of about a teaspoon, the size of Avogadro's number tells us two other things: (1) atoms and molecules are very small and (2) individually they do not weigh very much.

The relation between the atomic mass unit and the physical mass unit—that is, the gram—can now be stated numerically. That is, we are now in a position to calculate the mass of 1 amu in grams.

Because an Avogadro number of C-12 atoms has a mass of exactly 12 g, we can set up the following expression:

$$6.022 \times 10^{23} \ \cancel{C \ atoms} \left(\frac{12 \ amu}{\cancel{C \ atom}} \right) = 12 \ g$$

$$6.022 \times 10^{23} \ amu = 1 \ g$$

$$1 \ amu = \frac{1 \ g}{6.022 \times 10^{23}} = 1.661 \times 10^{-24} \ g$$

Now we can calculate the masses of individual atoms.

Example 4.6   Calculating atomic masses in grams

Calculate the mass, in grams, of the following atoms: hydrogen, oxygen, sulfur, sodium, phosphorus.

Solution

This problem is solved by listing the atoms and their relative atomic masses in atomic mass units and then multiplying each atomic mass by the mass, in grams, of 1 amu: 1.661×10^{-24} g.

Atom	Atomic mass (amu)	Grams per atomic mass unit	Atomic mass (grams per atom)
H	1.008	1.661×10^{-24}	1.674×10^{-24}
O	16.00	1.661×10^{-24}	2.658×10^{-23}
S	32.07	1.661×10^{-24}	5.325×10^{-23}
Na	22.99	1.661×10^{-24}	3.819×10^{-23}
P	30.97	1.661×10^{-24}	5.144×10^{-23}

Problem 4.6 Calculate the mass, in grams, of the following atoms: lithium, nitrogen, fluorine, calcium, carbon.

From the results calculated in Example 4.6, we can see that it is very difficult to weigh out a small number of atoms of any element. The weighing done in chemistry laboratories takes place on a much larger scale, with the use of a balance calibrated in units of 0.1000 g at the least. The following example shows how the weighing of many atoms of an element becomes a practical matter.

Example 4.7   Relating the numbers of oxygen atoms to mass

What number of oxygen atoms constitutes a practical amount of oxygen to weigh?

Solution

This number can best be illustrated by arranging numbers of atoms of oxygen and corresponding masses in grams in tabular form. (The mass of an oxygen atom was given in Example 4.6.)

Number of atoms of oxygen	Mass in grams
10	2.658×10^{-22}
100	2.658×10^{-21}
100,000	2.658×10^{-18}
1.0×10^{15}	2.658×10^{-8}
6.022×10^{20}	0.01600
6.022×10^{23}	16.00

The tabular arrangement shows that a very large number of atoms is required to obtain a practical mass of any substance. Only the last two numbers of atoms, 0.0010 mol and 1.000 mol, are practical for weighing.

Problem 4.7   Using the numbers of atoms in Example 4.7, calculate the corresponding masses, in grams, of calcium.

4.4   EMPIRICAL FORMULAS

Chemists did not use the periodic table to determine that the formula for calcium chloride was $CaCl_2$. Instead, it was the other way around. The formulas for compounds are first determined by quantitative analysis, specific identification methods that determine the identity and percent composition of each element in a sample of the compound. Box 4.1 describes an experimental method for the analysis of compounds containing carbon, hydrogen, and nitrogen. The calculation of percent composition of a compound was presented in Section 2.1. The relative mass of each atom is then used to determine the number of moles of each atom in the sample. From that information, we derive the simplest whole-number ratio of the constituents, which is called the **empirical formula.** The following example illustrates this approach.

Example 4.8   Finding the empirical formula: I

Quantitative analysis of calcium chloride reveals that the composition of the compound is 36.04% calcium and 63.96% chlorine. From those data, determine its empirical formula.

Solution

Because the data are expressed in percentages, it is convenient to assume that we have a 100.0 g sample of calcium chloride. This assumption means that, of the 100.0 g of calcium chloride, 36.04 g is calcium and 63.96 g is chlorine. The atomic mass of each kind of atom is used to calculate the number of moles of each:

4.1 CHEMISTRY IN DEPTH

Determining the Composition of a Compound

To determine the formula of a compound requires decomposing it into its component elements or converting each of the elements in it into compounds whose identity and mass are readily determined. The latter is the approach used for compounds consisting of carbon and hydrogen or of carbon, hydrogen, and nitrogen. When such a compound is fully combusted or burned in excess oxygen, the products will always be gaseous carbon dioxide and water, and, if the compound contained nitrogen, nitrogen dioxide.

The gases are collected by absorption and the increase in weight of the absorbent is measured, as illustrated in the accompanying diagram.

As oxygen flows through the combustion chamber (the furnace), all of the substance's carbon is converted into CO_2, its hydrogen into H_2O, and its nitrogen into NO_2. The CO_2 and H_2O are collected by absorbent materials whose increase in weight is then measured. After the mass of carbon and hydrogen in the original sample is accounted for, the mass of nitrogen is determined by difference. The percent composition of the compound is determined by the methods described in Sections 2.1 and 4.4.

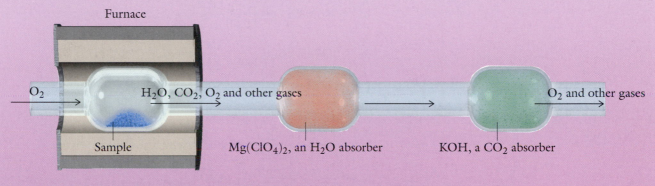

A combustion device used to analyze substances for carbon, hydrogen, and nitrogen.

$$36.04 \text{ g Ca} \left(\frac{1 \text{ mol Ca}}{40.08 \text{ g Ca}} \right) = 0.8992 \text{ mol Ca}$$

$$63.96 \text{ g Cl} \left(\frac{1 \text{ mol Cl}}{35.45 \text{ g Cl}} \right) = 1.804 \text{ mol Cl}$$

It is possible to have 0.8992 mol of calcium and 1.804 mol of chlorine but impossible to have 0.8992 atoms of calcium and 1.804 atoms of chlorine, because atoms are indivisible, cannot be present fractionally, and must be represented in the formula by small whole numbers. This apparent discrepancy is reconciled by dividing each of the experimentally derived values by the smallest numerical value of moles present. This procedure converts all the experimental values of moles into small whole numbers representing the relative numbers of atoms within the formula unit.

$$\frac{0.8992 \text{ mol Ca}}{0.8992} = 1.000 \text{ mol Ca}, \quad \text{which rounds to 1}$$

$$\frac{1.804 \text{ mol Cl}}{0.8992} = 2.006 \text{ mol Cl}, \quad \text{which rounds to 2}$$

Because these molar quantities are based on experimental values, it is unlikely that they will be exact whole numbers. In this type of analysis, it is acceptable to round 2.006 to 2. Therefore, the empirical formula of calcium chloride is $CaCl_2$. Example 4.10 presents another aspect of this procedure.

Problem 4.8 Quantitative analysis of magnesium oxide reveals that the composition of the compound is 60.31% magnesium and 39.69% oxygen. From those data, determine its empirical formula.

Example 4.9   Finding the empirical formula: II

Chemical analysis shows the composition of a compound containing potassium, hydrogen, phosphorus, and oxygen to be 44.66% potassium, 18.13% phosphorus, 36.63% oxygen, and 0.5785% hydrogen. Determine its empirical formula.

Solution

Convert percentages into grams by assuming that the sample contains 100.0 g of the compound. Then use the elements' atomic masses (as conversion factors) to determine the number of moles of each element in the sample:

$$44.66 \text{ g K} \left(\frac{1 \text{ mol K}}{39.10 \text{ g K}} \right) = 1.142 \text{ mol K}$$

$$18.13 \text{ g P} \left(\frac{1 \text{ mol P}}{30.97 \text{ g P}} \right) = 0.5854 \text{ mol P}$$

$$36.63 \text{ g O} \left(\frac{1 \text{ mol O}}{16.00 \text{ g O}} \right) = 2.289 \text{ mol O}$$

$$0.5785 \text{ g H} \left(\frac{1 \text{ mol H}}{1.008 \text{ g H}} \right) = 0.5739 \text{ mol H}$$

Again, the numbers of moles of each element must be represented by small whole numbers. This task is accomplished by dividing each of the preceding experimentally derived values by the smallest value.

$$\frac{1.142 \text{ mol K}}{0.5739 \text{ mol H}} = 1.990, \quad \text{which rounds to 2}$$

$$\frac{0.5854 \text{ mol P}}{0.5739 \text{ mol H}} = 1.020, \quad \text{which rounds to 1}$$

$$\frac{2.289 \text{ mol O}}{0.5739 \text{ mol H}} = 3.988, \quad \text{which rounds to 4}$$

$$\frac{0.5739 \text{ mol H}}{0.5739 \text{ mol H}} = 1.000, \quad \text{which rounds to 1}$$

These amounts now reveal the smallest whole-number ratios of the different elements in the compound. The reason that the calculated numbers of moles are not quite whole numbers is that the percentages were determined experimentally and were therefore subject to variability. The rounded values show the empirical formula of potassium monohydrogen phosphate to be K_2HPO_4.

Problem 4.9   Chemical analysis shows the composition of magnesium sulfate to be 20.20% magnesium, 26.60% sulfur, and 53.20% oxygen. Determine its empirical formula.

Example 4.10   Finding the empirical formula: III

Chemical analysis shows the composition of potassium dichromate to be 26.43% potassium, 38.09% oxygen, and 35.48% chromium. Determine its empirical formula.

Solution

Convert percentages into grams by assuming a sample of 100.0 g of the compound. Then determine the number of moles of each element by using the atomic mass of each element as a unit-conversion factor:

$$26.43 \text{ g K} \left(\frac{1.000 \text{ mol K}}{39.10 \text{ g K}} \right) = 0.6760 \text{ mol K}$$

$$38.09 \text{ g O} \left(\frac{1.000 \text{ mol O}}{16.00 \text{ g O}} \right) = 2.381 \text{ mol O}$$

$$35.48 \text{ g Cr} \left(\frac{1.000 \text{ mol Cr}}{52.00 \text{ g Cr}} \right) = 0.6823 \text{ mol Cr}$$

Dividing each value by the smallest one, we get

$$\frac{0.6760 \text{ mol K}}{0.6760 \text{ mol K}} = 1.000, \quad \text{which rounds to 1}$$

$$\frac{2.381 \text{ mol O}}{0.6760 \text{ mol K}} = 3.522, \quad \text{which rounds to 3.5}$$

$$\frac{0.6823 \text{ mol Cr}}{0.6760 \text{ mol K}} = 1.009, \quad \text{which rounds to 1}$$

The numbers just calculated are relative numbers of atoms. They are based on experimental fact. Therefore, the 3.522 atoms of oxygen should be rounded to 3.5. At the same time, because atoms are indivisible, only whole numbers of atoms are possible. A formula unit of this compound cannot have 3.5 atoms of oxygen. To solve the dilemma, we double the calculated values, which produces the smallest whole-number values and shows that the formula of potassium dichromate must be $K_2Cr_2O_7$.

Problem 4.10 By chemical analysis, the composition of an oxide of phosphorus was 43.66% phosphorus and 56.34% oxygen. Determine its empirical formula.

4.5 MOLECULAR FORMULAS

The empirical formula represents the basic combining ratios of a compound's elements. It does not reveal the actual numbers of each kind of atom in a molecule. That information is contained in the **molecular formula** of a compound. For example, the percent composition of fructose is carbon, 40.00%; hydrogen, 6.716%; and oxygen, 53.29%. From these data, the empirical formula can be calculated.

$$\text{Carbon:} \quad 40.00 \text{ g C} \left(\frac{1 \text{ mol C}}{12.00 \text{ g C}} \right) = 3.331 \text{ mol C}$$

$$\text{Hydrogen:} \quad 6.716 \text{ g H} \left(\frac{1 \text{ mol H}}{1.000 \text{ g H}} \right) = 6.663 \text{ mol H}$$

$$\text{Oxygen:} \quad 53.29 \text{ g O} \left(\frac{1 \text{ mol O}}{16.00 \text{ g O}} \right) = 3.331 \text{ mol O}$$

Dividing all numbers by 3.331, we find the smallest ratio of whole numbers to be $1:2:1$ and the empirical formula of fructose to be CH_2O. The molecular mass based on this empirical formula is approximately 30 amu.

If the molecular mass were 30 amu, then CH_2O would have to be accepted as the molecular formula, the presumed actual numbers of atoms of all kinds making up the molecule. However, the molecular mass of fructose, determined by an independent method, is known to be approximately 180 amu. This information in combination with the empirical formula allows us to determine the molecular formula. We know that the ratio of elements cannot be changed, only the actual numbers. Thus, if $1:2:1$ is the correct ratio of C:H:O, the molecular mass of 180 amu must be some whole-number multiple of the empirical formula mass of 30 amu.

The clearest way to derive the molecular formula is to construct a table of multiples of the mass of CH_2O, along with the corresponding molecular formulas. We can then select by inspection the correct molecular formula corresponding to the appropriate molecular mass.

Molecular formula	Molecular mass (amu)
CH_2O	30
$C_2H_4O_2$	60
$C_4H_8O_4$	120
$C_6H_{12}O_6$	180
$C_8H_{16}O_8$	240

The correct molecular formula corresponding to the molecular mass of 180 is $C_6H_{12}O_6$.

4.6 BALANCING CHEMICAL EQUATIONS

The atomic theory allows us to restate the Law of Conservation of Mass (Section 2.1) as the **Law of Conservation of Atoms:** In any chemical process, whatever atoms are present at the beginning of the reaction must be present at the end, albeit in new combinations. This idea is put into practice in the writing of balanced chemical equations. When you have mastered the skill of writing a chemical reaction as a balanced equation, you will also be able to predict the mass quantities of the products emerging from the reaction on the basis of the masses of the starting materials.

Chemists express the Law of Conservation of Mass in chemical shorthand by writing reactions as equations. An equal sign can be used, but most chemists use an arrow to separate components that react, called **reactants,** from components that result from the reaction, called **products.** Consider the reaction between sodium and chlorine. Remember that chlorine exists in elemental form as a diatomic molecule. A first attempt at writing an equation for that reaction might look like

$$Na + Cl_2 \longrightarrow NaCl$$

The components on the left-hand side of the equation, Na and Cl_2, are the reactants, and the component on the right-hand side is the product. The equation is not written correctly, however, because conservation of mass (or of atoms) requires that equal numbers of each kind of atom appear on both sides of the equation. The number of sodium atoms on the reactant side is the same as the number of sodium atoms on the product side, which means that the equation is **balanced** with respect to the sodium atoms. However, there are two chlorine atoms on the left but only one on the right, and the equation is therefore **unbalanced** with respect to the chlorine atoms. The chemical equation cannot be considered correct until the numbers of atoms of the elements on both sides of the equation are equal or balanced.

Balancing a reaction equation is accomplished by inserting numbers in front of individual molecular or ionic formulas. These numbers serve as multipliers. The numbers of chlorine atoms, for example, can be balanced by placing a 2 in front of the NaCl.

$$Na + Cl_2 \longrightarrow 2\,NaCl$$

The number 2 in front of the NaCl is called a balancing **coefficient.** Just as in an algebraic equation, the absence of a number in front of a formula is understood to indicate a balancing coefficient of 1: Na and Cl_2 can be read as 1 Na and 1 Cl_2.

Inserting a coefficient of 2 in front of NaCl increases the number of sodium atoms and chlorine atoms on the right to two each, but now sodium is unbalanced because there is only one sodium atom on the left. This imbalance is remedied by placing a balancing coefficient of 2 in front of the sodium on the left.

$$2\,Na + Cl_2 \longrightarrow 2\,NaCl$$

There are now equal numbers of each kind of atom on both sides of the equation, and the relation is called a **balanced chemical equation.**

Balancing can be accomplished only by inserting appropriate numbers before formulas, as multipliers, or balancing coefficients. Changing subscripts within a compound's formula is not permissible, because doing so changes the chemical identity of the substance. It might be mathematically correct to balance the equation by writing

$$Na + Cl_2 \longrightarrow NaCl_2$$

but it is not chemically correct, because $NaCl_2$ is not the same substance as NaCl and, in fact, does not exist. In NaCl, the element chlorine is no longer in the form of a gas; rather, it is in the form of chloride ions, and these ions combine with sodium ions in a $1:1$ ratio.

✓　To balance an equation, the correct formulas of reactants and products must be known.　　　　　　　　　　　　　　*Concept check*

Consider another reaction in which two elements react:

$$Al + O_2 \longrightarrow Al_2O_3$$

The key here is to recognize that there are two oxygen atoms on the left and three oxygen atoms on the right. By multiplying oxygen on the left-hand side by 3 and Al_2O_3 on the right-hand side by 2, the numbers of oxygen atoms on the left and right are made equal. Therefore, the first step in balancing this equation is

$$Al + 3\,O_2 \longrightarrow 2\,Al_2O_3$$

The result is six oxygen atoms on the left and six oxygen atoms on the right. However, there are now four aluminum atoms on the right; so a coefficient of 4 must be placed before the Al on the left. The final balanced equation is

$$4\,Al + 3\,O_2 \longrightarrow 2\,Al_2O_3$$

In chemical notation, a polyatomic ion is enclosed in parentheses when more than one of the ion is present in a formula. A subscript after the closing parenthesis indicates how many of that ion the formula unit contains. The compound $Ca_3(PO_4)_2$, for example, contains two phosphate ions in each formula unit. It is important to remember that the coefficients in a balanced chemical equation multiply the entire formula unit and therefore multiply the numbers of each atom in the formula—including each atom in each polyatomic ion in the formula.

Now consider the reaction between aluminum chloride, $AlCl_3$, and sodium phosphate, Na_3PO_4. Note that the PO_4^{3-} group is a polyatomic ion, which, as stated earlier, behaves as an individual ionic unit.

$$AlCl_3 + Na_3PO_4 \longrightarrow AlPO_4 + NaCl \quad \text{(unbalanced)}$$

At this point, the numbers of aluminum atoms and phosphate ions are the same on both sides of the equation, but only one-third the number of reacting sodium and chlorine atoms appear as products. To correct this imbalance, a 3 is placed before the formula for NaCl, as follows:

$$AlCl_3 + Na_3PO_4 \longrightarrow AlPO_4 + 3\,NaCl$$

There is no single, consistently successful method for balancing equations. It is an artful procedure, and each of us goes about it in his or her own way, developing effective methods as a result of diligent repetition. However, the following approach may prove helpful. This method of balancing equations focuses on one type of atom (or polyatomic ion) at a time, ignoring its partners in the compounds in which it exists. With that in mind, let's return to the reaction between aluminum chloride and sodium phosphate.

The most effective first step is to examine the fate of the cation having the largest charge—in this case, aluminum:

$$Al(X) \longrightarrow Al(Y)$$

X and Y are used to indicate that the compound containing aluminum on the reactant side is different from the compound containing aluminum on the product side. There are identical numbers of aluminum atoms on each side of the equation; therefore no multipliers or coefficients are required.

Phosphate is a polyatomic ion and can be considered a unit for balancing purposes. Examination of the preceding unbalanced equation reveals equal numbers of phosphate ion on both sides; so no further modification is suggested by the phosphate ion.

The next step is to examine the fate of sodium:

$$Na_3(X) \longrightarrow Na(Y)$$

In this case, a coefficient of 3 before the product, Na(Y), is required for balance.

$$Na_3(X) \longrightarrow 3\,Na(Y)$$

Now, consider the other consequence of inserting that coefficient. It balanced the sodium atoms, but it also multiplied the chloride in NaCl, resulting in three chloride ions on the right-hand side. Because there are also three chloride ions on the left-hand side of the arrow, the equation is now fully balanced:

$$AlCl_3 + Na_3PO_4 \longrightarrow AlPO_4 + 3\,NaCl$$

Let's try a more complex reaction, that of aluminum hydroxide with sulfuric acid, which results in the formation of the salt aluminum sulfate and water:

$$Al(OH)_3 + H_2SO_4 \longrightarrow Al_2(SO_4)_3 + H_2O$$

Notice that, in these compounds, the parentheses emphasize that there are three polyatomic OH^- ions per unit of aluminum hydroxide and three polyatomic SO_4^{2-} ions per unit of aluminum sulfate.

Again, we will consider one type of atom at a time, ignoring its partners in the compounds in which it exists. We focus first on aluminum, the cation of greatest charge, Al^{3+}:

$$Al(X) \longrightarrow Al_2(Y)$$

Because two atoms of Al appear in the product, there must be two atoms in the reactant. This imbalance requires a coefficient of 2 before the reactant:

$$2\,Al(X) \longrightarrow Al_2(Y)$$

Now we turn to the polyatomic ion of greatest charge, SO_4^{2-}, counting its numbers in product and reactant before balancing:

$$(X)SO_4 \longrightarrow (Y)(SO_4)_3$$

This ion can be balanced by inserting a coefficient of 3 in front of the reactant that contains the sulfate ion:

$$3\,(X)SO_4 \longrightarrow (Y)(SO_4)_3$$

The equation is now balanced with respect to Al and SO_4, but it is not balanced with respect to H_2O:

$$2\,Al(OH)_3 + 3\,H_2SO_4 \longrightarrow Al_2(SO_4)_3 + H_2O$$

The formula for water, H_2O, may also be written as HOH. The formula for water written in this way makes it easier to visualize the balancing coefficients needed to complete the procedure. There are 6 OH^- ions and 6 H atoms on the left-hand side of the arrow. A coefficient of 6 inserted in front of H_2O on the product side balances them out:

$$6\,(OH) + 6\,H \longrightarrow 6\,HOH \longrightarrow 6\,H_2O$$

The final, balanced equation is

$$2\,Al(OH)_3 + 3\,H_2SO_4 \longrightarrow Al_2(SO_4)_3 + 6\,H_2O$$

Example 4.11 **Balancing a chemical equation**

Balance the following equation:

$$K_3PO_4 + BaCl_2 \longrightarrow Ba_3(PO_4)_2 + KCl$$

Solution

Focus on the cation of greatest charge, Ba^{2+}:

$$Ba(X) \longrightarrow Ba_3(Y)$$

The barium atoms can be balanced, provisionally, by placing a 3 before the Ba on the left-hand side.

$$3\,Ba(X) \longrightarrow Ba_3(Y)$$

The consequence of inserting this balancing coefficient is that there are now six chlorine atoms on the left-hand side. Inserting a coefficient of 6 in front of the Cl-containing compound on the right-hand side increases the number of potassium atoms in the product to six. To balance those 6 K, we must now place a coefficient of 2 in front of the reactant K_3PO_4. The final balanced equation is

$$2\,K_3PO_4 + 3\,BaCl_2 \longrightarrow Ba_3(PO_4)_2 + 6\,KCl$$

Problem 4.11 Balance the following equation:

$$Ca(OH)_2 + H_3PO_4 \longrightarrow Ca_3(PO_4)_2 + H_2O$$

4.7 OXIDATION–REDUCTION REACTIONS

Chemical reactions in which changes in the valence shell of the reaction participants take place are called **oxidation–reduction** or **redox reactions.** The earliest oxidation reactions were considered to be those in which there was an increase in the oxygen content of the new compound. These compounds were called **oxides.** It was found that oxygen could be displaced from oxides by heating with hydrogen. An increase in a compound's oxygen content or a decrease in its hydrogen content or both is still considered to be an oxidation. Conversely, a decrease in a compound's oxygen content or an increase in its hydrogen content or both is considered to be a reduction. However, we now know that these reactions are a subset of reactions in which **oxidation** refers to the loss of electrons and **reduction** refers to the gain in electrons. The best generality is that, in an oxidation–reduction reaction, there are changes in the valence shells of atoms as a result of a chemical reaction.

Loss and Gain of Electrons

In Chapter 3, the reaction of sodium with chlorine was written as two separate processes called **half-reactions** in which electrons lost by one of the components were gained by the other. The loss of electrons was an oxidation and the gain in electrons a reduction. The practice of separating the complete reaction equation into equations showing only reduction or oxidation allows balancing with respect to electrons as well as atoms.

$$2\,Na \longrightarrow Na^+ + 2\,e^-$$
$$Cl_2 + 2\,e^- \longrightarrow 2\,Cl^-$$

Balancing redox reactions requires not only the application of the Law of Conservation of Mass—that is, equating the number of atoms of each type on the reactant side with the number of atoms of each type on the product side—but also the conservation of electrons. In a redox reaction, the number of electrons lost by the **reductant** must be equal to the number of electrons gained by the **oxidant.** To decide when a redox reaction has taken place, we assign a number to an atom that defines its **oxidation state,** called its **oxidation number.** If that number undergoes a change as a result of a chemical reaction, the

A PICTURE OF HEALTH

Examples of Stoichiometry in Nutritional Intake

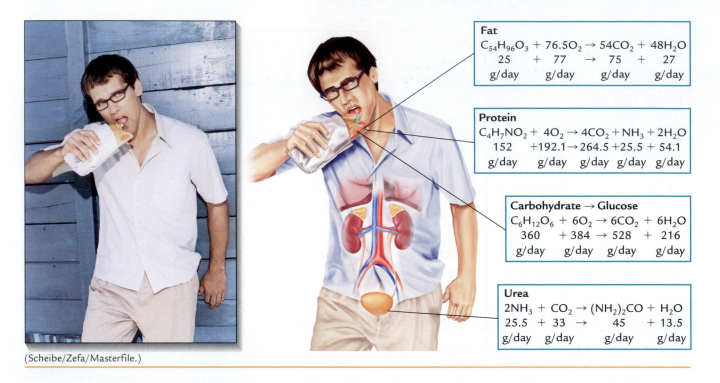

Fat
$$C_{54}H_{96}O_3 + 76.5O_2 \rightarrow 54CO_2 + 48H_2O$$
| 25 | + | 77 | → | 75 | + | 27 |
| g/day | | g/day | | g/day | | g/day |

Protein
$$C_4H_7NO_2 + 4O_2 \rightarrow 4CO_2 + NH_3 + 2H_2O$$
| 152 | +192.1→ | 264.5 | +25.5 | + 54.1 |
| g/day | g/day | g/day | g/day | g/day |

Carbohydrate → Glucose
$$C_6H_{12}O_6 + 6O_2 \rightarrow 6CO_2 + 6H_2O$$
| 360 | + 384 → | 528 | + 216 |
| g/day | g/day | g/day | g/day |

Urea
$$2NH_3 + CO_2 \rightarrow (NH_2)_2CO + H_2O$$
| 25.5 | + 33 | → | 45 | + 13.5 |
| g/day | g/day | | g/day | g/day |

(Scheibe/Zefa/Masterfile.)

reaction is a redox reaction. Furthermore, the changes in oxidation number of reductant and oxidant are used to balance redox reaction equations.

Oxidation States

The following set of rules will allow you to assign an oxidation number to atoms whether in ionic or covalent compounds. They are listed in priority of application; that is, rule 1 takes priority over rule 2, which takes priority over rule 3, and so on:

1. Free elements are assigned an oxidation number of zero. The modern usage is to call an oxidation number an oxidation state; so free elements have an oxidation state of zero.

2. The sum of oxidation states of all the atoms in a species (compound or polyatomic ion) must be equal to the net charge of the species.

3. In compounds, the Group I metals, such as Na and K, are assigned an oxidation state of $+1$. Note that the charge of an ion is written as $1+$ or $2+$. The oxidation states are written $+1$, $+2$, and so on, to differentiate them from ionic charge.

4. In its compounds, fluorine is always assigned an oxidation state of -1, because it is the most electronegative element.

5. The Group II metals, such as Ca and Mg, are always assigned an oxidation state of $+2$, and Group III ions, such as Al and Ga, are assigned an oxidation state of $+3$. These states are identical with the ionic charges of these elements in compounds.

6. Hydrogen in compounds is assigned an oxidation state of $+1$.

7. Oxygen in compounds is assigned an oxidation state of -2.

These rules were developed by considering numbers of electrons and relative electronegativities so that ions such as Na^+ or Ca^{2+} have ionic charges and oxidation states that are identical. However, in covalent compounds such as NO_2 or CH_4, this is not the case. Even though the combined atoms have no real charge, they are assigned oxidation states that imply an electrical charge.

Example 4.12 | **Determining oxidation states of all atoms in a compound**

Determine the oxidation states of each atom in the following compounds:
(a) $CaCl_2$; (b) NaH; (c) SO_2; (d) H_2O_2.

Solution
(a) Ca is assigned an oxidation state of +2 (rule 5). Because $CaCl_2$ is a neutral species, the oxidation state of Cl must be −1. This is established by using rule 2 and setting up an algebraic equation with one unknown quantity to solve for the oxidation state of Cl in $CaCl_2$:

$$(+2) + (2\ Cl) = 0$$
$$Cl = -1$$

The oxidation state of Cl in $CaCl_2$ is −1.
(b) Na is assigned an oxidation state of +1 (rule 3), and rule 2 informs us that, in NaH, H must have an oxidation state of −1. Rule 6 is not violated, because it is of lower priority than rule 3. Hydrogen forms compounds with metals of Groups I and II called hydrides.
(c) Oxygen is assigned an oxidation state of −2 (rule 7). The oxidation state of sulfur (rule 2) is therefore

$$(S) + (2 \times -2) = 0$$
$$S = +4$$

The oxidation state of sulfur in SO_2 is +4.
(d) Rule 6 requires H to have an oxidation state of +1; therefore, by rule 2,

$$(2 \times +1) + 2\ O = 0$$
$$O = -1$$

The oxidation state of oxygen in H_2O_2 is −1, typical of peroxides. There is no conflict, because rule 6 has priority over rule 7.

Problem 4.12 Complete and balance the following equation. The reaction takes place in acidic solution.

$$N_2H_4 + I^- \longrightarrow NH_4^+ + I_2$$

As mentioned earlier, the reaction of sodium with chlorine was written as two separate processes in Chapter 3 to emphasize the loss of electrons by sodium with the simultaneous gain of electrons by chlorine:

$$(1)\ 2\ Na \longrightarrow Na^+ + 2\ e^-$$
$$(2)\ Cl_2 + 2\ e^- \longrightarrow 2\ Cl^-$$

Reaction 1 is an oxidation, and reaction 2 is a reduction. The rules tell us that the oxidation state of sodium changed from 0 to +1 and that of chlorine changed from 0 to −1. We can generalize these observations to say that any change in oxidation state to a more-positive state is an oxidation and that any change in oxidation state to a less-positive (more-negative) state is a reduction.

Example 4.13 | **Determining oxidation states of carbon in different compounds**

What is the oxidation state of carbon in each of the following compounds, and how are these compounds related to one another in reference to their oxidation states? (a) CH_4; (b) CH_3OH; (c) CH_2O; (d) HCOOH; (e) CO_2.

Solution

(a) By rule 6, H is assigned an oxidation state of +1. With the use of rule 2,

$$(C) + (4 \times +1) = 0$$
$$C = -4$$

In methane, carbon has an oxidation state of -4.

(b) Rule 6 assigns an oxidation state of +1 to H, and rule 7 assigns an oxidation state of -2 to O. With the use of rule 2,

$$(C) + (4 \times +1) + (-2) = 0$$
$$C = -2$$

In methyl alcohol, carbon has an oxidation state of -2.

(c) By rule 2, with the use of the preceding assignments of oxidation states to hydrogen and oxygen,

$$(C) + (2 \times +1) + (-2) = 0$$
$$C = 0$$

In formaldehyde, carbon has an oxidation state of 0.

(d) By rule 2, with the use of the preceding assignments of oxidation states to hydrogen and oxygen,

$$(C) + (2 \times +1) + (2 \times -2) = 0$$
$$C = +2$$

In formic acid, carbon has an oxidation state of $+2$.

(e) By rule 2, with the use of the preceding assignments of oxidation state to oxygen,

$$(C) + (2 \times -2) = 0$$
$$C = +4$$

In carbon dioxide, carbon has an oxidation state of $+4$.

Problem 4.13   Assign an oxidation state to chlorine in each of the following compounds.

(a) Cl_2O; (b) Cl_2O_3; (c) ClO_2; (d) Cl_2O_6.

The series of compounds in Example 4.13 represents the range of oxidation states of carbon in compounds with hydrogen and oxygen. Carbon is in its maximally reduced state in methane and maximally oxidized state in carbon dioxide. This range of oxidation states will be of more significance when you begin your study of organic chemistry.

Balancing Redox Reactions

Let's reexamine the reaction between sodium and chlorine but rewrite it as one reaction:

$$Na + Cl_2 \longrightarrow Na^+ + 2\,Cl^-$$

This equation is not balanced but, by inspection, it is easily balanced:

$$2\,Na + Cl_2 \longrightarrow 2\,Na^+ + 2\,Cl^-$$

When we write the reaction as two half-reactions, we will see that not only are the atoms balanced but the electrons transferred also are balanced.

$$(1)\ 2\,Na \longrightarrow 2\,Na^+ + 2\,e^-$$
$$(2)\ Cl_2 + 2\,e^- \longrightarrow 2\,Cl^-$$

Balancing this redox reaction is deceptively simple, because the atom balance and electron balance are the same. Guessing the balancing coefficients for the typical and more-complex redox reactions can be a frustrating experience; so we will demonstrate a systematic procedure that will permit a straightforward approach to solving these problems. It is called the ion-electron or half-reaction method and it appears in Box 4.2.

4.2 CHEMISTRY IN DEPTH

Balancing Oxidation–Reduction Reactions by the Ion-Electron Method

The ion-electron method of balancing redox reactions requires the balance of mass (atoms) and electrons. This process is accomplished by the following series of steps.

- **Step 1.** Separate the oxidation and reduction half-reactions.
- **Step 2.** Make certain that both half-reaction equations are balanced with respect to all elements other than oxygen and hydrogen.
- **Step 3.** Balance the oxygen atoms by adding water.
- **Step 4.** Balance the equations with respect to hydrogen by using protons.
- **Step 5.** Balance the equations with respect to charge.
- **Step 6.** Obtain the complete and balanced equation by adding the two half-reactions and canceling any terms that appear on both sides.

Balancing the reaction equation sometimes used for the roadside detection of alcohol on the breath is an illustration of the ion-electron method. In this reaction, ethanol, C_2H_5OH, is oxidized to acetaldehyde, C_2H_4O, by the orange dichromate ion, $Cr_2O_7{}^{2-}$, which is reduced to the green chromium(III) ion, Cr^{3+}. The unbalanced net ionic equation is

$$H^+ + Cr_2O_7{}^{2-} + C_2H_5OH \longrightarrow$$
$$Cr^{3+} + C_2H_4O + H_2O$$

The first step in the solution of this problem is to separate the oxidation half-reaction from the reduction half-reaction:

Step 1. The two half-reactions are

$$Cr_2O_7{}^{2-} \longrightarrow Cr^{3+}$$
$$C_2H_5OH \longrightarrow C_2H_4O$$

Step 2. Make certain that both equations are balanced with respect to all elements other than oxygen and hydrogen. In this case, the oxidation half-reaction is balanced, but the dichromate reduction requires a coefficient of 2 before the product, chromium(III) ion:

$$Cr_2O_7{}^{2-} \longrightarrow 2\,Cr^{3+}$$
$$C_2H_5OH \longrightarrow C_2H_4O$$

Step 3. Balance the oxygen atoms. This balancing is done by noting that the oxygen of $Cr_2O_7{}^{2-}$ is lost on its conversion into Cr^{3+}. The lost oxygen is assumed to have reacted with H^+ to form H_2O. In this case, only the reduction half-reaction includes changes in oxygen. There are seven oxygen atoms on the left and zero on the right; so the oxygen atoms are balanced by adding 14 H^+ ions to the left-hand side of the dichromate half-reaction, which results in the formation of 7 H_2O:

$$Cr_2O_7{}^{2-} + 14\,H^+ \longrightarrow 2\,Cr^{3+} + 7\,H_2O$$
$$C_2H_5OH \longrightarrow C_2H_4O$$

Step 4. Balance with respect to hydrogen. There are six hydrogen atoms in ethanol on the left-hand side of the oxidation half-reaction and four hydrogen atoms in acetaldehyde on the right-hand side. Therefore two hydrogen atoms in the form of H^+ ions must be added to the right-hand side of that half-reaction:

$$Cr_2O_7{}^{2-} + 14\,H^+ \longrightarrow 2\,Cr^{3+} + 7\,H_2O$$
$$C_2H_5OH \longrightarrow C_2H_4O + 2\,H^+$$

The two half-reactions are now individually balanced with respect to atoms. Next, they must be balanced with respect to charge.

Step 5. Balancing with respect to charge means that the charges on both sides of the arrow must be equal to each other but not necessarily 0. Each half-reaction must be charge balanced by adding electrons to the side with excess positive charge. The dichromate half-reaction has a total charge of +12 on the left and +6 on the right; so six electrons must be added to the left-hand side. This addition of electrons makes the charge of +6 on the left equal to a +6 charge on the right:

$$Cr_2O_7{}^{2-} + 14\,H^+ + 6\,e^- \longrightarrow 2\,Cr^{3+} + 7\,H_2O$$

The alcohol half-reaction has a charge of 0 on the left-hand side and +2 on the right-hand side; so two electrons must be added to the right-hand side. This addition of electrons makes the charge of 0 on the left equal to the charge of 0 on the right:

$$C_2H_5OH \longrightarrow C_2H_4O + 2\,H^+ + 2\,e^-$$

It is now necessary to recognize that, in a redox reaction, the number of electrons consumed in the reduction step must be equal to the number of electrons produced in the oxidation step. The reduction of 1 mol of dichromate requires 6 mol of electrons. The oxidation of 1 mol of ethanol produces 2 mol of electrons. In this case, the oxidation half-reaction must be multiplied by 3 to provide the 6 mol of electrons required by the reduction reaction. When that is done, the equations are balanced with respect to both atoms and electrons.

Step 6. The complete and balanced equation is obtained by adding the two half-reactions and canceling any terms that appear on both sides:

$$Cr_2O_7{}^{2-} + 14\,H^+ + 6\,e^- \longrightarrow 2\,Cr^{3+} + 7\,H_2O$$
$$3\,C_2H_5OH \longrightarrow 3\,C_2H_4O + 6\,H^+ + 6\,e^-$$

$$Cr_2O_7{}^{2-} + 3\,C_2H_5OH + 8\,H^+ \longrightarrow$$
$$2\,Cr^{3+} + 3\,C_2H_4O + 7\,H_2O$$

In the complete equation, the electrons do not appear. They were equal on both sides of the equation and therefore canceled by summation. Electrons should never appear in the completely balanced equation.

Direct Reaction with Oxygen

Metals react directly with oxygen to form oxides: for example,

$$2\,Mg + O_2 \longrightarrow 2\,MgO$$
$$4\,Fe + 3\,O_2 \longrightarrow 2\,Fe_2O_3$$
$$4\,Al + 3\,O_2 \longrightarrow 2\,Al_2O_3$$

Each of these reactions is a redox reaction in which the metals are oxidized and oxygen is reduced.

An example of a reaction in which a compound loses oxygen is the reaction of an ore (in this case, iron ore) with a reducing agent (in this case, carbon):

$$2\,Fe_2O_3 + 3\,C \longrightarrow 4\,Fe + 3\,CO_2$$

Most metals are obtained in elemental form by the reduction of ores that, in many cases, are oxides of metals.

When the direct reaction of an element or compound with oxygen is accompanied by the familiar sight of burning, the reaction is called a **combustion reaction.** Hydrogen, sulfur, and nitrogen, for example, each burn in excess oxygen to produce H_2O, SO_2, and NO_2, respectively.

Compounds containing carbon react with oxygen to produce CO_2 if sufficient oxygen is present. For example, if an alkane such as C_3H_8 is burned, the products will be CO_2 and H_2O. The energy-yielding reactions of metabolism are equivalent to slow, controlled combustion reactions.

If less than sufficient O_2 is available, the combustion product of an alkane will be CO (carbon monoxide). A significant number of deaths occur every year from the carbon monoxide generated when fossil fuels are burned in heating systems that have insufficient oxygen intake. For our purposes, assume that there is sufficient oxygen to produce CO_2 unless otherwise indicated.

Example 4.14   Balancing a combustion equation: I

Write the balanced equation for the combustion of the alkane C_5H_{12}.

Solution

The products of the alkane's combustion will be water and carbon dioxide:

$$C_5H_{12} + O_2 \longrightarrow CO_2 + H_2O \quad \text{(unbalanced)}$$

All the carbon is contained in one product, and all the hydrogen in the other, and so the coefficients of the products are quickly determined:

$$C_5H_{12} + O_2 \longrightarrow 5\,CO_2 + 6\,H_2O \quad \text{(unbalanced)}$$

The sum of O atoms on the right is 16; therefore, the coefficient of O_2 is 8.

$$C_5H_{12} + 8\,O_2 \longrightarrow 5\,CO_2 + 6\,H_2O$$

Problem 4.14   Write the balanced equation for the combustion of the alkane C_4H_{10}. To avoid fractional balancing coefficients, multiply all coefficients by an appropriate whole number—for example, 2 or 3—to change the fractions to whole numbers.

Example 4.15   Balancing a combustion equation: II

Write the balanced equation for the combustion of the carbohydrate glucose ($C_6H_{12}O_6$).

Solution

The elements of which glucose (and any other carbohydrate) is composed are carbon, hydrogen, and oxygen. Therefore the products of the combustion of glucose will be water and carbon dioxide:

$$C_6H_{12}O_6 + O_2 \longrightarrow CO_2 + H_2O \quad \text{(unbalanced)}$$

All the carbon is contained in one product and all the hydrogen in the other, and so the coefficients of the products are quickly determined:

$$C_6H_{12}O_6 + O_2 \longrightarrow 6\,CO_2 + 6\,H_2O \quad \text{(unbalanced)}$$

The sum of O atoms on the right is 18, but, in this case, both reactants contribute O atoms. Six oxygen atoms are contributed by glucose. Therefore oxygen need contribute only 12, and its coefficient is 6:

$$C_6H_{12}O_6 + 6\,O_2 \longrightarrow 6\,CO_2 + 6\,H_2O$$

Although the metabolic oxidation of glucose in living cells is far more complex, it, too, is represented by this balanced equation.

Problem 4.15 Write the balanced equation for the combustion of lactic acid ($C_3H_6O_3$), a product of carbohydrate metabolism.

Dehydrogenation

One of the reactions through which the cell extracts energy from food is the oxidation of malic acid to oxaloacetic acid. This reaction is an example of oxidation by hydrogen loss, also called a **dehydrogenation** reaction. The hydrogen is transferred as a hydride ion ($H:^-$) to an intermediary compound called nicotinamide adenine dinucleotide, abbreviated NAD^+, which is reduced to form NADH in an enzyme-catalyzed reaction. The hydrogen eventually emerges in the form of water. The oxidation can be written as

$$C_4H_6O_5 + NAD^+ \longrightarrow C_4H_4O_5 + NADH + H^+$$

In this reaction, one hydrogen atom in the form of a hydride ion is transferred to NAD^+ along with two electrons, and the other hydrogen atom enters solution as a proton.

By using structural formulas, we write the reaction as follows:

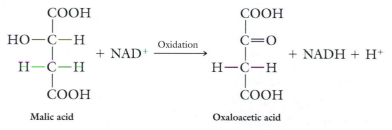

The reverse of this reaction, a reduction, in which the oxaloacetic acid gains hydrogen to form malic acid also takes place in biological systems.

✓ Oxidation takes place when an atom's oxidation state becomes more positive.

✓ Reduction takes place when an atom's oxidation state becomes more negative.

✓ Oxidation takes place when a compound's oxygen content increases or its hydrogen content decreases or both.

✓ Reduction takes place when a compound's oxygen content decreases or its hydrogen content increases or both.

The following list summarizes the process of balancing chemical equations.

• Write a chemical reaction as an equation with an arrow separating the reacting components, or reactants, from the resulting components, or products.

• To balance an equation, the correct formulas of reactants and products must be known.

- Changing the subscripts in a compound's formula in an effort to balance an equation is not permissible, because altering subscripts changes the chemical identity of the substance.

- An equation can be balanced only by inserting whole numbers, called balancing coefficients, in front of the formulas of both reactants and products. The coefficient multiplies the entire formula unit and therefore the numbers of all the atoms in it.

- When a final check shows that there are equal numbers of each kind of atom on both sides of the equation, it is balanced.

4.8 STOICHIOMETRY

The coefficients, or multipliers, in balanced equations can now be used to connect the Law of Conservation of Atoms to the Law of Conservation of Mass. In Section 4.7, the emphasis was on the balancing of equations with respect to number of atoms. Note, however, that, if we multiply a balancing coefficient by Avogadro's number, we are in effect converting it into Avogadro numbers of formula units, or into moles. With this conversion, we can relate the balancing coefficients to mass. The quantitative mass relations of chemical reactions define an aspect of chemistry called **stoichiometry.**

When we use the mole (the Avogadro number of formula units), the balanced equation

$$2\,Al(OH)_3 + 3\,H_2SO_4 \longrightarrow Al_2(SO_4)_3 + 3\,H_2O$$

can be interpreted to mean that 2 mol of $Al(OH)_3$ will react with 3 mol of H_2SO_4 to produce 1 mol of $Al_2(SO_4)_3$ and 3 mol of H_2O. These relations can be expressed as a series of equivalences:

$$2\,Al(OH)_3 \Leftrightarrow 1\,Al_2(SO_4)_3$$
$$2\,Al(OH)_3 \Leftrightarrow 3\,H_2SO_4$$
$$2\,Al(OH)_3 \Leftrightarrow 3\,H_2O$$
$$3\,H_2SO_4 \Leftrightarrow 1\,Al_2(SO_4)_3$$
$$3\,H_2SO_4 \Leftrightarrow 3\,H_2O$$
$$1\,Al_2(SO_4)_3 \Leftrightarrow 3\,H_2O$$

The notation $\Leftrightarrow$ is an equivalence sign. In this context, equivalence means, for example, that 2 mol of $Al(OH)_3$ in this reaction will produce 1 mol of Al_2SO_4. It does not mean that 2 mol of $Al(OH)_3$ is equal to 1 mol of Al_2SO_4.

Relations among these balancing coefficients provide conversion factors that allow us to calculate how much product will be formed by the reaction of

Stoichiometric calculations in chemistry are analogous to multiplying or dividing the amounts of ingredients in a bread recipe in order to produce a smaller or larger number of loaves. (Richard Hutchings/PhotoEdit.)

given amounts of starting materials. In the more explicit example that follows, a balanced chemical equation is used to construct unit-conversion factors relating moles of one component to moles of any other component.

Example 4.16 **Using stoichiometry to create conversion factors: I**

Derive all possible unit-conversion factors from the following balanced equation:

$$2\,Al(OH)_3 + 3\,H_2SO_4 \longrightarrow Al_2(SO_4)_3 + 3\,H_2O$$

Solution

The following 6 unit-conversion factors and their reciprocals add up to a total of 12 different unit-conversion factors from this chemical equation.

$$\frac{2\ \text{mol Al(OH)}_3}{1\ \text{mol Al}_2(\text{SO}_4)_3} \qquad \frac{2\ \text{mol Al(OH)}_3}{3\ \text{mol H}_2\text{SO}_4} \qquad \frac{2\ \text{mol Al(OH)}_3}{3\ \text{mol H}_2\text{O}}$$

$$\frac{3\ \text{mol H}_2\text{SO}_4}{1\ \text{mol Al}_2(\text{SO}_4)_3} \qquad \frac{3\ \text{mol H}_2\text{SO}_4}{3\ \text{mol H}_2\text{O}} \qquad \frac{1\ \text{mol Al}_2(\text{SO}_4)_3}{3\ \text{mol H}_2\text{O}}$$

Problem 4.16 Derive all possible unit-conversion factors from the following balanced equation:

$$3\,Mg(OH)_2 + 2\,FeCl_3 \longrightarrow 2\,Fe(OH)_3 + 3\,MgCl_2$$

Example 4.17 **Using stoichiometry to create conversion factors: II**

In the reaction of aluminum hydroxide with sulfuric acid, how many moles of aluminum sulfate can be produced if the reaction is begun with 1.500 mol of $Al(OH)_3$?

Solution

This problem can be solved with the unit-conversion method, by first deciding on the required conversion factor. Starting with some number of moles of $Al(OH)_3$, how many moles of $Al_2(SO_4)_3$ can we prepare?

$$\text{Moles of Al(OH)}_3 \times \text{conversion factor} = \text{moles of Al}_2(\text{SO}_4)_3$$

$$\text{Conversion factor} = \frac{\text{moles of Al}_2(\text{SO}_4)_3}{\text{moles of Al(OH)}_3}$$

Substituting the appropriate numerical values from the balanced equation, we have

$$1.500\ \text{mol Al(OH)}_3 \left(\frac{1\ \text{mol Al}_2(\text{SO}_4)_3}{2\ \text{mol Al(OH)}_3} \right) = 0.7500\ \text{mol Al}_2(\text{SO}_4)_3$$

Problem 4.17 In the reaction of aluminum hydroxide with sulfuric acid, how many moles of aluminum hydroxide are required to produce 3.200 mol of $Al_2(SO_4)_3$?

We must again point out that the mathematical conventions followed in Example 4.17 might seem to suggest that, in the reaction of $Al(OH)_3$ with H_2SO_4, 1.500 mol of $Al(OH)_3$ is equal to 0.7500 mol of $Al_2(SO_4)_3$. That is not quite correct; $Al(OH)_3$ is not $Al_2(SO_4)_3$. A more precise meaning is that 1.500 mol of $Al(OH)_3$ is equivalent to 0.7500 mol of $Al_2(SO_4)_3$ in that particular reaction. It is important to note that conversion factors of the kind used in Example 4.17 allow us to express the numerical equivalences between moles of different materials. Such numerical values are one thing, but the identity of the substances is a separate matter entirely. To clarify this idea, a few more examples will be presented.

Using stoichiometry to create conversion factors: III

In the reaction

$$N_2 + 3H_2 \longrightarrow 2NH_3$$

how many moles of hydrogen are equivalent to 0.5000 mol of nitrogen?

Solution

We require a factor that will convert moles of nitrogen into their equivalence in moles of hydrogen, in regard to this particular balanced equation.

$$0.5000 \text{ mol } N_2 \left(\frac{3 \text{ mol } H_2}{1 \text{ mol } N_2} \right) = 1.500 \text{ mol } H_2$$

This calculation shows that 0.5000 mol N_2 is equivalent to but certainly not equal to 1.500 mol H_2 in the reaction of nitrogen with hydrogen to form ammonia.

Problem 4.18   How many moles of ammonia will be formed from 0.5000 mol of N_2 and 1.500 mol of H_2?

In the laboratory, chemists weigh grams of substances, not moles of substances. But balanced chemical equations relate moles of reactant to moles of products. To predict grams of product from grams of reactant, we therefore must convert grams of reactant into moles, then use the balanced equation to predict the number of moles of product, and, finally, convert the number of moles of product into grams of product. The detailed procedure is as follows:

- First, balance the equation describing the reaction.
- Using the formula mass as a unit-conversion factor, convert nitrogen's mass in grams into its equivalent number of moles.
- Using the balancing coefficients as unit-conversion factors, calculate the number of moles of hydrogen required to react with the calculated number of moles of nitrogen.
- Using the formula mass as a unit-conversion factor, convert the calculated number of moles of hydrogen into its mass in grams.

Figure 4.2 is a flow diagram of the procedure for calculating either mass or moles from a balanced chemical equation. The procedure is based on the use of the balancing coefficients to convert moles of reactants into moles of products or vice versa. The methods of stoichiometry are based on the fact that reactions can be interpreted on a molecular, a molar, or a mass basis, and these interpretations are summarized in Figure 4.3. Example 4.19 illustrates the method.

Calculating the mass relations in a chemical reaction

How many grams of hydrogen are required for the complete conversion of 42 g of nitrogen into ammonia, and how many grams of ammonia can be formed from that quantity of nitrogen?

Solution

Step 1.   Balance the equation:

$$N_2 + 3H_2 \longrightarrow 2NH_3$$

Step 2.   Convert the mass of nitrogen into moles of nitrogen:

$$42.00 \text{ g } N_2 \left(\frac{1 \text{ mol } N_2}{28.02 \text{ g } N_2} \right) = 1.499 \text{ mol } N_2$$

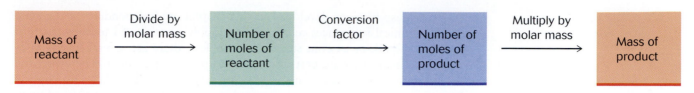

Figure 4.2 The procedure for calculating either mass or moles from balanced chemical equations. The central feature of the procedure is the use of balancing coefficients to convert moles of reactants into moles of product or vice versa.

Step 3. Calculate the numbers of moles of hydrogen and ammonia equivalent to 1.499 mol of N_2 by using the balancing coefficients as unit-conversion factors:

$$1.499 \text{ mol } N_2 \left(\frac{2 \text{ mol } NH_3}{1 \text{ mol } N_2} \right) = 2.998 \text{ mol } NH_3$$

$$1.499 \text{ mol } N_2 \left(\frac{3 \text{ mol } H_2}{1 \text{ mol } N_2} \right) = 4.497 \text{ mol } H_2$$

Step 4. Convert moles of H_2 and NH_3 into their corresponding mass in grams by using their formula masses as unit-conversion factors:

$$2.998 \text{ mol } NH_3 \left(\frac{17.03 \text{ g } NH_3}{1 \text{ mol } NH_3} \right) = 51.06 \text{ g } NH_3$$

$$4.497 \text{ mol } H_2 \left(\frac{2.016 \text{ g } H_2}{1 \text{ mol } H_2} \right) = 9.066 \text{ g } H_2$$

Add up the masses of the reactants and see if they equal the mass(es) of the product(s) to check for conservation of mass:

$$42.00 \text{ g } N_2 + 9.066 \text{ g } H_2 = 51.07 \text{ g } NH_3$$

Problem 4.19 Given the reaction

$$K_3PO_4 + BaCl_2 \longrightarrow Ba_3(PO_4)_2 + KCl,$$

calculate the number of grams of $BaCl_2$ needed to react with 42.4 g of K_3PO_4, and calculate the number of grams of $Ba_3(PO_4)_2$ produced.

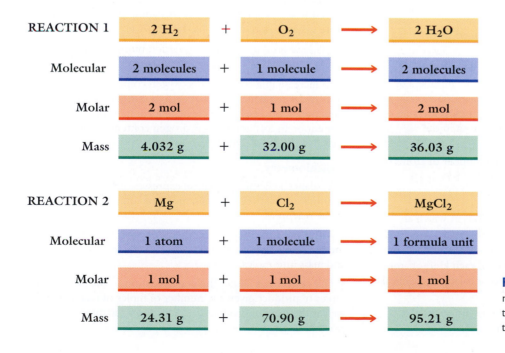

Figure 4.3 Three interpretations—molecular, molar, and mass—of the stoichiometric relations in two chemical reactions.

The mass relations developed in this chapter are independent of the state of the chemical substances reacting. They apply equally well to substances in the solid, liquid, or gaseous states. Nevertheless, there are several specific uses of stoichiometry that are best discussed in a particular context. Therefore, the determination of molecular mass will be presented in Chapter 5, "The Physical Properties of Gases," and more detailed aspects of the stoichiometry of reactions in solution will be delayed until Chapters 7 and 9.

Summary

Chemical Formulas and Formula Masses

• A chemical compound's formula tells us how many atoms of each element are in a formula unit.

• The sum of the relative masses of the atoms of the formula unit is called the formula mass.

• A compound's formula mass can be calculated only if its formula is known.

The Mole

• The atomic mass (in grams) of an element or the formula mass (in grams) of a compound is called the mole.

• A mole of any element or compound contains the same number of atoms or formula units as a mole of any other element or compound.

• The mole is the factor for converting mass into number of formula units and number of formula units to mass.

Avogadro's Number

• The number of fundamental units in a mole has been determined by experiment to be 6.022×10^{23} and is called Avogadro's number.

Empirical and Molecular Formulas

• The formulas for compounds are determined by quantitative analysis.

• If the numbers of moles obtained by quantitative analysis are fractional, they are transformed into whole numbers by dividing each fractional number by the smallest fractional value. This transformation produces the empirical formula, which gives the number of moles of each atom in the compound.

• The molecular formula gives the actual number of atoms of each element in a molecule.

Balancing Chemical Equations

• Chemical reactions are written as equations in which an arrow separates components that react, called reactants, from components resulting from the reaction, called products.

• When an equation has equal numbers of each kind of atom on both sides of the equation, it is called a balanced chemical equation.

• To balance a chemical equation, the correct formulas of reactants and products must be known.

• A chemical equation can be balanced only by inserting whole numbers, called balancing coefficients, in front of the formulas of both reactants and products.

• Each coefficient in a balanced chemical equation multiplies the entire formula unit that it precedes and, therefore, the numbers of all the atoms in it.

Oxidation–Reduction Reactions

• Chemical reactions in which electrons are lost by one of the reactants and gained by another are called oxidation–reduction or redox reactions.

• The reactions are written as two separate processes called half-reactions in which electrons are lost in one of the half-reactions and gained in the other.

• In a redox reaction, the number of electrons lost by a compound (the reductant) must be equal to the number of electrons gained by the other (the oxidant).

• To decide when a redox reaction has taken place, we assign a number to an atom that defines its oxidation state, called its oxidation number.

• If the oxidation number becomes more positive, an oxidation has taken place. If the oxidation number becomes more negative, a reduction has taken place.

• Rules are used to assign an oxidation number to atoms whether in ionic or in covalent compounds.

• An increase in a compound's oxygen content or a decrease in its hydrogen content or both also is considered to be an oxidation. Conversely, a decrease in a compound's oxygen content or an increase in its hydrogen content or both is considered to be a reduction.

Stoichiometry

• A balanced chemical equation expresses the quantitative relations between moles of components in the reaction.

• With the use of these relations and the formula masses of the reaction components, moles can be converted into mass and mass into moles.

• This conversion allows the determination of the number of moles of product given the number of moles of reactant, as well as the mass of product in grams given the mass of reactant in grams.

Key Words

Exercises

Chemical Formulas and Formula Masses

4.1 Calculate the formula masses, in grams, of the following compounds: (a) $CaCrO_4$; (b) $Mg(OH)_2$; (c) $TiCl_4$; (d) $Na_2Cr_2O_7$; (e) C_3H_7OH (propyl alcohol).

4.2 Calculate the formula masses, in grams, of the following compounds: (a) $Zn_3(PO_4)_2$; (b) $Ba(C_2H_3O_2)_2$; (c) Hg_2Cl_2; (d) $Sr(NO_3)_2$; (e) BF_3.

4.3 Calculate the formula masses, in grams, of the following compounds: (a) $Zn_3(AsO_4)_2$; (b) $Al(C_2H_3O_2)_3$; (c) $HgCl_2$; (d) $Sr(ClO_3)_2$; (e) BI_3.

4.4 Calculate the formula masses, in grams, of the following compounds: (a) Ag_2CO_3; (b) Cr_2O_3; (c) Hg_2Br_2; (d) $ZnCl_2$; (e) B_3N_2.

The Mole

4.5 How many moles are there in 24.67 g of sodium chloride?

4.6 How many moles are there in 20.27 g of sodium thiocyanate, NaSCN?

4.7 How many grams of sodium tripolyphosphate, $Na_5P_3O_{10}$, are there in 0.275 mol of that compound?

4.8 How many grams are there in 0.8750 mol of nickel iodate, $Ni(IO_3)_2$?

Avogadro's Number

4.9 How many zinc ions are present in 30.5 g of zinc pyrophosphate, $Zn_2P_2O_7$?

4.10 How many H atoms are there in 5.40 g of water?

4.11 What is the mass, in grams, of 10.54×10^{24} molecules of CH_4?

4.12 How many moles of water are there in 2.3×10^{24} molecules of water?

Empirical Formulas

4.13 By chemical analysis, the composition of zinc chromate is 36.05% zinc, 28.67% chromium, and 35.28% oxygen. Calculate its empirical formula.

4.14 By chemical analysis, the composition of calcium phosphate is 38.71% calcium, 20.00% phosphorus, and 41.29% oxygen. Determine its empirical formula.

Balancing Chemical Equations

4.15 Balance the following equations:
(a) $N_2 + Br_2 \longrightarrow NBr_3$
(b) $HNO_3 + Ba(OH)_2 \longrightarrow Ba(NO_3)_2 + H_2O$
(c) $HgCl_2 + H_2S \longrightarrow HgS + HCl$

4.16 Balance the following equations:
(a) $Mg(OH)_2 + FeCl_2 \longrightarrow Fe(OH)_2 + MgCl_2$
(b) $KBr + Cl_2 \longrightarrow KCl + Br_2$
(c) $Li + H_2O \longrightarrow LiOH + H_2$

4.17 Balance the following equations:
(a) $P + O_2 \longrightarrow P_2O_5$
(b) $FeCl_2 + K_2SO_4 \longrightarrow FeSO_4 + KCl$
(c) $HgCl + NaOH + NH_4Cl \longrightarrow$
$\qquad\qquad Hg(NH_3)_2Cl + NaCl + H_2O$

4.18 Balance the following equations:
(a) $BrCl + NH_3 \longrightarrow NBr_3 + NH_4Cl$
(b) $Cu(NO_3)_2 + NaI \longrightarrow I_2 + CuI + NaNO_3$
(c) $TiCl_4 + NaH \longrightarrow Ti + NaCl + H_2$

Oxidation–Reduction Reactions

4.19 Complete and balance the following combustion reaction:
$$C_6H_6 + O_2 \longrightarrow$$

4.20 Complete and balance the following combustion reaction:
$$C_6H_{12}O_6 + O_2 \longrightarrow$$

4.21 Complete and balance the following equations:
(a) $C_5H_{12} + O_2 \longrightarrow$
(b) $C_8H_{18} + O_2 \longrightarrow$
(c) $C_{10}H_{20} + O_2 \longrightarrow$

4.22 Complete and balance the following equations:
(a) $C_2H_6O + O_2 \longrightarrow$
(b) $C_4H_8O_2 + O_2 \longrightarrow$
(c) $C_{12}H_{22}O_{11} + O_2 \longrightarrow$

4.23 Write the balanced equation for the combustion of methane, CH_4.

4.24 Complete and balance the following equation:
$$C_3H_8O_3 + O_2 \longrightarrow$$

Stoichiometry

4.25 What unit-conversion factors can be derived from the following balanced equation?
$$2\,Al(OH)_3 + 3\,H_2SO_4 \longrightarrow Al_2(SO_4)_3 + 6\,H_2O$$

4.26 What unit-conversion factors can be derived from the following balanced equation?
$$Ca(NO_3)_2 + Na_2SO_4 \longrightarrow CaSO_4 + 2\,NaNO_3$$

4.27 How many moles of $Al_2(SO_4)_3$ can be prepared from 2.5 mol of H_2SO_4 by using sufficient $Al(OH)_3$?

4.28 How many moles of $Al(OH)_3$ are required to completely react with 2.75 mol of H_2SO_4?

4.29 How many moles of H_2SO_4 must be used to produce 6.5 mol of water?

4.30 How many moles of $Al_2(SO_4)_3$ can be prepared from 245 g of H_2SO_4 by using sufficient $Al(OH)_3$?

4.31 How many grams of $Al_2(SO_4)_3$ can be prepared from 46.0 g of H_2SO_4 by using sufficient $Al(OH)_3$?

4.32 How many grams of $Al(OH)_3$ are required to completely convert 144 g of H_2SO_4 into $Al_2(SO_4)_3$?

4.33 Given the following balanced equation,

$$2\,Cu(NO_3)_2 + 4\,NaI \longrightarrow I_2 + 2\,CuI + 4\,NaNO_3$$

(a) How many moles of $Cu(NO_3)_2$ will react with 0.87 mol of NaI?

(b) How many moles of NaI must react to produce 1.45 mol of CuI?

(c) How many grams of $Cu(NO_3)_2$ will react with 15.0 g of NaI?

(d) How many grams of $NaNO_3$ will be produced from 15.00 g of NaI?

4.34 Given the following balanced equation,

$$3\,BaCl_2 + 2\,Na_3PO_4 \longrightarrow Ba_3(PO_4)_2 + 6\,NaCl$$

(a) How many moles of Na_3PO_4 will react with 0.45 mol of $BaCl_2$?

(b) How many moles of $Ba_3(PO_4)_2$ can be produced from 33.3 g of $BaCl_2$?

(c) How many grams of NaCl can be produced from 59.0 g of Na_3PO_4?

(d) How many grams of $Ba_3(PO_4)_2$ can be produced from 0.550 mol of Na_3PO_4?

Unclassified Exercises

4.35 Balance the following equations:

(a) $Ca_3(PO_4)_2 + H_3PO_4 \longrightarrow Ca(H_2PO_4)_2$

(b) $FeCl_2 + (NH_4)_2S \longrightarrow FeS + NH_4Cl$

(c) $KClO_3 \longrightarrow KCl + O_2$

(d) $O_2 \longrightarrow O_3$

(e) $C_5H_6 + O_2 \longrightarrow CO_2 + H_2O$

4.36 How many grams are there in 1.50 mol of Na_3PO_4?

4.37 Calculate the number of moles contained in the given amounts of the following compounds: (a) 102.6 g of $BaCO_3$; (b) 60.75 g of HBr; (c) 148.5 g of CuCl; (d) 50.4 g of HNO_3; (e) 65 g of $C_6H_{12}O_6$.

4.38 Calculate the number of moles contained in the given amounts of the following compounds: (a) 45 g of H_2SO_4; (b) 133 g of $Ca(OH)_2$; (c) 35 g of C_2H_6O; (d) 110 g of C_4H_{10}; (e) 210 g of $FeCl_2$.

4.39 How many formula units are there in 133 g of $Ca(OH)_2$?

4.40 How many molecules are there in 35 g of C_2H_6O (ethyl alcohol)?

4.41 How many electrons are lost by 42 g of sodium when its atoms are oxidized to ions?

4.42 How many molecules are there in 1.000 kg of water?

4.43 By chemical analysis, silver chloride was found to be 75.26% silver and 24.74% chlorine. Determine its empirical formula.

4.44 By chemical analysis, an oxide of copper was found to be 88.81% copper and 11.19% oxygen. Determine its empirical formula.

4.45 Chemical analysis of methyl ether showed it to be composed of 52.17% carbon, 13.05% hydrogen, and 34.78% oxygen. Determine its empirical formula.

4.46 The empirical formula of glycerin is $C_3H_8O_3$. Determine its percent composition.

4.47 Complete and balance the following equations:

(a) $AgNO_3 + KCl \longrightarrow AgCl +$

(b) $Ba(NO_3)_2 + Na_2SO_4 \longrightarrow BaSO_4 +$

(c) $(NH_4)_3PO_4 + Ca(NO_3)_2 \longrightarrow Ca_3(PO_4)_2 +$

(d) $Mg(OH)_2 + H_3PO_4 \longrightarrow Mg_3(PO_4)_2 +$

4.48 Calculate the formula masses of the following compounds: (a) $Hg_3(PO_4)_2$; (b) $BaSO_4$; (c) $C_{12}H_{22}O_{11}$; (d) $MgC_8H_4O_4$; (e) CuCl.

4.49 What is the mass, in grams, of 2.000 mol of $Al_2(SO_4)_3$?

4.50 Balance the following equations:

(a) $P + O_2 \longrightarrow P_2O_5$

(b) $Ti + O_2 \longrightarrow Ti_2O_3$

Expand Your Knowledge

Note: The icons ▲▲▲ denote exercises based on material in boxes.

4.51 The chemical formula for cocaine is $C_{17}H_{21}NO_4$. Calculate its percent composition.

4.52 How many electrons are lost by 60 g of calcium when its atoms are oxidized to ions?

4.53 Calculate the number of grams contained in the given amounts of the following compounds: (a) 0.46 mol H_2SO_4; (b) 1.80 mol $Ca(OH)_2$; (c) 0.76 mol C_2H_6O; (d) 1.89 mol C_4H_{10}; (e) 1.66 mol $FeCl_2$.

4.54 When aluminum is heated with an element from Group VI of the periodic table, Al forms an ionic compound. On one occasion, the Group VI element was unknown, but the compound was 35.93% Al by weight. Identify the unknown element.

4.55 Analysis of an oxide of phosphorus showed it to be 43.64% phosphorus and 56.36% oxygen. Its molecular mass was shown to be 282 amu. Calculate its molecular formula.

4.56 Calculate the number of moles in the recommended daily allowance of the following substances: (a) 11 mg of zinc;

(b) 30 mg of vitamin C, $C_6H_8O_6$; (c) 8 mg of iron; (d) 2.4 μg of vitamin B_{12}, $C_{63}H_{88}CoN_{14}O_{14}P$.

4.57 Calculate the number of moles in (a) 75.0 g of aluminum sulfate, $Al_2(SO_4)_3$: (b) 25.0 g of potassium dichromate, $K_2Cr_2O_7$.

4.58 Calculate the number of moles in (a) 11.5 g of parathion, $C_{10}H_{14}NO_5PS$; (b) 500 mg of aspirin, $C_9H_8O_4$.

4.59 The world's population is about 6.4×10^9. Suppose Avogadro's number of dollars were distributed equally among the world's population. How many dollars would each person receive?

4.60 Calculate, in grams, the mass of (a) one methane (CH_4) molecule; (b) one formula unit of titanium chloride ($TiCl_4$).

4.61 Calculate, in grams, the mass of (a) one sucrose ($C_{12}H_{22}O_{12}$) molecule; (b) one formula unit of iron(II) sulfate ($FeSO_4$).

4.62 Acetylene is produced when water is added to calcium carbide. For many years, acetylene was burned in lanterns to produce light in mines and caves. By chemical analysis, calcium carbide is 62.5% calcium by mass and 37.5% carbon by mass. Determine its empirical formula.

4.63 A 10.0-g sample of cobalt underwent reaction with an excess of gaseous chlorine. The mass of the compound formed was 22.0 g. Determine the compound's empirical formula.

4.64 Given the mass percentages of the following compounds, determine their empirical formulas. (a) 59.78% Li, 40.22% N; (b) 14.17% Li, 85.83% N.

4.65 A 2.006-g sample of a metal was heated in excess oxygen to form an oxide whose formula was MO and whose mass was 2.166 g. Determine the atomic mass of the metal.

4.66 By chemical analysis (see Box 4.1), glucose consists of 40.0% carbon, 6.71% hydrogen, and 53.3% oxygen. Its molecular mass, determined by an independent method, is 180.2 amu. Determine its molecular formula.

4.67 The protein myoglobin is known to contain one molecule of iron per molecule of protein. A sample of the protein was found to contain 0.349% of iron by mass. Determine the molecular mass of the protein.

4.68 Aspirin ($C_9H_8O_4$) is synthesized by the reaction of acetic anhydride ($C_4H_6O_3$) and salicylic acid ($C_7H_6O_3$).

(a) Balance the following unbalanced equation:

$$C_7H_6O_3 + C_4H_6O_3 \longrightarrow C_9H_8O_4 + H_2O$$

(b) What mass of acetic anhydride will react with 10 g of salicylic acid?
(c) What mass of aspirin will be produced in the reaction?

4.69 A mixture of citric acid and sodium hydrogen carbonate is often used to produce a "fizz."

(a) Balance the following unbalanced equation:

$$NaHCO_3 + C_6H_8O_7 \longrightarrow CO_2 + H_2O + Na_3C_6H_5O_7$$

(b) What mass of $C_6H_8O_7$ should be used for 200 mg of $NaHCO_3$?
(c) What mass of CO_2 would be produced from this mixture?

4.70 Bromine, Br_2, is produced by reacting seawater containing bromide ion with chlorine, Cl_2, according to the following equation:

$$Br^- + Cl_2 \longrightarrow Br_2 + Cl^-$$

(a) Balance the equation.
(b) What mass of chlorine will produce 1000 kg of bromine?

4.71 Balance the following equation using the method described in Box 4.2:

$$Fe^{2+} + MnO_4^- \xrightarrow{H^+} Fe^{3+} + Mn^{2+}$$

4.72 Pyruvic acid, a product of glucose metabolism, has the formula

$$O=C-OH$$
$$|$$
$$C=O$$
$$|$$
$$CH_3$$

What are the oxidation states (numbers) for each of the carbon atoms?

4.73 Identify the atoms undergoing oxidation and reduction in the following reactions:

(a) $I_2 + OH^- \longrightarrow IO_3^- + I^-$
(b) $Al + OH^- \longrightarrow Al(OH)_4^- + H_2$

4.74 Assign oxidation states to the elements in the following compounds.

(a) C_6H_6; (b) $NaC_2H_3O_2$; (c) $Na_2S_2O_3$.

CHAPTER 5

THE PHYSICAL PROPERTIES OF GASES

(Michael DeMocker/Visuals Unlimited.)

Chemistry in Your Future

While scuba diving with some friends, you surface to find that one of the party is barely conscious and appears to be in pain. The group had been exploring a reef 150 feet below the surface and didn't notice when this person surfaced much earlier than the rest. Three of you immediately force this diver back down 100 feet and slowly bring him up in stages of 50 feet, staying at each depth for about 30 minutes. At the end of the process, he is fully recovered. The certified divers in the group knew how to handle the emergency, but you are the only one who studied chemistry and can explain what happened. It has to do with the material in Chapter 5.

For more information on this topic and others in this chapter, go to www.whfreeman.com/bleiodian2e

Learning Objectives

- Define gas pressure and its units, and describe how it is measured.
- Summarize the gas laws' quantitative descriptions of the physical behavior of gases.
- Apply the appropriate gas laws to particular experimental conditions.
- Describe the properties of mixtures of gases.
- Use the gas laws to determine molecular mass.
- Determine the amount of gas dissolved in a liquid.

About 60% of a person's body mass is skeletal muscle, a tissue that often functions quite effectively for significant periods of time in the absence of oxygen. However, the heart and brain cannot be deprived of oxygen for even a few minutes without serious long-term consequences. To maintain homeostasis—that is, a healthy physiological steady state—the body's blood-oxygen concentration must remain high and the carbon dioxide concentration low. Understanding the behavior of gases is vital in this regard. The respirators and barometric chambers used in inhalation therapy are only a few of the life-saving tools in every hospital. These machines function according to a short list of basic principles that chemists have discovered about gases. We will be exploring these principles in this chapter.

Experiments have shown that all gases exhibit virtually the same behavior in response to temperature and pressure changes, no matter what the gaseous compound's formula mass or shape. To illustrate this universal behavior, Table 5.1 lists several gases, differing in number and types of atom per molecule and in formula mass. At the same temperature and pressure (defined in Section 5.1), each of these gases occupies about the same amount of space—close to 22.4 L.

The space around each individual particle in these samples of gas may be a thousand or more times the actual volume of the gas molecule itself. In other words, the volume of a gas molecule is insignificant compared with the space surrounding it. That is the basic reason why both the quantitative and the qualitative ideas presented in this chapter apply to all gases without regard to their chemical nature.

5.1 GAS PRESSURE

This chapter explores the physical properties of gases. These properties include two that we have studied before—volume and temperature—and a new one, called pressure is one of the most important of the physical properties with which a gas can be characterized. Differences in pressure cause the movement of gases from one place to another. This phenomenon explains the transport of oxygen from the lung airspace into the blood, the transport of carbon dioxide from the blood into the lung airspace, and the transport of oxygen from a respirator into the lung airspace.

Pressure is defined as a force applied over a unit area, and it is measured in units such as pounds per square inch, kilograms per square meter, or grams per square centimeter. In mathematical terms,

$$\text{Pressure} = \frac{\text{force}}{\text{area}}$$

Thus, pressure will increase if the force increases but the area remains constant or if the force stays the same but the area decreases. For example, the same force applied over a smaller and smaller area will produce a larger and larger

TABLE 5.1 | **Volume Occupied by 1 mol of Several Different Gases at 0°C and 1 atm Pressure**

Gas	Formula	Formula mass (amu)	Volume (L)*
hydrogen	H_2	2.016	22.43
helium	He	4.003	22.42
nitrogen	N_2	28.02	22.38
carbon monoxide	CO	28.01	22.38
oxygen	O_2	32.00	22.40

*The volumes are expressed to four significant figures to show the variability that accompanied these experimentally determined values.

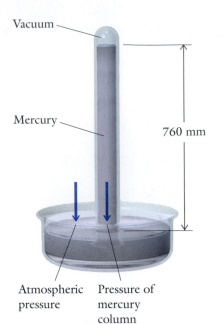

Vacuum

Mercury

760 mm

Atmospheric   Pressure of
pressure   mercury
               column

Figure 5.1 A mercury barometer. The mass of the mercury column is balanced by the pressure of the atmosphere. The atmospheric pressure is measured by the height of the mercury column. The small space at the top of the column is a vacuum and does not contribute to any pressure inside the barometer tube.

pressure. Compare how it would feel to have your toe stepped on by an infant or by a pro football linebacker. It is not difficult to imagine the pressure exerted by someone stepping on your toe. But a gas cannot step on your toe; so how does it exert pressure?

The first clue can be found by examining the meaning of atmospheric pressure, a gas pressure that is routinely mentioned in weather reports. Low pressure usually means bad weather on the way, and high pressure signals a pleasant day. The atmosphere extends to heights in excess of 20 miles and is a mixture of about 20% oxygen, 79% nitrogen, and 1% other gases, including variable amounts of water vapor. These gases have mass, and **atmospheric pressure** is the result of their mass pressing against Earth's surface. The variations in atmospheric pressure are caused by variation in the mass of the atmospheric gas column over a particular location. Cold air is more dense than hot air; therefore it is heavier and leads to a higher local atmospheric pressure. The higher the elevation, the shorter the air column above that location and the lower the average atmospheric pressure.

Atmospheric pressure is often measured by means of a **barometer** (Figure 5.1). The barometer consists of a mercury-filled, closed-end tube inverted in a bath of mercury, whose surface is exposed to the atmosphere. The mass of the atmosphere presses down against the exposed mercury surface and supports the column of mercury in the inverted tube to about 760 mm high. The small space above the mercury column contains no gas. It is a vacuum that is created as the mercury recedes slightly from the end of the tube because of its own weight.

Because atmospheric pressure varies from place to place and from time to time, a standardized unit of pressure independent of location and time has been created. It is called the **atmosphere** (abbreviated **atm**) and is defined as the pressure that will support a vertical column of mercury to a height of exactly 760 mm at 0°C. (Remember, an exact number has an infinite number of significant figures.)

Pressure can be expressed either in atmospheres, millimeters of mercury (mm Hg), or torr. The unit called **torr,** which is equal to **1 mm Hg,** is named in honor of Evangelista Torricelli, who invented the first barometer. In English units, 760 mm Hg is equivalent to 14.7 lb per square foot. In this book, pressure will be expressed in units of torr or atmospheres. To summarize,

$$1 \text{ atm} = 760 \text{ torr} = 760 \text{ mm Hg}$$

The effect of gas pressure on a physiological state is described in Box 5.1.

5.1   CHEMISTRY WITHIN US

Diving Time and Gas Pressure

In a manometer using water rather than mercury as the column of fluid, 1 atmosphere of pressure would raise the liquid's level to a height of more than 33 feet. Compare this height with that of a mercury column at that pressure, about 30 inches. The reason for the difference is that mercury has a density of 13.6 g/cm³ compared with water's density of 1.0 g/cm³. Nevertheless, water's density is so much greater than that of the air that marine divers are subjected to one additional atmosphere of pressure for every 33 feet that they descend. This effect translates to 4 atm at a depth of about 100 feet.

Novice divers describe the experience of breathing at that depth and pressure as quite unsettling. Some say

that the sensation is almost like "drinking" air. They must make a significant physical effort to maintain gas flow into and out of their lungs. This effort is required because an increase in gas pressure is accompanied by a corresponding increase in gas viscosity. (Viscosity is a property of gases and liquids that describes their resistance to flow. For example, glycerol, a liquid whose viscosity is about the same as that of maple syrup, is more than 1000 times as viscous as water.) The divers' problem is alleviated somewhat by breathing-tube design, but the increased physical effort nevertheless leads rapidly to fatigue and requires close attention to diving time. Divers who forget to check their wristwatches pay a heavy price.

Example 5.1 Converting units of pressure

Calculate (a) the pressure, in atmospheres, that a gas exerts if it supports a 385-mm column of mercury and (b) the equivalent of 0.750 atm in millimeters of mercury and in torr.

Solution

To solve these problems, we will use conversion factors based on the equivalence 1 atm = 760 mm Hg = 760 torr.

(a) $385 \text{ mm Hg} \left(\dfrac{1 \text{ atm}}{760 \text{ mm Hg}} \right) = 0.507 \text{ atm}$

(b) $0.750 \text{ atm} \left(\dfrac{760 \text{ mm Hg}}{1 \text{ atm}} \right) = 570 \text{ mm Hg} = 570 \text{ torr}$

Problem 5.1 Calculate (a) the pressure, in atmospheres, that a gas exerts if it supports a 563-mm column of mercury and (b) the equivalent of 0.930 atm in millimeters of mercury and in torr.

To measure the pressure of a gas in a closed container, a device called a **manometer** is used. Figure 5.2 shows a common type of manometer known as a differential manometer, so called because it measures pressure by revealing the difference between two pressures. It consists of a glass U-tube that is open at both ends and filled with a fluid, typically mercury. However, other fluids will serve as well (for example, as described in Box 5.1), as long as the gas does not dissolve in the fluid. In Figure 5.2a, the liquid reaches the same height in each arm of the tube because each surface of the liquid is in contact with the atmosphere, which exerts the same force on each surface.

In Figure 5.2b, the left arm of the manometer is connected to a gas source, and the other arm is open to the atmosphere. The difference in the levels of the liquids on each side is caused by a difference in pressure. The pressure of the gas is reported as the difference between the heights of the mercury columns in each arm in millimeters of mercury, or torr.

Three possible conditions affect the positions of the fluid in a differential manometer. When gas pressure is greater than atmospheric pressure, the

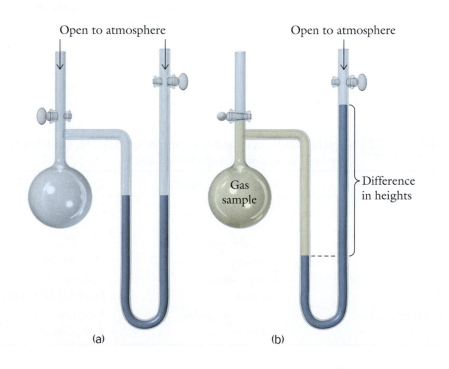

(a) (b)

Figure 5.2 Differential manometers. (a) Both columns are at the same height because both sides are exposed to the atmosphere. (b) The stopcock on the left is closed, and the stopcock on the right is open to the atmosphere. The difference in heights is a direct measure of the difference in pressure between the flask on the left and the atmospheric pressure.

Figure 5.3 A gas-pressure gauge on a tank of oxygen used in respiratory therapy. (Mary Kate Denny/PhotoEdit.)

≫ The influence of pressure on physical states and on the properties of solutions is developed in Chapters 6 and 7.

mercury column will be higher on the atmospheric side of the U-tube, as illustrated in Figure 5.2b. When gas pressure is less than atmospheric pressure, the mercury column will be higher on the side of the gas source. When gas pressure is equal to atmospheric pressure, the mercury column will be the same on both sides of the manometer.

The differential manometer is not a sturdy device, and a more practical meter must be used in a setting such as a hospital. The gas gauge on an oxygen tank (Figure 5.3) employs a membrane to separate the gas, not from the atmosphere but from a springlike mechanical device. The device exerts a force opposing the gas pressure that is calibrated to read in pressure units of pounds per square inch or millimeters of mercury.

The pressure of a gas in a container is the result of the gas molecules colliding with the walls of their container. The greater the collision rate—that is, the more collisions per unit time—the greater the pressure. The rate of collisions can be increased in two ways. The first is to increase the number of molecules in a fixed-volume container. The second is to increase the temperature of the gas. The latter causes the molecules to move faster and therefore collide more often with the container walls. Imagine sitting in a shed with a metal roof at the beginning of a rainstorm. The individual drops on the roof can be heard as the rain begins. But, as the intensity of the storm increases, the sounds of the individual drops begin to merge and, finally, become a constant roar. That "roar" of a gas as it hits the walls of a container is what we detect when its pressure is measured. A practical use of manometry in medicine is described in Box 5.2.

5.2 THE GAS LAWS

The physical properties of gases depend on only four variables: pressure (P), volume (V), Kelvin temperature (T), and number of moles (n). All four variables can be changed simultaneously, but experimenting with only two at a time while keeping the other two constant results in some simple but important mathematical descriptions of gas behavior.

We will first consider a group of laws that describe changes in the state of a gas by changes in only two of the four possible gas variables: pressure and volume, volume and temperature, and, finally, pressure and temperature. The laws are listed in Table 5.2 and are named after their discoverers: Boyle's law, Charles's law, Gay-Lussac's law, and Avogadro's law.

The effects of physical changes on the properties of gases can be visualized by imagining a system consisting of a cylinder with a movable piston (see Figures 5.4 to 5.7). The gas cannot escape, so the number of moles in the cylinder must remain constant. If the cylinder is not insulated, any temperature

TABLE 5.2 | **Variable and Constant Quantities of Boyle's, Charles's, Gay-Lussac's, and Avogadro's Laws**

Law	Variable quantities	Constant quantities
Boyle's law	Pressure Volume	Temperature (K) Number of moles
Charles's law	Temperature (K) Volume	Pressure Number of moles
Gay-Lussac's law	Temperature (K) Pressure	Volume Number of moles
Avogadro's law	Number of moles Volume	Pressure Temperature (K)

5.2 CHEMISTRY AROUND US

Manometry and Blood Pressure

Blood pressure is measured by means of a mercury-filled manometer called a sphygmomanometer. The sphygmomanometer is a manometer in which the column of mercury is connected to both a hand-held air pump and a balloon in the form of a cuff on the other side. The cuff is designed to be wrapped around the upper arm above the elbow. It is then inflated with air until its pressure is great enough to stop the blood flow through the artery to the lower arm. This pressure is transmitted to the manometer and can be read as the height of the mercury column in millimeters. A stethoscope is placed just below the cuff, and the air in the cuff is slowly released. When the cuff pressure falls to a point just below what is known as the systolic pressure (the peak pressure developed by the pumping of the heart), the artery begins to open under the pressure developed by the heart's contractions. At that point, the blood begins to flow with great velocity through a narrower than normal opening, which causes turbulence and the production of vibrations indicated by the soft tapping sounds (heard through the stethoscope).

As the cuff pressure falls further, the tapping sounds are separated by longer intervals, because the arteries remain open longer during each cycle of cardiac contractions. The tapping sounds also become louder. At the point at which the cuff pressure equals what is known as diastolic pressure, the artery remains open, allowing continuous turbulent flow. The sounds heard through the stethoscope become dull and muffled. Just below diastolic pressure, the blood flow becomes continuous, nonturbulent, and silent because the artery is now completely open.

Systolic pressure is recorded as the cuff pressure at which sounds first appear, and diastolic pressure is recorded as the cuff pressure at which sounds disappear. Normal systolic pressures are in the range from 120 torr to 150 torr. Normal diastolic pressures are in the range from 70 torr to 90 torr. Diastolic pressures above that range indicate difficulty in pumping blood through the capillary bed because of either vasoconstriction or vascular obstruction. This capillary narrowing may cause systolic pressures to rise high enough to cause small blood vessels to burst. Such events in the central nervous system are called strokes.

(Jose L. Palaez/The Stock Market.)

change in the gas will be followed by the loss or gain of heat to or from the environment, effectively keeping the gas temperature constant. If the cylinder were insulated, such a loss or gain of heat would be prevented.

This model allows us to predict qualitative changes in gas properties when its pressure, temperature, volume, and mass are varied. Remember that a gas takes the shape of its container and occupies all of it, which means that 1 L of a gas is a sample of that gas enclosed in a 1-L container. Furthermore, the pressure of the gas within the cylinder is defined by the external force on the piston. If the piston does not move, the pressure of the gas within the cylinder must be equal to that external pressure. For example, a 10-kg weight on a piston compressing a gas in a cylinder of 10-cm^2 area exerts an external pressure of 1 kg/cm^2, and that is also the pressure of the gas contained in the cylinder.

On the following pages, we will use our cylinder–piston system to illustrate the four fundamental principles formulated in the laws of Boyle, Charles, Gay-Lussac, and Avogadro.

5.3 BOYLE'S LAW

Figure 5.4 on the following page shows three "snapshots" of a gas confined in our model of a cylinder with a movable piston. We can increase the pressure of the gas in the system by increasing the external force on the piston. In Figure 5.4, this force is represented by a weight sitting on the piston's handle. Note that, as the pressure increases, the gas volume decreases.

Figure 5.4 A gas confined in a cylinder at three different pressures illustrates Boyle's law. Doubling the pressure on the gas halves its volume. The diagram shows a doubling of the force on the piston from 1 kg to 2 kg and again from 2 kg to 4 kg.

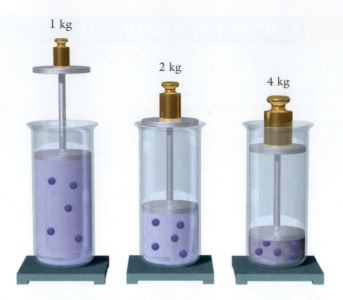

Robert Boyle (1627–1691). While studying at Oxford, Boyle discovered the properties of gases that later came to be known as Boyle's law. He was a cofounder of the Royal Society of London. In his book titled *The Skeptical Chymist,* he attacked alchemical principles regarding substances and proposed an atomistic view of the elements that was a forerunner of the modern concept of matter. (Science Photo Library/Photo Researchers.)

Figure 5.4 also shows that, as the volume decreases, the gas particles become more and more crowded. This crowding causes more collisions with the container walls per unit time and accounts for the increase in gas pressure.

The relation between gas pressure and volume is described by **Boyle's law,** which states that the product of a gas's pressure multiplied by its volume is a constant. (This law means that multiplying the pressure of any sample of gas by its volume will always yield the same numerical value.) Increasing the pressure on a gas will decrease its volume, and vice versa. This principle underlies the mechanics of breathing presented in Box 5.3.

The data listed in Table 5.3 illustrate the results of an experiment in which a gas's pressure or volume were varied while the amount of gas and the temperature (0°C) were held constant. Table 5.3 reveals a number of things:

- As the pressure decreases, the volume increases; or, as the pressure increases, the volume decreases.
- The $P \times V$ product is a constant, where the P is the gas pressure, and V is the volume.

If the condition of the gas at state 1 is compared with that of the gas after it has been changed to state 2, then

$$P_1 \times V_1 = 2.00 \text{ atm} \times \text{L} = P_2 \times V_2$$

or, because both are equal to the same constant, Boyle's law at constant temperature and number of moles becomes

$$P_1 \times V_1 = P_2 \times V_2$$

This equality can be solved in two ways, representing two experimental situations. It can be used to tell us (1) what happens to the volume when pressure is varied and (2) what happens to the pressure when volume is varied.

TABLE 5.3	Variation in Gas Pressure and Volume When Temperature and Number of Moles of Gas Are Held Constant		
State	Pressure (atm)	Volume (L)	$P \times V$ (atm $\times$ L)
1	5.31	0.377	2.00
2	2.67	0.750	2.00
3	1.20	1.67	2.00
4	0.663	3.02	2.00

5.3 | **CHEMISTRY WITHIN US**

Breathing and the Gas Laws

All fluids, whether gas or liquid, will flow when subjected to a gradient in pressure. In other words, a gas will flow from a region of higher pressure to a region of lower pressure. Breathing, a process in which air must flow into and then out of two elastic sacs called lungs, takes advantage of this universal principle. Air will flow into the lungs if the pressure within them is lower than the pressure outside the body, and it will flow out of the lungs if the pressure within them is higher than the pressure outside. The necessary pressure changes are brought about by changes in lung volume caused by the action of a number of powerful muscles within the chest. Breathing is Boyle's law in action.

The figure below shows the lungs located in the airtight thoracic cavity, a space formed by the rib cage and its muscles and a muscle at the base of the cavity called the diaphragm. When relaxed, the diaphragm muscle assumes an upwardly convex shape. The increase in lung volume is caused by expansion of the rib cage and a flattening of the diaphragm. Lung volume decreases on relaxation of the diaphragm upward accompanied by relaxation of the muscles of the rib cage. Boyle's law describes the interaction: As lung volume increases, gas pressure within the lung will decrease and external air will flow into the lungs. As lung volume decreases, gas pressure within the lung will increase and air within the lungs will flow out.

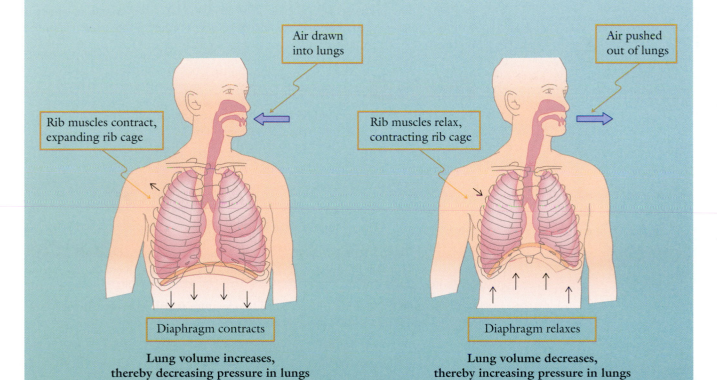

Air drawn into lungs

Rib muscles contract, expanding rib cage

Diaphragm contracts

Lung volume increases, thereby decreasing pressure in lungs

Air pushed out of lungs

Rib muscles relax, contracting rib cage

Diaphragm relaxes

Lung volume decreases, thereby increasing pressure in lungs

The equation can be solved directly by substitution. If you know any three of the variables, the fourth is obtained algebraically. It can also be solved as a unit-conversion problem.

For situation 1, when pressure is varied,

$$V_2 = V_1\left(\frac{P_1}{P_2}\right)$$

This is a unit-conversion calculation in which the conversion factor is a ratio of pressures. If the pressure on the gas is increased [that is, $P_2 > P_1$ (P_2 is greater than P_1)], then the conversion factor must be smaller than 1.0 and the gas volume must decrease, as predicted by Boyle's law. If, instead, P_2 is less than P_1, the reverse will occur.

For situation 2, when volume is varied,

$$P_2 = P_1\left(\frac{V_1}{V_2}\right)$$

This also is a unit-conversion calculation, but now the conversion factor is a ratio of volumes. If the volume of the gas increases—that is, $V_2 > V_1$—the conversion factor must be smaller than 1.0 and the pressure must decrease, as Boyle's law predicts. If, instead, V_2 is less than V_1, the conversion factor will be greater than 1.0 and, as Boyle's law predicts, the pressure must increase.

Example 5.2   Using Boyle's law: I

A 712-mL sample of gas at 505 torr is compressed at constant temperature until its final pressure is 825 torr. What is its final volume?

Solution
First, tabulate the conditions:

	State 1	State 2
Pressure	505 torr	825 torr
Volume	712 mL	?
Temperature	constant	constant
Number of moles	constant	constant

Because the amount of gas and its temperature remain constant, Boyle's law applies and is solved for V_2, the new volume:

$$V_2 = V_1\left(\frac{P_1}{P_2}\right)$$

Because the pressure on the gas increased, the volume must have decreased, and the conversion factor must be smaller than 1.0. Substituting the tabulated values gives

$$V_2 = 712 \text{ mL}\left(\frac{505 \text{ torr}}{825 \text{ torr}}\right) = 436 \text{ mL}$$

The smaller value for the final volume makes sense because the gas was compressed. The pressure units cancel because they are in a ratio, and the final volume units are the same as the original units. Always make certain that the pressure and volume units are clearly stated.

Problem 5.2   An 862-mL sample of gas at 425 torr is compressed at constant temperature until its final pressure is 901 torr. What is its final volume?

Example 5.3 | Using Boyle's law: II

A 2.00-L sample of gas at 0.800 atm must be compressed to 1.60 L at constant temperature. What pressure in atmospheres must be exerted to bring it to that volume?

Solution

First, tabulate the conditions:

	State 1	State 2
Pressure	0.800 atm	?
Volume	2.00 L	1.60 L
Temperature	constant	constant
Number of moles	constant	constant

Because the amount of gas and its temperature remain constant, Boyle's law applies and is solved for P_2.

$$P_2 = P_1 \left(\frac{V_1}{V_2} \right)$$

The volume of the gas decreased from state 1 to state 2, and so the pressure must have increased. Therefore, the ratio of volumes must be greater than 1.0. Substituting the tabulated values gives

$$P_2 = 0.800 \text{ atm} \left(\frac{2.00 \text{ L}}{1.60 \text{ L}} \right) = 1.00 \text{ atm}$$

Here, again, the volume units in the ratio cancel, and the final pressure unit is the same as the original unit. If asked to calculate the final pressure in torr, use the conversion factor 1 atm = 760 torr to change the final answer to torr.

Problem 5.3 A 1.80-L sample of gas at 0.739 atm must be compressed to 1.40 L at constant temperature. What pressure in atmospheres must be exerted to bring it to that volume?

5.4 CHARLES'S LAW

If the temperature of a gas is increased and the external pressure (the weight on the piston) and amount of gas are kept constant, then the volume of the gas will increase. Raising the temperature of a gas increases the velocity of the gas molecules. They undergo more collisions with the container walls per unit time, creating a consequent rise in pressure. Because the external pressure on the gas is not increased, the gas expands. This effect is illustrated in Figure 5.5.

The quantitative relation between gas volume and temperature, called **Charles's law,** relates the volume of a gas to the Kelvin temperature. Mathematically, Charles's law is

$$V = \text{constant} \times T$$

This form of the law tells us, for example, that, if the Kelvin temperature is doubled, the volume doubles and, when it is halved, the volume also is halved.

Table 5.4 on the following page lists data from an experiment in which temperature changes led to volume changes for a fixed amount of gas kept at a constant pressure. The table indicates that

- As the temperature increases, the volume increases.
- The ratio V/T is a constant, but the ratio V/t (t representing the Celsius temperature) is not constant.

Figure 5.5 Equal quantities of a gas confined in insulated cylinders at two temperatures illustrate Charles's law. When the Kelvin temperature is doubled and the pressure is held constant, the gas volume doubles.

TABLE 5.4	Variation in Gas Volume with Change in Temperature at Constant Pressure				
State	$t(°C)$	$T(K)$	Volume (L)	$V/T(L/K)$	$V/t(L/°C)$
1	−23.0	250	5.50	0.0220	−0.239
2	27.0	300	6.60	0.0220	0.244
3	77.0	350	7.70	0.0220	0.100
4	152	425	9.35	0.0220	0.0620

When the temperature is expressed on the Kelvin scale, the ratio of volume to temperature is a constant that takes the same value at all temperatures if the quantity of gas and its pressure remain constant.

The volumes of a sample of gas at two different temperatures can be compared by noting that

$$\frac{V_1}{T_1} = \text{constant} = \frac{V_2}{T_2}$$

Because the two ratios of volume to Kelvin temperature are both equal to the same constant, they are also equal to each other. The final form of Charles's law is

$$\frac{V_1}{T_1} = \frac{V_2}{T_2}$$

Two experimental situations can be solved with Charles's law: one situation determines what happens to the volume of a gas when the temperature is changed, and the other determines what happens to the temperature of a gas when the volume is changed.

For the first experimental situation, the equation is solved for V_2:

$$V_2 = V_1\left(\frac{T_2}{T_1}\right)$$

This is a unit-conversion calculation, in which the conversion factor is a ratio of Kelvin temperatures. Charles's law predicts that the volume of the gas will increase if the temperature increases. Because $T_2 > T_1$, the ratio of temperatures is greater than 1.0, and the volume must increase.

Example 5.4   Using Charles's law: I

A 512-mL sample of a gas, in a cylinder with a movable piston, at 0.000°C is heated at a constant pressure of 0.800 atm to 41.0°C. What is its final volume?

Solution

To use Charles's law, it is necessary to convert Celsius temperature into Kelvin temperature. The conversion uses the relation

Kelvin temperature = Celsius temperature + 273

Tabulate the conditions:

	State 1	State 2
Pressure	0.800 atm	0.800 atm
Volume	512 mL	? mL
Number of moles	constant	constant
Celsius temperature	0.000°C	41.0°C
Kelvin temperature	273 K	314 K

The pressure and number of moles are constant, and so Charles's law applies. Solving for V_2 and introducing the Kelvin temperatures into Charles's law gives

$$V_2 = 512 \text{ mL} \left(\frac{314 \text{ K}}{273 \text{ K}} \right) = 589 \text{ mL}$$

Charles's law informs us that, if the temperature increased, the volume had to increase. Therefore, it is necessary to multiply the initial volume by a fraction, formed by the initial and final temperatures, that is greater than 1.0, or 314/273.

Problem 5.4 A 755-mL sample of a gas, in a cylinder with a movable piston, at 3.00°C is heated at a constant pressure of 0.800 atm to 142°C. What is its final volume?

Example 5.5   Using Charles's law: II

A 1.00-L sample of gas at 27.0°C is heated so that it expands at constant pressure to a final volume of 1.50 L. Calculate its final Kelvin temperature.

Solution
Tabulate the conditions:

	State 1	State 2
Pressure	constant	constant
Volume	1.00 L	1.50 L
Number of moles	constant	constant
Celsius temperature	27.0°C	? °C
Kelvin temperature	300 K	? K

The pressure and amount of gas are constant, and so Charles's law applies. Again, it is necessary to convert Celsius temperature into kelvins. Solving for T_2 and making the appropriate substitutions gives

$$T_2 = 300 \text{ K} \left(\frac{1.50 \text{ L}}{1.00 \text{ L}} \right) = 450 \text{ K}$$

Charles's law specifies that, if the volume increased, the temperature had to increase. Therefore the initial temperature is multiplied by a fraction, formed by the initial and final volumes, greater than 1.0, or 1.5/1.0.

Problem 5.5 A 0.630-L sample of gas at 16.0°C is heated so that it expands at constant pressure to a final volume of 1.35 L. Calculate its final temperature.

5.5   GAY-LUSSAC'S LAW

Gay-Lussac's law states that, if the Kelvin temperature of a fixed amount of gas kept at a constant volume is increased, the pressure will increase. In addition, if the pressure on a gas is increased at constant volume, the temperature will increase. Figure 5.6 indicates that, to keep the volume of a fixed amount of gas constant when the Kelvin temperature is doubled, the external pressure also must be doubled. Gay-Lussac's law is expressed mathematically as

$$P = \text{constant} \times T$$

The data from an experiment that measured the variation in pressure of a fixed amount of gas with change in temperature at a constant volume are found in Table 5.5 on the following page. The table shows that

- As the temperature increases, the pressure increases.
- The ratio P/T is a constant, but the ratio P/t is not constant.

$T = 200 \text{ K}$ $T = 400 \text{ K}$

Figure 5.6 Illustration of Gay-Lussac's law. The volume of a gas confined in an insulated cylinder will remain the same when its Kelvin temperature is doubled if, at the same time, the pressure also is doubled.

TABLE 5.5   Variation in Gas Pressure and Change in Temperature at Constant Volume

State	$t(°C)$	$T(K)$	Pressure (atm)	P/T(atm/K)	P/t(atm/°C)
1	−23.0	250	1.50	0.00600	−0.0652
2	27.0	300	1.80	0.00600	0.0667
3	77.0	350	2.10	0.00600	0.0273
4	152	425	2.55	0.00600	0.0168

When the temperature is expressed on the Kelvin scale, the ratio of pressure to temperature is a constant that takes the same numerical value at all temperatures if the quantity of gas and its volume remain constant.

The pressures of a sample of gas at two different temperatures can be compared by noting that

$$\frac{P_1}{T_1} = \text{constant} = \frac{P_2}{T_2}$$

Because the two ratios of pressure to Kelvin temperature are both equal to the same constant, they are also equal to each other, and the final form of Gay-Lussac's law is

$$\frac{P_1}{T_1} = \frac{P_2}{T_2}$$

It is important to remember that both the amount of gas—that is, the number of moles of gas—and the volume remain constant.

Gay-Lussac's law is applicable in experimental situations when we want to know (1) what happens to the pressure of a gas when the temperature is changed and (2) what happens to the temperature of a gas when the pressure is changed.

For experimental situation 1, we solve the equation for P_2:

$$P_2 = P_1 \left(\frac{T_2}{T_1} \right)$$

This is a unit-conversion calculation, with the conversion factor in the form of a ratio of Kelvin temperatures. Gay-Lussac's law predicts that the pressure of the gas increases if the temperature increases. Accordingly, because $T_2 > T_1$, the ratio of temperatures is greater than 1.0, and the pressure must increase. Gay-Lussac's law is the basis for the operation of the autoclave, described in Box 5.4. It also explains automobile tire blowout at high speeds.

Example 5.6   Using Gay-Lussac's law: I

Calculate the final pressure of a sample of gas in a fixed-volume cylinder at 0.500 atm and 0.000°C when the gas is heated to 41.0°C.

Solution
Tabulate the conditions:

	State 1	State 2
Pressure	0.500 atm	? atm
Volume	constant	constant
Number of moles	constant	constant
Celsius temperature	0.000°C	41.0°C
Kelvin temperature	273 K	314 K

5.4 CHEMISTRY AROUND US

The Autoclave and Gay-Lussac's Law

A number of different approaches have been developed for using heat to inactivate microorganisms. The choice of method depends on the materials being sterilized. Items such as scalpels or glass syringes, characterized by hard surfaces, can withstand harsh treatment at high temperatures, but heat-sensitive materials must be sterilized by more gentle methods. The fact that steam can be heated to high temperatures is particularly useful because bacteria such as *Salmonella* are killed in only 7 to 12 min when exposed to steam at 125°C but require much longer exposures when subjected to dry heat at the same temperature.

At atmospheric pressures, liquid water and its accompanying vapor can never reach temperatures above its boiling point of 100°C, but the temperature of steam can be raised to much higher values. Because steam is gaseous water, it is subject to the gas laws. When heated but allowed to expand, its temperature will not rise. However, if it is heated at constant volume, its temperature will increase—the situation that is quantitatively described by the Gay-Lussac law. The pressure developed in converting steam at 100°C into steam at 125°C at a constant volume can be calculated as follows:

$$760 \text{ torr} \left(\frac{498 \text{ K}}{373 \text{ K}} \right) = 1015 \text{ torr} = 1.34 \text{ atm}$$

Hospital and clinical laboratories are equipped with a device called an autoclave for producing steam at high temperatures. An autoclave is shown in the photograph above, at right. Essentially a double-walled, airtight,

(Ulrich Sapountsis/Okapia/Photo Researchers.)

constant-volume kettle, the autoclave resists pressures significantly greater than 2 atm. Steam is produced in a boiler and is introduced at higher temperature, and consequently under high pressure, into the autoclave's hollow wall. When the temperature of the inner chamber reaches that within the hollow wall, the high-temperature steam is injected into the chamber and comes into contact with the instruments and other objects placed there for sterilization. From 15 to 20 min is sufficient for complete sterilization. Gay-Lussac's law indicates that higher temperatures could be attained if desired, but temperatures near 125°C are sufficient for all practical purposes.

The volume and number of moles are constant, and so Gay-Lussac's law applies. To use Gay-Lussac's law, Celsius temperature must be converted into Kelvin temperature. The conversion uses the relation

$$\text{Kelvin temperature} = \text{Celsius temperature} + 273$$

Solving for P_2 and introducing the Kelvin temperatures into Gay-Lussac's law gives

$$P_2 = 0.500 \text{ atm} \left(\frac{314 \text{ K}}{273 \text{ K}} \right) = 0.575 \text{ atm}$$

According to Gay-Lussac's law, if the gas temperature increased, the pressure also had to increase. Therefore, the initial pressure had to be multiplied by a fraction that is formed by the initial and final temperatures and is greater than 1.0, or 314/273.

Problem 5.6 Calculate the final pressure of a sample of gas in a fixed-volume cylinder at 1.25 atm and 14.0°C when the gas is heated to 125°C.

Example 5.7 Using Gay-Lussac's law: II

A constant-volume sample of gas at 27.0°C and 1.00 atm is heated so that its pressure increases to a final value of 1.50 atm. What is its final temperature?

Solution

Tabulate the conditions:

	State 1	State 2
Pressure	1.00 atm	1.50 atm
Volume	constant	constant
Number of moles	constant	constant
Celsius temperature	27.0°C	? °C
Kelvin temperature	300 K	? K

The volume and amount of gas are constant, and so Gay-Lussac's law applies. Again, it is necessary to convert Celsius temperature into Kelvin temperature. Solving for T_2 and making the appropriate substitutions gives

$$T_2 = 300 \text{ K} \left(\frac{1.50 \text{ atm}}{1.00 \text{ atm}} \right) = 450 \text{ K}$$

Gay-Lussac's law informs us that, if the pressure increased, the temperature had to increase. Therefore the initial temperature is multiplied by a fraction formed by the initial and final pressures and is greater than 1.0, or 1.5/1.0.

Problem 5.7   A constant-volume sample of gas at 3.00°C and 0.650 atm is heated so that its pressure increases to a final value of 2.15 atm. What is its final temperature?

5.6   AVOGADRO'S LAW

Avogadro's law, illustrated by the data presented in Table 5.1 and in Figure 5.7, states that equal volumes (V) of gases at the same temperature and pressure contain equal numbers of molecules (n). The molecules may be monatomic like helium, diatomic like oxygen or chlorine, triatomic like carbon dioxide, or tetratomic like ammonia, or they may be of any size that can exist in the gaseous state. Expressed mathematically,

$$V = \text{constant} \times n$$

Therefore, the volumes of gas samples containing different numbers of moles can be compared by noting that

$$\frac{V_1}{n_1} = \text{constant} = \frac{V_2}{n_2}$$

Because the two ratios of volume to number of moles are both equal to the same constant, they are also equal to each other, and the final form of Avogadro's law is

$$\frac{V_1}{n_1} = \frac{V_2}{n_2}$$

It is important to remember that application of this law requires that both the temperature of the gas sample and its pressure remain constant.

Avogadro's law can be used to answer the questions, (1) What happens to the volume of a gas when the number of moles changes? and (2) What number of moles would be contained in a new and different volume?

The equation expressing Avogadro's law can be solved, for example, for V_2:

$$V_2 = V_1 \left(\frac{n_2}{n_1} \right)$$

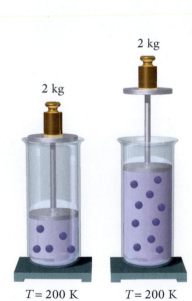

2 kg

2 kg

$T = 200$ K $T = 200$ K

Figure 5.7 Illustration of Avogadro's law. Doubling the amount of a gas at the same Kelvin temperature and pressure will double its volume. Equal volumes of gases at the same Kelvin temperature and pressure contain the same number of molecules.

This is a unit-conversion calculation in which the conversion factor is a ratio of moles. Avogadro's law predicts that the volume of the gas increases if the number of moles increases. If $n_2 > n_1$, the ratio of moles is greater than 1.0, and the volume must increase.

In another form, when $n_1 = n_2$, then $V_1 = V_2$. This is a quantitative expression of the most important consequence of Avogadro's law:

- Equal volumes of gases at the same pressure and temperature contain the same number of molecules.

This idea is the basis for relating the measured mass of a gas to the relative mass of its molecules, thus establishing its molar mass (see Example 5.11) and, in addition, for many of the relations considered in Section 4.6.

5.7 THE COMBINED GAS LAW

The gas laws that we've considered so far have applied to a situation in which, for example, we measure the effect of a change in gas volume on its pressure. Now we will consider situations in which two gas variables—for example, pressure and temperature—are changed simultaneously and will determine the effect of those changes on the gas volume.

Suppose that we use our model cylinder for an experiment in which a fixed quantity of gas is compressed but a constant temperature is not maintained. According to Boyle's law, if the pressure on a gas is increased, its volume will decrease; but, according to Gay-Lussac's law, if the pressure goes up, the temperature will increase. The proposed experiment takes place in one step, but we can think of it as being composed of two separate but simultaneous events and describe it by a combination of the two laws:

$$\frac{PV}{T} = \text{constant}$$

The equation can be written to describe two different states of the same sample of gas:

$$\frac{P_1 V_1}{T_1} = \text{constant} = \frac{P_2 V_2}{T_2}$$

Because each expression is equal to the same constant, they are equal to each other:

$$\frac{P_1 V_1}{T_1} = \frac{P_2 V_2}{T_2}$$

This relation is called the **combined gas law.**

The combined gas law can be used to calculate the answers to three kinds of questions about gases:

1. What is the new pressure of a gas when both the temperature and the volume are simultaneously changed?
2. What is the new temperature of a gas when both the pressure and the volume are simultaneously changed?
3. What is the new volume of a gas when both the temperature and the pressure are simultaneously changed?

Example 5.8 Using the combined gas law

A 1.20-L sample of gas at 27.0°C and 1.00 atm pressure is heated to 177°C and 1.50 atm final pressure. What is its final volume?

Solution

Tabulate the conditions:

	State 1	State 2
Pressure	1.00 atm	1.50 atm
Volume	1.20 L	? L
Number of moles	constant	constant
Celsius temperature	27.0°C	177°C
Kelvin temperature	300 K	450 K

Use the combined gas law, $\dfrac{P_1 V_1}{T_1} = \dfrac{P_2 V_2}{T_2}$, and solve it for V_2:

$$V_2 = V_1 \left(\frac{T_2}{T_1}\right)\left(\frac{P_1}{P_2}\right)$$

This expression has the form of a unit-conversion calculation in which two conversion factors multiply the initial volume. Because the volume increases as the temperature increases, the temperature ratio is T_2/T_1; and, because the volume decreases as the pressure increases, the pressure ratio is P_1/P_2.

Substituting appropriate values gives

$$V_2 = 1.20 \text{ L} \left(\frac{450 \text{ K}}{300 \text{ K}}\right)\left(\frac{1.00 \text{ atm}}{1.50 \text{ atm}}\right) = 1.20 \text{ L}$$

Notice that increasing the temperature should increase the volume, but the increase in pressure was just enough to keep the volume constant.

Problem 5.8 A 1.76-L sample of gas at 6.00°C and 0.920 atm pressure is heated to 254°C and 2.75 atm final pressure. What is its final volume?

Concept check

✓ The combined gas law enables us to calculate the new value of one of the variables when both of the others are changed simultaneously.

5.8 THE IDEAL GAS LAW

All four gas variables, P, V, T, and n, can be combined into the following single mathematical expression:

$$\frac{PV}{nT} = \text{constant}$$

The constant, given the symbol **R,** is called the **universal gas constant.** The equation is more commonly written as

$$PV = nRT$$

which is called the **ideal gas law.**

At room temperature and atmospheric pressure, all gases, regardless of type or molecular mass, follow this relation quite closely. The law is universally applicable, and its use is necessary to solve problems concerning moles or masses of gas.

The value of the universal gas constant R was determined by experimentally establishing that, at 1.00 atm and 273 K, 1.00 mol of any gas occupies 22.4 L. Avogadro predicted that all gases would occupy the same volume under the same conditions, but the precise value of the volume could not be predicted and had to be determined by experiment. Because the volumes of gases depend on temperature and pressure, the only way to compare volumes of different gases is to measure them at the same temperature and pressure. The specific conditions adopted by scientists were 1.00 atm and 273 K and were designated as the **standard temperature and pressure (STP)** of a gas.

Inserting the experimentally determined gas volume and STP values into the ideal gas equation and solving for R, we have

$$R = \left(\frac{1.00 \text{ atm}}{1.00 \text{ mol}}\right)\left(\frac{22.4 \text{ L}}{273 \text{ K}}\right) = 0.0821 \frac{\text{L} \times \text{atm}}{\text{K} \times \text{mol}}$$

To avoid expressing the units of R as a fraction, in this chapter R will be written as follows: $R = 0.0821$ L·atm·K^{-1}·mol^{-1}. A centered dot (·) is used to denote multiplication in R, which is a derived unit. Take note of these units: pressure is in atmospheres, temperature in kelvins, volume in liters, and number (n) in moles. To use this equation for solving gas-law problems, you must be careful to express all measurements in these units.

Example 5.9   Using the ideal gas law

Inhaled air contains O_2 at a pressure of 160 torr. The capacity of a human lung is about 3.00 L. Body temperature is 37.0°C. Calculate the mass of O_2 in a lung after inspiration.

Solution

The variables given are P, T, and V, and the requirement is to find mass, so the universal gas law must be used. However, pressure must be converted from torr into atmospheres, and temperature from degrees Celsius into kelvins before the universal gas law can be used.

$$160 \text{ torr} \left(\frac{1.00 \text{ atm}}{760 \text{ torr}}\right) = 0.211 \text{ atm}$$

$$\text{Kelvin temperature} = \text{Celsius temperature} + 273$$

$$37°C + 273°C = 310 \text{ K}$$

Solving the universal gas law for n, we have

$$n = \frac{PV}{RT} = \frac{0.211 \text{ atm} \times 3.00 \text{ L}}{0.0821 \text{ L·atm·K}^{-1}\text{·mol}^{-1} \times 310 \text{ K}} = 0.0249 \text{ mol}$$

Finally,

$$\text{Grams } O_2 = 0.0249 \text{ mol} \left(\frac{32.0 \text{ g } O_2}{1.00 \text{ mol } O_2}\right) = 0.797 \text{ g}$$

Problem 5.9   Exhaled air contains CO_2 at a pressure of 28.0 torr. The capacity of a human lung is about 3.00 L. Body temperature is 37.0°C. Calculate the mass of CO_2 in a lung before expiration.

5.9   THE IDEAL GAS LAW AND MOLAR MASS

Substances in the gaseous state consist of molecules. Application of the universal gas law allows us to determine the mass of a mole of molecules—called, appropriately, **molar mass.**

The number of moles in any mass of a substance is calculated by dividing the mass in grams by the formula mass. In a sample of a gas, the number of moles, n, is equal to the mass of the sample in grams (g) divided by the molar mass (M). That is,

$$n = \frac{g}{M}$$

This relation is combined with the ideal gas law in the following equations:

$$n = \frac{g}{M} = \frac{PV}{RT}$$

Solving for M gives

$$M = \frac{gRT}{PV} \quad \text{or} \quad M = \left(\frac{g}{V}\right)\frac{RT}{P}$$

Note that the factor (g/V) is in fact the gas density in grams per liter. When correct units are used for R, T, and P, the units of molar mass are grams per mole.

Example 5.10   Calculating molar mass with the ideal gas law

The density of CO_2 at 25.0°C and 1.00 atm is 1.80 g/L. Calculate its molar mass.

Solution

The density of the gas is the factor g/V in the preceding equation:

$$M = \left(\frac{1.80 \text{ g}}{1.00 \text{ L}}\right)\left(\frac{0.0821 \text{ L}\cdot\text{atm}\cdot\text{K}^{-1}\cdot\text{mol}^{-1} \times 298 \text{ K}}{1.00 \text{ atm}}\right)$$

$$M = 44.0 \text{ g/mol}$$

Problem 5.10   The density of O_2 at 15.0°C and 1.00 atm is 1.36 g/L. Calculate its molar mass.

Avogadro's law provides another method for using gas densities to establish molar mass. Because 1.00 mol of any gas occupies 22.4 L at STP, all we need to do is weigh some volume of a gas and, using the combined gas equation, calculate its equivalent mass at 22.4 L—the volume of 1.00 mol at STP.

Example 5.11   Using Avogadro's law and gas density to determine molar mass

Determine the molar mass of methane, 6.60 L of which, collected at 745 torr and 25.0°C, weighs 4.24 g.

Solution

Because 1.00 mol of a gas occupies 22.4 L at STP, the 6.60 L of methane at 25.0°C and 745 torr must be changed into its equivalent volume at STP:

$$6.60 \text{ L}\left(\frac{273 \text{ K}}{298 \text{ K}}\right)\left(\frac{745 \text{ torr}}{760 \text{ torr}}\right) = 5.93 \text{ L}$$

The next step is to calculate the mass of gas that would occupy 22.4 L. This calculation requires a conversion factor defined by the fact that 1.00 mol of any gas = 22.4 L.

$$\left(\frac{4.24 \text{ g}}{5.93 \text{ L}}\right)\left(\frac{22.4 \text{ L}}{1.00 \text{ mol}}\right) = 16.0 \frac{\text{g}}{\text{mol}}$$

Problem 5.11   Determine the molar mass of argon, 14.10 g of which occupies 7.890 L at STP.

The experimental determination of molar mass also provides the value of molecular mass. The two values give the same number but in different units, **g/mol** for molar mass and **amu** for molecular mass (see Section 4.2).

5.10   DALTON'S LAW OF PARTIAL PRESSURES

Mixtures of gases are described by **Dalton's law of partial pressures.** This law states that the total pressure of a mixture of gases is equal to the sum of the individual pressures of all the individual gases in the mixture. These individual pressures are called **partial pressures,** defined as the pressure that each gas would exert if it were present alone instead of in a mixture. Expressed mathematically for a mixture of two gases, A and B,

$$P_{total} = P_A + P_B$$

The chief characteristic of gases is their physical similarity to one another. Gaseous substances composed of widely differing molecules nevertheless display virtually identical physical characteristics. Dalton's law merely states this characteristic in a new way—namely, that 0.10 mol of oxygen and 0.10 mol of carbon dioxide in a 10-L container will have the same pressure as 0.20 mol of oxygen or 0.20 mol of any other gas in the same container under the same conditions of temperature and pressure. Each molecule of a gas is independent of every other molecule present, whether it is the same substance or a different one because molecules in the gaseous state do not interact with one another. Therefore, the total pressure of a gas mixture is the sum of the partial pressures of each member of the mixture.

In Example 5.9, the pressure of oxygen in air was specified as 160 torr. Because air is a mixture mainly of O_2 and N_2 and the total pressure of air is 760 torr, 160 torr is the partial pressure of O_2. Medical oxygen from a tank at 760 torr therefore has a pressure that is five times the pressure of oxygen in air. The result of this difference is that oxygen from the tank is transported into the blood from the lung airspace at about five times the rate of transport from air.

Example 5.12   Calculating partial pressures: I

As just stated, the partial pressure of O_2 in air is 160 torr. What is the partial pressure of N_2 in air?

Solution
Dalton's law of partial pressures can be written as

$$P_{air} = P_{O_2} + P_{N_2}$$

Atmospheric pressure (pressure of air) is usually about 760 torr. Assuming that value and rearranging the equation to solve for P_{N_2} gives

$$P_{N_2} = 760 \text{ torr} - 160 \text{ torr} = 600 \text{ torr}$$

Problem 5.12   Water can be decomposed by an electric current into oxygen and hydrogen whose collected volumes are in the ratio of 1 to 2, respectively. The total pressure of these two dry gases is 741 torr. What are their partial pressures?

Example 5.13   Calculating partial pressures: II

Gases are collected in the laboratory by bubbling them into a bottle filled with water and then inverted under water contained in a vessel open to the atmosphere (Figure 5.8 on the following page). Water in the bottle is then displaced by the entering gas. This experimental arrangement can keep the water's level inside the beaker even with the level of water surrounding it, ensuring that the trapped gas is present at atmospheric pressure. However, because the gas is in contact with the water, some water in the form of gas, or vapor, is mixed with the collected gas. If the atmospheric pressure is 755 torr and the partial pressure of water is 23.0 torr, what is the pressure of a gas collected under these conditions?

Figure 5.8 Collection of a gas over water. A gas produced by chemical reaction in the test tube is conducted by tubing under and into an inverted beaker filled with water. The gas displaces the water and is trapped above the water surface.

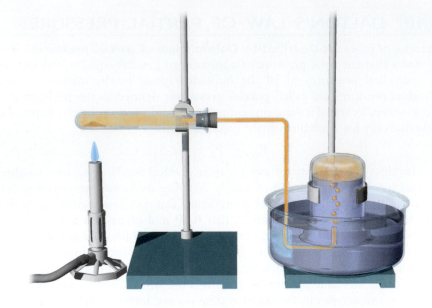

Solution

$$P_{atmosphere} = P_{gas} + P_{water}$$
$$755 \text{ torr} = P_{gas} + 23.0 \text{ torr}$$
$$P_{gas} = 755 \text{ torr} - 23.0 \text{ torr} = 732 \text{ torr}$$

Problem 5.13 Hydrogen was collected over water at 25.0°C and a total pressure of 755 torr. The vapor pressure of water at that temperature is 23.8 torr. The amount of hydrogen collected was 0.320 mol. What was the volume of the gas?

5.11 GASES DISSOLVE IN LIQUIDS

The amount of gas that will dissolve in a liquid such as water depends strongly on the gas pressure. As the gas pressure increases, more gas will dissolve.

- The general rule is: The greater a gas's partial pressure in a gas mixture, the greater the extent to which that gas will dissolve in any liquid present.

Gas solubility is described by **Henry's law,** a statement of the relation between the amount of gas dissolved in a liquid (at a fixed temperature) and the gas pressure (or partial pressure for a mixture of gases). Solubility is a quantitative measurement of one substance's ability to dissolve in another. We will explore the concept of solubility in detail in Chapters 6 and 7 but introduce it here to briefly consider the extent to which gases dissolve in liquids. Henry's law for a pure gas takes the form

$$\left(\frac{\text{Amount of gas dissolved}}{\text{Unit volume of solvent}} \right) = C_H \times \text{gas pressure}$$

and, for a mixture of gases,

$$\left(\frac{\text{Amount of gas dissolved}}{\text{Unit volume of solvent}} \right) = C_H \times \text{gas partial pressure}$$

C_H is called the **Henry's law constant.** It has a unique value for a particular gas at a particular temperature. The Henry's law constant for CO_2 at 25°C is different from that for O_2 at 25°C and from that for CO_2 at 30°C. A practical definition for the units of the Henry's law constant is obtained by solving for C_H in Henry's law for a pure gas:

A PICTURE OF HEALTH
Solubilities of O_2 and CO_2 in Body Fluids

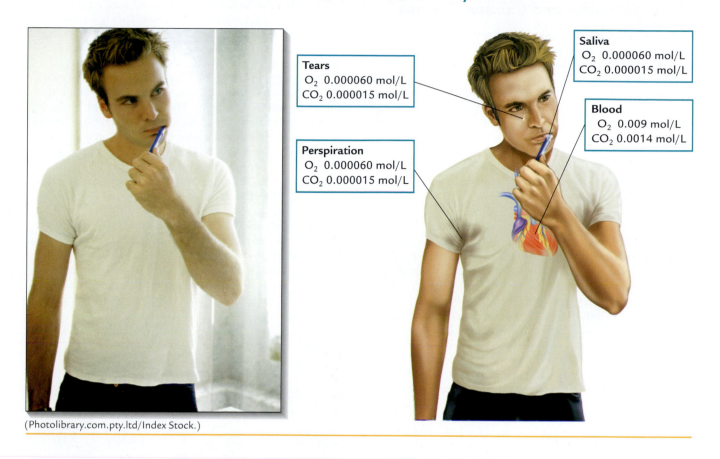

Tears
O_2 0.000060 mol/L
CO_2 0.000015 mol/L

Perspiration
O_2 0.000060 mol/L
CO_2 0.000015 mol/L

Saliva
O_2 0.000060 mol/L
CO_2 0.000015 mol/L

Blood
O_2 0.009 mol/L
CO_2 0.0014 mol/L

(Photolibrary.com.pty.ltd/Index Stock.)

$$C_H = \frac{\left(\begin{array}{c}\text{amount of gas dissolved} \\ \text{within a unit volume of solvent}\end{array}\right)}{\text{gas pressure}}$$

$$C_H = \frac{\text{mL gas}}{\text{mL solvent} \times \text{atm}}$$

The units of gas dissolved and the volume of solvent are given in milliliters, and the gas pressure is given in atmospheres. Gas pressure may also be given in torr. To use Henry's law to calculate the solubility of a gas, we need to know the Henry's law constant for that gas or be able to calculate it from the data in hand. Values of Henry's law constants for many gases can be found in chemical and biological handbooks.

Because gas solubility depends on the pressure of the gas in contact with the liquid, the amounts dissolved can be described in terms of gas pressure alone. For example, the solubility of carbon dioxide in arterial blood is expressed as 46 torr. This means that the amount of carbon dioxide dissolved in arterial blood is the amount that dissolves when the partial pressure of carbon dioxide gas in contact with arterial blood is 46 torr. Physiologists and anesthesiologists define this method of expressing gas solubility as the **gas tension.** The amounts of gas dissolved in blood plasma are critical in anesthesiology and in respiratory therapy. Remember, the general rule is: The greater a gas's partial pressure in the gas mixture, the greater the extent to which that gas will dissolve in any liquid present.

Air consists mostly of nitrogen. When blood comes into contact with air in the lungs, nitrogen, as well as oxygen, dissolves in it. Nitrogen is not used up

5.5   CHEMISTRY WITHIN US

Gas Solubility and Caisson Disease

Caisson disease, more commonly known as "the bends," was first encountered in the United States in 1867, in the course of the construction of the Eads Bridge over the Mississippi River at St. Louis, Missouri. The Eads was the first steel bridge ever built, and its construction required piers to be sunk down to bedrock some 100 feet below the river's surface. The construction workers accomplished the necessary digging by descending to the river bottom inside giant upright steel tubes, or caissons. The caissons, resembling a wide-mouthed jar with open end down, were filled with air under high pressure that kept water and mud out of the working area.

As a result of working under these novel conditions, 119 men of a labor force of 600 suffered intense pains in their muscles and joints, often with significant damage to the tissues. Similar problems arose in the construction of the Brooklyn Bridge in New York City in 1886. Although the disease was still not fully understood, the builders recognized that it developed when the caisson workers left the high pressure within the caisson and returned to the surface.

The fundamental cause of caisson disease is the increase in gas solubility as the result of an increase in gas pressure. Because nitrogen is the major component of air, large amounts of nitrogen as well as oxygen dissolved in the construction workers' blood under the high-gas-pressure conditions that they experienced inside the caissons. Oxygen is used up in metabolism, but nitrogen is not and therefore accumulates in large amounts in the blood. As long as the workers remained under those conditions, they suffered no symptoms. However, when they were too rapidly elevated to the surface, the dissolved gases literally bubbled out of solution, just as CO_2 escapes from warm soda when the bottle cap is suddenly removed.

Caisson disease is virtually unknown among tunnel and bridge workers today because adequate precautions are enforced in those industries. Workers exposed to high air pressures must decompress very slowly. They are returned to surface pressures in gradual stages over time, allowing the dissolved gases to leave their blood and tissues at a rate that does no damage.

One of the by-products of the decompression technology developed to prevent or treat the bends is the introduction of hyperbaric chambers. These chambers are airtight and can be large enough to hold several people at a time. The pressure inside the chamber can be regulated so that, for example, a diver with the bends can quickly be provided with a high-gas-pressure environment to reverse the bubbling of nitrogen from the blood and relieve the painful symptoms. Then the pressure can be lowered slowly and gradually until the danger is over.

In the increased total pressure within the hyperbaric chamber, the partial pressure of oxygen also increases. Therefore, hyperbaric chambers have proved useful in other situations, including the prevention and reversal of infections caused by anaerobic pathogens (such as gangrene) and the treatment of fire fighters who have been exposed to carbon monoxide.

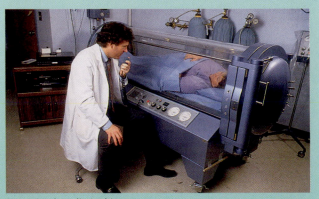

(SIU/Visuals Unlimited.)

in metabolism, and, for that reason, in contrast with oxygen, the blood is saturated with nitrogen at all times. Under special circumstances, such as under water at a significant depth, the total gas pressure of inhaled air is much greater than at Earth's surface. For every 10 m of depth in water, this pressure increases by 1 atm. Consequently, the amounts of all gases dissolving in the blood will also increase. If the external gas pressure is then suddenly decreased, as occurs when a person rises rapidly from deep immersion, the gases will quickly leave solution in the form of bubbles. When gas bubbles form within the body, a result is a potentially lethal condition called the bends (also called caisson disease), which is considered in more detail in Box 5.5.

Example 5.14   Using Henry's law for pure gases

The Henry's law constant for pure oxygen at 20.0°C is 0.0310 mL O_2/mL H_2O at 1.00 atm total pressure. Calculate the oxygen content of water at 20.0°C in contact with pure oxygen at 1.00 atm O_2 pressure.

Solution

The Henry's law expression for the solubility of O_2 in water under the stated conditions is

$$1.00 \text{ atm} \left(\frac{0.0310 \text{ mL } O_2}{\text{mL } H_2O \times 1.00 \text{ atm}} \right) = 0.0310 \text{ mL } O_2/\text{mL } H_2O$$

Problem 5.14   The Henry's law constant for pure nitrogen in water at 20.0°C and 1.00 atm pressure of N_2 is 0.0152 mL N_2/mL H_2O. Calculate its solubility in water at 20.0°C and a pressure of 1.00 atm.

Example 5.15   Using Henry's law for a mixture of gases

Calculate the oxygen content of pure water in the presence of air at 20.0°C.

Solution

As given in Example 5.14, the Henry's law constant for pure oxygen at 20.0°C is 0.0310 mL O_2/mL H_2O at 1.00 atm total pressure. Air consists of about 20% O_2, which means that, at 1.00 atm air pressure, the O_2 content of air can be written as 0.200 atm. Therefore, the Henry's law expression for the solubility of O_2 in water in the presence of air is

$$0.200 \text{ atm} \left(\frac{0.0310 \text{ mL } O_2}{1.00 \text{ mL } H_2O \times 1.00 \text{ atm}} \right) = 0.00620 \text{ mL } O_2/\text{mL } H_2O$$

Problem 5.15   The solubility of pure nitrogen in water at 20.0°C and 1.00 atm pressure is 0.0152 ml N_2/mL H_2O. It constitutes about 80% of air. Calculate its solubility when air is in contact with water at 20.0°C and a pressure of 1.00 atm.

The solubility of any gas decreases as temperature increases. Gas molecules dissolve in liquids by occupying spaces between the liquid's molecules. As the temperature increases, the liquid's molecules move about more and more chaotically, with the result that the number of spaces available to gas molecules decrease—and so does the gas's solubility. The loss of "zip" as a glass of cold carbonated beverage warms to room temperature is a good example of this behavior. Figure 5.9 shows how the solubility of gases in water decreases as the temperature increases.

The inverse of this phenomenon is that, as temperature decreases, more gas will dissolve. A fortunate consequence is the number of people—particularly children, with their low body masses—who have been able to survive near

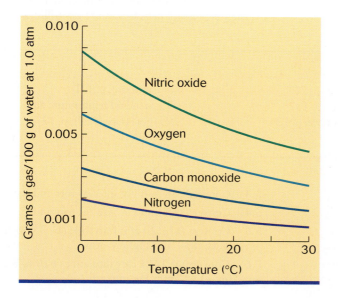

Figure 5.9 The solubility of gases in water decreases with increasing temperature.

drowning in water at freezing temperatures. The amounts of oxygen dissolved in cold body fluids can be sufficient to support metabolism of the brain and other vital organs until the person is revived. This phenomenon also explains why cryogenic surgery is useful under certain conditions.

Quantitative relations to remember	
• Boyle's law:	pressure × volume = constant
• Charles's law:	volume = temperature (K) × constant
• Gay-Lussac's law:	pressure = temperature (K) × constant
• Avogadro's law:	volume = number of moles × constant
• Combined gas law:	pressure × volume = temperature (K) × constant
• Ideal gas law:	pressure × volume = number of moles × temperature (K) × constant
• Henry's law:	volume of gas dissolved / volume of liquid = C_H × gas pressure

Summary

Properties of Gases

• The physical properties of all gases depend on only four variables: pressure (P), volume (V), temperature (T), and number of moles (n).

Quantitative Description of Gas Behavior

• Boyle's law is a mathematical description of the relation between gas pressure and volume when the temperature and number of moles of gas remain constant.

• Charles's law is a mathematical description of the relation between gas volume and temperature when the gas's pressure and the number of moles of gas remain constant.

• Gay-Lussac's law is a mathematical description of the relation between gas temperature and pressure when the gas's volume and number of moles remain constant.

• Avogadro's law is a mathematical description of the relation between gas volume and number of moles when the gas's temperature and pressure remain constant.

The Combined Gas Law

• Three of the gas laws (Boyle's, Charles's, and Guy-Lussac's) can be combined to predict results when two variables are changed simultaneously.

The Ideal Gas Law

• All four gas variables, P, V, T, and n, can be combined into the ideal gas law. It is used to predict the behavior of gases when mass is a consideration and to determine the molar mass of gaseous substances.

Mixtures of Gases

• The pressures of different gases in a mixture are called partial pressures. Dalton's law of partial pressures states that the total pressure of a gas mixture is equal to the sum of the partial pressures.

Gases in Liquids

• The amount of a gas that will dissolve in a liquid depends chiefly on the gas pressure and is described by Henry's law. As the gas pressure increases, more gas will dissolve.

Key Words

atmospheric pressure, p. 122	manometer, p. 123	pressure, p. 121
barometer, p. 122	molar mass, p. 137	standard temperature and
gas tension, p. 141	partial pressure, p. 139	pressure (STP), p. 137

Exercises

Gas Pressure

5.1 Convert the following gas pressures into torr: (a) 364 mm Hg; (b) 483 mm Hg; (c) 675 mm Hg; (d) 735 mm Hg.

5.2 Convert the following gas pressures into atm: (a) 825 torr; (b) 375 torr; (c) 680 torr; (d) 745 torr.

5.3 Convert 735 mm Hg into (a) torr and (b) atm.

5.4 Convert 160 mm Hg into (a) torr and (b) atm.

5.5 Convert (a) 1.20 atm into torr; (b) 0.450 atm into mm Hg; (c) 850 mm Hg into atm.

5.6 Convert (a) 1421 torr into atm; (b) 0.675 atm into mm Hg; (c) 927 mm Hg into atm.

The Gas Laws

5.7   In applications of Boyle's law, which of the four gas variables, P, V, T, and n, are held constant?

5.8   In applications of Charles's law, which of the four gas variables, P, V, T, and n, are held constant?

5.9   In applications of Gay-Lussac's law, which of the four gas variables, P, V, T, and n, are held constant?

5.10   In applications of Avogadro's law, which of the four gas variables, P, V, T, and n, are held constant?

5.11   Calculate the new volume of 2.0 L of a gas when its pressure is changed from 1.0 atm to 0.80 atm at constant temperature.

5.12   While the pressure of 1.40 L of a gas is kept constant, its temperature is increased from 27.0°C to 177°C. Calculate the new volume.

5.13   The initial state of 1.00 L of a gas is $P = 0.800$ atm, $t = 25.0$°C. The conditions were changed to $P = 1.00$ atm, $t = 100$°C. Calculate the new volume.

5.14   The initial state of a gas is $P = 0.80$ atm, $t = 25$°C, $V = 1.2$ L. The conditions were changed to $P = 1.6$ atm, $t = 25$°C. What is the final volume?

5.15   The initial state of a gas is $P = 0.800$ atm, $t = 25$°C, $V = 1.20$ L. The conditions were changed to $P = 1.00$ atm, $V = 2.00$ L. Find the new temperature.

5.16   The volume of a 3.60-L sample of gas at 27.0°C is reduced at constant pressure to 3.00 L. What is its final temperature?

5.17   The initial state of a gas is $P = 1.50$ atm, $t = 25.0$°C, $V = 1.20$ L. The conditions were changed to $V = 0.600$ L, $t = 50.0$°C. What is the final pressure?

5.18   The initial state of a gas is $P = 1.25$ atm, $t = 27.0$°C, $V = 1.50$ L. The conditions were changed to $V = 1.00$ L, $t = 77.0$°C. What is the final pressure?

5.19   The pressure on a 3.00-L sample of gas is doubled and its absolute temperature (K) is increased by 50%. What is its final volume?

5.20   What is the final volume of a gas if the volume of a 4.00-L sample of the gas is first doubled and then its absolute temperature (K) is halved while its pressure is held constant during both changes?

5.21   A sample of gas at 0.830 atm and 25.0°C has a volume of 2.00 L. What will its volume be at STP?

5.22   Ammonia gas occupies a volume of 5.00 L at 4.00°C and 760 torr. Find its volume at 77.0°C and 800 torr.

5.23   Calculate the value of R at STP when volume is in milliliters and pressure is in torr.

5.24   Calculate the value of R at STP when the volume is in milliliters and the pressure is in atm.

5.25   Calculate the new volume of 1.35 L of hydrogen gas at STP when 0.100 mol of hydrogen at STP is added to its expandable container.

5.26   How many moles of helium are required to fill a 3.50-L tank at 25.0°C to a final pressure of 8.74 atm?

5.27   Calculate the volume occupied by 8.00 g of oxygen at STP.

5.28   Compare the volumes occupied by 4.00 g of helium and 44.0 g of carbon dioxide at STP.

5.29   The volumes of hydrogen and ammonia samples at STP are 11.2 L and 5.60 L, respectively. What masses are contained in those volumes?

5.30   A 6.13-g sample of a hydrocarbon occupied a 5.00-L container at 25.0°C and 1.00 atm. Calculate its molar mass.

Law of Partial Pressures

5.31   The total pressure in a vessel containing oxygen collected over water at 25.0°C was 723 torr. The vapor pressure of water at that temperature is 23.8 torr. What was the pressure of the oxygen in atmospheres?

5.32   The total pressure in a vessel containing hydrogen collected over water at 25.0°C was 0.8 atm. The vapor pressure of water at that temperature is 23.8 torr. What was the pressure of the oxygen in torr?

5.33   The volume of a sample of hydrogen collected over water at 25.0°C was 6.00 L. The total pressure was 752 torr. How many moles of hydrogen were collected?

5.34   The volume of a sample of methane (CH_4) collected over water at 25.0°C was 7.00 L. The total pressure was 688 torr. How many moles of methane were collected?

5.35   In different containers, you have 18.0 g of H_2O, 2.02 g of H_2, and 44.0 g of CO_2, all under standard conditions of temperature and pressure. Which occupies the largest volume?

5.36   In different containers, you have 9.0 g of H_2O, 16 g of CH_4, and 44.0 g of CO_2, all under standard conditions of temperature and pressure. Which occupies the smallest volume?

5.37   A mixture of 0.250 mol of oxygen and 0.750 mol of nitrogen is prepared in a container such that the total pressure is 800 torr. Calculate the partial pressure of oxygen in the mixture.

5.38   A mixture of 1.25 mol of argon and 3.75 mol of nitrogen is prepared in a container such that the total pressure is 760 torr. Calculate the partial pressure in torr of argon in the mixture.

Unclassified Exercises

5.39   What volume in liters does a 1.50-mol sample of H_2 occupy at 55.0°C and 1.00 atm pressure?

5.40   The pressure of a sample of N_2 was changed from 1.00 atm to 2.75 atm. As a result, its volume changed from 725 mL to 450 mL, and its final temperature was 235°C. Calculate its initial temperature in Celsius degrees.

5.41   The pressure of a sample of He at 25.0°C was changed from 565 torr to 760 torr. As a result, its volume changed from 800 mL to 850 mL. Calculate its final temperature.

5.42   The volume of a sample of O_2 at 0.75 atm was changed from 2.0 L to 3.5 L. As a result, its temperature changed from 25°C to 77°C. Calculate its final pressure.

5.43   The temperature of a 2.00-L sample of Ar at 600 torr was changed from 25.0°C to 125°C. As a result, its pressure changed from 600 torr to 783 torr. Calculate its final volume.

5.44 The temperature of a sample of CO_2 at 600 torr was changed from 25.0°C to 125°C. As a result, its volume changed from 1200 mL to 1440 mL. Calculate its final pressure.

5.45 The final volume of a sample of methane (CH_4) compressed from 550 torr to 850 torr was 1550 mL. In the process, its temperature changed from 50.0°C to 587°C. Calculate its initial volume.

5.46 How many moles of CO_2 are contained in a 1.5-L vessel at 50°C and 0.86 atm?

5.47 What is the density of H_2 contained in a 2.75-L vessel at 850 torr and 30.0°C?

5.48 The density of benzene vapor at 100°C and 650 torr is 2.18 g/L. Calculate its molar mass.

5.49 In the reaction $Zn + 2 HCl \rightarrow ZnCl_2 + H_2$, 13.1 g of Zn was consumed. At STP, how many milliliters of H_2 were formed?

5.50 The Henry's law constant for nitrogen at 20.0°C and 1.00 atm total pressure is 0.0152 mL N_2/mL H_2O. How many milliliters of nitrogen are dissolved in a milliliter of water at 20.0°C at a nitrogen partial pressure of 456 torr?

5.51 Convert the following gas pressures into torr: (a) 0.750 atm; (b) 1.23 atm; (c) 0.950 atm; (d) 1.750 atm.

5.52 What is the meaning of the expression standard temperature and pressure (STP)?

5.53 Fill in the correct values in the empty places in the table below.

$P(atm)$	$V(L)$	$T(K)$	$n(mol)$
(a) 0.800	_____	300	0.900
(b) _____	1.60	225	2.00
(c) 0.900	15.0	_____	0.500
(d) 4.00	10.0	490	_____

5.54 A 2.50-L sample of a gas at 1.00 atm is heated from 50.0°C to 375°C. Calculate the pressure required to keep the volume constant at 2.50 L.

Expand Your Knowledge

Note: The icons ▲▲▲ denote exercises based on material in boxes.

5.55 Medical oxygen contained in a 20.0-L steel tank was removed at a constant temperature of 27.0°C for a period of 65 min, during which time the pressure in the tank decreased from 14.0 atm to 11.5 atm. How much oxygen, in moles, was removed from the tank?

5.56 If the oxygen removed from the tank in Exercise 5.55 were at STP, what would be the rate of delivery of the gas in liters per minute?

5.57 The partial pressure of CO_2 in contact with human arterial blood is 46 torr. The Henry's law constant for CO_2 in blood plasma at 37°C is 0.51 mL CO_2/mL plasma at 1.0 atm of CO_2. Calculate the amount of carbon dioxide, in milliliters, dissolved in a milliliter of arterial blood.

5.58 A helium weather balloon filled at ground level and atmospheric pressure of 1.0 atm contains 50 L of gas at 20°C. When it rises to the stratosphere, the external pressure will be 0.40 atm, and the temperature will be −50°C. What should the capacity of the balloon be at ground level? Hint: Have you noticed how limp stratospheric balloons appear as they are released from ground level?

5.59 Methane, a greenhouse gas generated by anaerobic bacterial metabolism, appears as bubbles at the surface of ponds and lakes. A bubble of methane at the bottom of a lake has a volume of 0.0850 mL, where the pressure is 4.25 atm. What is the volume of that bubble at the lake's surface, where the pressure is 1.00 atm?

5.60 Methane burns according to the equation:

$$CH_4 + 2 O_2 \longrightarrow CO_2 + 2 H_2O$$

Calculate the volume of air that is required to burn 10.0 L of methane when both are at the same temperature and pressure. Assume that air is 20.0 percent oxygen by volume.

5.61 Calculate the number of grams of propane C_3H_8, in a 50.00-L container at a pressure of 7.500 atm at a temperature of 25.00°C.

5.62 (a) Calculate the pressure exerted by 18.0 g of steam confined to a volume of 18.0 L at 373 K. (b) What volume would water occupy if the steam were converted to liquid water at 298 K? Assume the density of liquid water is 1.00 g/mL at 298 K.

5.63 Ultraviolet radiation from the sun is absorbed by ozone molecules in the stratosphere. The temperature of the stratosphere is 250 K. Assume that the pressure due to the ozone is 4.00×10^{-7} atm. Calculate the number of ozone molecules present in 1.00 mL.

5.64 Vacuum pumps are used in the laboratory to obtain low pressures. Typical low pressures of 0.000100 torr are readily obtained. Calculate the number of molecules in 1.00 ml of gas at that pressure and 293 K.

5.65 The density of liquid water at 373 K and 1.00 atm is 0.958 g/ml. What is its density in the gas phase under the same conditions?

5.66 A 0.271 g-sample of an unknown vapor occupies 294 ml at 373 K and 765 torr. The simplest formula of the compound is CH_2. What is the molecular formula of the compound?

5.67 Chemical analysis of a gaseous hydrocarbon shows it to contain 88.82% carbon and 11.18% hydrogen by mass (see Box 4.1). A 62.6 mg sample of the compound occupies 34.9 ml at 772 torr and 373 K. Determine the molecular formula of the hydrocarbon.

5.68 A 3.00-L sample of hydrogen at 1.00 atm, a 9.00 L sample of nitrogen at 2.00 atm, and a 6.00 L sample of argon at 0.500 atm are all transferred to a single 10.0 L container. Calculate the partial pressure of each gas and the total final pressure of the mixture. Assume constant temperature conditions.

5.69 A 65.0-L tank of carbon dioxide at 8.00 atm pressure and 298 K was used for one month. At the end of the month, the pressure had dropped to 5.10 atm. Calculate the number of grams of carbon dioxide used during the month.

5.70 You are in a restaurant and notice that a patron at the next table begins turning blue and cannot breathe. One of his companions jumps up, gets behind him, wraps his arms around him, and squeezes suddenly and quite hard. A piece of meat is ejected from the choker's mouth and he can breathe again. You recognize the Heimlich maneuver. Explain this procedure using the gas laws.

5.71 A mixture of oxygen and methane has a density of 1.071 g/L at 273 K and 760 torr. Compute the ratio of the number of moles of oxygen to the number of moles of methane in the mixture.

5.72 Two containers are connected by a valve. One of the containers has a volume of 550.0 mL and is occupied by methane at 750.0 torr. The other has a volume of 800.0 mL and is occupied by oxygen at 650.0 torr. The valve is opened and the two gases mix. Calculate the total pressure and partial pressures of methane and oxygen in the resulting mixture.

5.73 One of the instructions in the National Association of Underwater Instructors (NAUI) dive tables is to always make your deepest dive first when making a series of dives. Then make each of your repetitive dives to a shallower depth than your previous dive. What is the reason for this instruction?

CHAPTER 6

INTERACTIONS BETWEEN MOLECULES

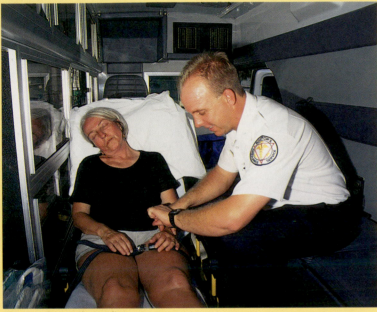

(Spencer Grant/PhotoEdit.)

Chemistry in Your Future

Having recently moved from Phoenix to Chicago, you start your new job at Northwestern Memorial Hospital's emergency room in the midst of a record-breaking heat wave. The temperatures are not any higher than the temperatures typical for this time of year in Phoenix, and so you are perplexed when you encounter many more, and far more serious, cases of heat prostration than you ever did back home. The phrase "it's not the heat, it's the humidity" runs through your mind. Could that explain the riddle? Chapter 6 reveals the physical basis of this phenomenon and of other physiological problems as well.

For more information on this topic and others in this chapter, go to www.whfreeman.com/bleiodian2e

Learning Objectives

- Use molecular concepts to explain the properties of the three states of matter and transitions between them.
- Describe secondary forces, and correlate chemical structure with types of secondary forces.
- Correlate physical properties such as vapor pressure, normal boiling point, melting point, and surface tension with types of secondary forces.
- Understand the molecular characteristics necessary for solution formation.
- Define dynamic equilibrium.

So far, we have concentrated on the forces that hold atoms together to form chemical compounds—covalent and ionic bonds. In this chapter, we will study a new group of forces that operate not within molecules but between them. Walking on water—when it is frozen, of course—is a good reminder that it can exist as a solid (ice) or a liquid or a gas (steam). Water is not unique in this regard; all substances may exist in any of the three states of matter, depending on the conditions of temperature and pressure.

Substances exist as solids and liquids because of attractive forces between the fundamental particles. This concept is easy to explain when the substance is an ionic compound, such as sodium chloride or potassium nitrate, because the component particles are positive and negative ions that are strongly attracted to each other. But the molecules of water, carbon dioxide, ammonia, and many other substances are electrically neutral, yet these compounds can exist as liquids or solids. The forces that cause water and other neutral molecules to attract one another are much weaker than ionic or covalent chemical bonds, but they underlie the formation of solids and liquids, as well as many physiological phenomena. These phenomena include the structures and properties of biomolecules such as proteins, nucleic acids, carbohydrates, and lipids.

Many biological processes are based on brief weak interactions between different molecules. The conduction of nerve impulses, for example, depends on a small molecule released from one nerve cell interacting weakly with large molecules embedded in the surface of an adjacent nerve cell. The weak interaction guarantees that the interactions formed are very short lived. If the attractions were strong, a nerve impulse would never cease. You would feel pain, taste sourness, or see red endlessly.

Our primary interest in this chapter is in the weak interactions between neutral molecules. We begin with a description of the properties of the three states of matter and the transitions between them. Next, we consider the nature of the attractive forces between molecules. Then we explore some properties of liquids that are basic to many biochemical and physiological processes. A brief concluding section looks at some of the properties of solids.

6.1 THE THREE STATES OF MATTER AND TRANSITIONS BETWEEN THEM

A **solid** has a fixed volume and shape, both of which are independent of its container. A **liquid** also has a fixed volume but not a fixed shape. It can be poured from one container to another and takes the shape of its container. The volume of a gas depends on the volume of its container. A gas always occupies all of its container. The volumes of both liquids and solids change very little when pressure is applied. They are said to be incompressible. On the other hand, a gas is easily compressed into a smaller volume. Figure 6.1 is a sketch depicting the arrangements of molecules in each of the three states of matter. There is little increase in volume when a solid melts, but, when a liquid is

Solid   Liquid   Gas

Figure 6.1 A molecular view of the three states of matter. The molecules in both the solid and the liquid states are in contact, but those in the liquid state are somewhat disordered. Molecules in the gaseous state are very far apart and totally disordered.

vaporized, there is a large change in volume. The volume of 1 mol of liquid water at 100°C is 18.8 mL, and the volume of 1 mol of gaseous water (steam) at the same temperature and pressure is about 3.1×10^4 mL.

A solid cannot be compressed, because the constituent molecules are in contact, as close to one another as they can be. A solid holds its shape because strong attractive forces hold the molecules in fixed positions from which they cannot escape.

The compressibility of liquids is similar to that of solids, and so the molecules are assumed to be as close to one another as they can be. The fact that a liquid flows and cannot hold its shape is explained by assuming that the attractive forces between molecules in the liquid state are weaker than those existing in solids. The internal structure of liquids is less orderly than that of a typical solid.

In a gas, the particles move at high speeds and change directions only when they collide with the walls of their container or with other particles. Because the molecules in gases are so far apart, they have virtually no opportunity to interact, and so attractive forces between molecules play a small role in gases. Because the molecules of gases are not subject to such forces, their physical properties are independent of the nature of the individual molecules. For example, in the gaseous state, there is virtually no difference in the physical behavior of hydrogen, methane, or any other gaseous compound. As a consequence, all gases can be quantitatively described by the few simple mathematical laws presented in Chapter 5.

The energy of a moving body, or energy of motion, is called **kinetic energy (KE).** It is equivalent to the work necessary to bring that body to rest or to impart motion to a resting body. Kinetic energy at the molecular level depends only on temperature. As the temperature increases, so does kinetic energy, and, as a result, molecules move faster. In hot systems, molecules move about much faster than in cold systems. Kinetic energy tends to oppose the attractive forces between molecules. For example, if two molecules collide energetically, the attractive forces between them may not be strong enough to prevent the molecules from simply bouncing off each other and proceeding in new directions.

Most solids melt and are transformed into liquids on the addition of sufficient heat and can revert back to the solid state if the same amount of heat is removed. The transition from solid to liquid is called **melting.** The reverse process, transforming liquid into solid, is called **freezing.** The **melting point** and **freezing point** of a pure solid are identical. The transition from liquid to gas also requires heat and is called **vaporization.** The reverse of this process is called **condensation.** Continuous vaporization from the surface of a liquid into the atmosphere is called evaporation, which is considered further in Section 6.7.

The changes from one physical state into another as a result of the addition or removal of heat are called **phase transitions.** Phase transitions occur in either direction; that is, they are **reversible.** Heat added to a system in transition from solid to liquid or from liquid to gas does not cause the temperature to increase. Instead, this heat is used to overcome the attractive forces between molecules. Only when the phase transition is complete—that is, when all the molecules of a solid, for example, have been converted into the liquid state—will the added heat cause the temperature to increase. The temperature of a sample of liquid water remains at 100°C until all of it is converted into steam at 100°C. Any heat then added to the steam will raise its temperature. The temperature at which gas bubbles form throughout the liquid in a container open to the atmosphere is called the **normal boiling point.**

The amount of heat required to melt a solid and then to vaporize it is a unique physical property distinguishing one substance from another. The heat per mole of solid required for melting is called the **molar heat of fusion,**

» Chapter 8 describes the role of kinetic energy in a chemical reaction.

« Phase transitions are physical changes (see Section 1.1).

ΔH_f, in kilocalories per mole (kcal/mol). The heat per mole of liquid required for vaporization is called the **molar heat of vaporization, ΔH_v**, in kcal/mol. The molar heat of vaporization is always considerably greater than the molar heat of fusion because, in the transition from liquid to gas, the attractive forces between molecules must be completely overcome. Melting requires only a loosening of the solid structure, and the attractive forces are largely still at work.

In the next section, we will examine the nature of the forces that operate between molecules.

6.2 ATTRACTIVE FORCES BETWEEN MOLECULES

The attractive forces between molecules are often called **intermolecular forces** or **van der Waals forces.** In this book, we will use the term **secondary forces** because it includes both intermolecular and intramolecular forces. Intermolecular forces operate between molecules. In very large molecules such as proteins, parts of the molecules can fold back on themselves. Therefore, attractive forces can operate within molecules and are called intramolecular forces. You will see that these forces are central to the properties of biologically important molecules such as proteins, polysaccharides, and DNA. It is important to note that secondary forces are far weaker than chemical bonds—perhaps at most from 10% to 20% of the strength of a covalent or ionic bond—and must not be confused with a true chemical bond. Even though the energies of these secondary forces are much smaller than those of chemical bonds, they underlie not only the physical properties of solids and liquids, but many physiological phenomena as well. To better understand these physical properties, we must explore the origin and nature of secondary forces.

Dipole–Dipole Interactions

Secondary forces cause gases to condense into liquids and liquids to freeze into solids. These forces are electrical and arise from molecular polarity. Permanent molecular polarity, characterized by a dipole moment, was described in Chapter 3. Molecules with permanent dipole moments, or **polar molecules,** attract not only each other but also other, different molecules possessing dipole moments.

Figure 6.2 shows the interaction of polar molecules in the liquid state. The oppositely charged parts of the polar molecules attract, and the similarly charged parts repel each other. Attractive forces between polar molecules—for example, methyl chloride, CH_3Cl, or chloroform, $CHCl_3$, or mixtures of the two—are called **dipole–dipole forces.** Substances composed of polar molecules have higher melting and boiling points than do those composed of molecules possessing no permanent dipoles, or **nonpolar molecules.** Methane, CH_4, has no permanent dipole moment and, at room temperature, is a gas. However, water, H_2O, a molecule of about the same mass, has a large dipole moment and is a liquid at the same temperature.

London Forces of Interaction

Even in molecules such as methane that have no permanent polarity, temporary dipoles can exist for brief moments. Such temporary dipoles are caused by the erratic motions of electrons that result in uneven distributions of electrical charge within the molecule at any given time. For example, for a brief instant, perhaps 1×10^{-12} s, most of the electrons in a molecule may be at one end. The attractive force resulting from this temporary dipole is called a **London force.** This phenomenon occurs in all molecules. The strength of the London force can range from being many times weaker than dipole–dipole forces to being about equal to them. London forces exist between all molecules, but

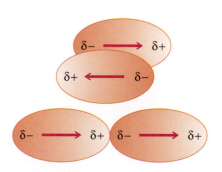

Figure 6.2 Polar molecules attract each other by the interaction of their electric dipoles (represented by the arrows). Two relative orientations are shown: end-to-end and side-by-side. (Adapted from P. Atkins and L. Jones, *Chemistry: Molecules, Matter, and Change,* 3d ed., New York, W. H. Freeman and Company, 1997.)

TABLE 6.1 Boiling Points Characteristic of the Three Types of Secondary Forces

Secondary force	Substance	Boiling point (°C)
London force	helium (He)	−269
	hydrogen (H_2)	−253
	methane (CH_4)	−164
	butane (C_4H_{10})	−0.5
	hexane (C_6H_{14})	69
dipole–dipole	methyl fluoride (CH_3F)	−78
	fluoroform (CHF_3)	−82
	formaldehyde (HCHO)	−21
hydrogen bond	methanol (CH_3OH)	65
	water (H_2O or HOH)	100
	ethylene glycol (HOC_2H_4OH)	198

their contribution to physical properties is particularly significant in those molecules that do not possess permanent dipoles.

For example, nonpolar substances such as methane or helium condense to liquids only because of London forces. Compare the boiling points of nonpolar substances with those possessing permanent dipoles in Table 6.1. Methane, a nonpolar molecule, condenses to a liquid at −164°C compared with formaldehyde, a polar molecule that condenses at −21°C.

Because London forces arise from uneven distributions of electrons within molecules, molecules with large numbers of electrons exert larger London forces. Therefore, nonpolar molecules of large molecular mass can exert greater attractive force than can nonpolar or even polar molecules of smaller molecular mass. Compare methane's normal boiling point with that of hydrogen and methanol's with that of hexane in Table 6.1.

Example 6.1 **Using structural formulas to predict polarity**

Which of the following substances are characterized by polar secondary forces?

$$F_2 \qquad CH_4 \qquad HF \qquad CH_3Cl$$

Solution
The structural formulas for these compounds were developed in Section 3.7 on Lewis formulas. They are:

$$
\begin{array}{ccccc}
 & & H & & H \\
 & & | & & | \\
F{-}F & H{-}C{-}H & & H{-}F & H{-}C{-}Cl \\
 & & | & & | \\
 & & H & & H
\end{array}
$$

F_2 and CH_4 have electrically symmetrical structures and therefore possess no dipole moments. They cannot interact through polar secondary forces. HF and CH_3Cl have electrically asymmetrical structures. Therefore they possess permanent dipole moments and interact through polar secondary forces.

Problem 6.1 Which of the following compounds condense to the liquid phase from the gas phase through polar secondary forces?

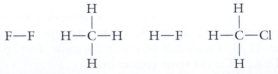

(a) H—H (b) Cl—C—Cl (with Cl above and Cl below the central C) (c) H—C≡N (d) I—I

6.3 THE HYDROGEN BOND

A covalent bond between hydrogen and any one of the three elements oxygen, nitrogen, and fluorine possesses a unique character. The differences in electronegativity between hydrogen and these elements, and therefore the polarity within such a bond, are very large. This polarity leads to a unique and very strong attractive force, called **hydrogen bonding,** between molecules such as H_2O, NH_3, and HF.

Figure 6.3 is a sketch of hydrogen bonding between two water molecules drawn with appropriate bond angles. The hydrogen bond between molecules appears as a dashed line to differentiate it from the covalent bonds within the water molecules. A hydrogen atom in one of the water molecules, with its intense partial positive charge (indicated by δ^+), is able to interact strongly with a lone electron pair of the oxygen atom in another water molecule. In Figure 6.3, the partial charges on all other atoms in the diagram are ignored so as to focus on this intermolecular hydrogen–oxygen interaction.

A hydrogen bond, sometimes abbreviated H-bond, is a special case of a dipole–dipole attraction because it is significantly stronger than any other dipole–dipole secondary force. It forms not only between identical molecules, as occurs in liquid water between molecules of H_2O or in ammonia between molecules of NH_3, but also between different molecules, allowing mixtures, for example, of water and ammonia.

The hydrogen bond allows the hydrogen atom to act as "glue" between two different molecules; that is, a hydrogen atom of one molecule binds to the oxygen, nitrogen, or fluorine atom of another molecule. The bond, though strong, is weaker than a covalent bond and is therefore depicted as a longer bond than the covalent bonds.

Water molecules in the solid state, ice, are completely hydrogen bonded. Figure 6.4 shows that each molecule in ice forms four hydrogen bonds and

INSIGHT INTO PROPERTIES

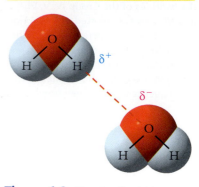

Figure 6.3 Sketch of hydrogen bonding between two water molecules. The hydrogen bond is a dashed line to differentiate it from the covalent bonds (solid lines) of the water molecules. The hydrogen bond explains why water is a liquid.

www.whfreeman.com/bleiodian2e

INSIGHT INTO PROPERTIES

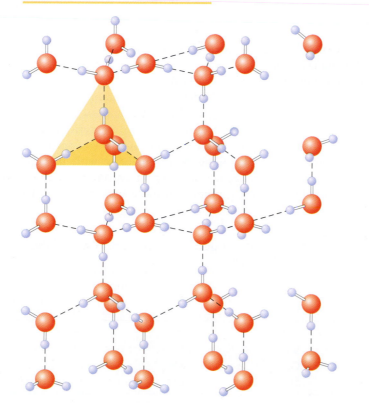

Figure 6.4 The crystal structure of ice. Every oxygen (shown in red) is located at the center of a tetrahedron formed by four other oxygen atoms. Hydrogen bonds (dashed lines) link all oxygen atoms together. To form this number of hydrogen bonds, the resulting structure requires a greater volume than does the liquid and is therefore of lower density than liquid water. That explains why ice floats in liquid water.

Figure 6.5 The boiling points of the hydrides of the Group IV, V, VI, and VII elements plotted as a function of their row in the periodic table. The anomalously high boiling points of the hydrides of the three elements nitrogen, oxygen, and fluorine are the consequences of hydrogen bonding.

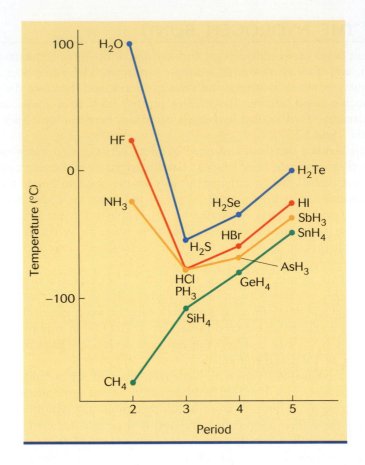

that every oxygen atom is located at the center of a tetrahedron formed by four other oxygen atoms. The entire solid structure is held together by hydrogen bonds, which forces the molecules in the solid to take up as much room as possible—to form an open structure. When ice melts, some 15 to 20% of the hydrogen bonds are broken, and the open structure collapses a bit. The result is that the density of liquid water is greater than that of ice, and solid water floats on its liquid form.

Binary (two-element) molecular compounds formed between hydrogen and one other element are called **hydrides.** Water, ammonia, and hydrogen fluoride are hydrides. In Figure 6.5, the boiling points of the binary molecular hydrides are plotted as a function of their row in the periodic table. Table 6.2 identifies these hydrides with respect to group and row of the periodic table.

For the most part, the boiling points increase with increasing mass, as would be expected for molecules interacting through London forces. The linear relation between boiling point and molecular structure of the hydrides

TABLE 6.2   **Hydrides Arranged by Group and Row of the Periodic Table**

	Group			
Row	IV	V	VI	VII
2	CH_4	NH_3	H_2O	HF
3	SiH_4	PH_3	H_2S	HCl
4	GeH_4	AsH_3	H_2Se	HBr
5	SnH_4	SbH_4	H_2Te	HI

from period 3 through period 5 of the periodic table implies that the boiling points of H_2O, NH_3, and HF should be below those of the hydrides in rows 3, 4, and 5. However, the behaviors of H_2O, NH_3, and HF break the pattern—their boiling points are actually higher than any of the others. The reason is that oxygen, nitrogen, and fluorine atoms are strongly electronegative, and this electronegativity causes extensive hydrogen bonding in H_2O, NH_3, and HF. The special behavior of the hydrides of only the three elements nitrogen, oxygen, and fluorine but not the other hydrides shows that hydrogen bonding cannot just occur between all molecules that contain a hydrogen atom. A hydrogen bond can form only under three circumstances: when a hydrogen atom is covalently bound to (1) a fluorine atom or (2) an oxygen atom or (3) a nitrogen atom.

✓   Look for hydrogen bonding when a molecule contains a hydrogen atom covalently bonded to either oxygen, nitrogen, or fluorine.

Concept check

Example 6.2   **Recognizing molecules that can form hydrogen bonds**

Which of the following molecules can form hydrogen bonds?

$$H_2 \qquad CH_4 \qquad NH_3 \qquad HF \qquad CH_3OH$$

Solution

A hydrogen bond can form if compounds possess a hydrogen atom covalently bonded to either oxygen, nitrogen, or fluorine. Three of the preceding compounds fit that criterion: NH_3, ammonia; HF, hydrogen fluoride; and CH_3OH, methyl alcohol. Neither H_2 (hydrogen) nor CH_4 (methane) can form hydrogen bonds.

Problem 6.2   Which of the following molecules can form hydrogen bonds? (a) SiH_4; (b) CH_4; (c) CH_3NH_2; (d) H_2Te; (e) C_2H_5OH.

The importance of water in biological systems developed in large part because of its hydrogen-bonding properties. The anomalous behavior of water's density and its effect on aquatic life were presented in Box 1.4. Hydrogen bonding also has a key role in the regulation of body temperature, which will be considered in Section 6.8. The characteristics of many important biological substances—proteins, nucleic acids, and carbohydrates—strongly depend on their ability to form hydrogen bonds within their own molecules, as well as external hydrogen bonds with other molecules, particularly water. Hydrogen bonds are responsible in large part for the properties of biomolecules such as the protein hemoglobin and the double-helical DNA of chromosomes.

6.4   SECONDARY FORCES AND PHYSICAL PROPERTIES

As mentioned earlier, Table 6.1 lists normal boiling points for compounds whose liquid states are dominated by different kinds of secondary forces: London forces, dipole–dipole forces, and hydrogen bonds. The boiling points are indicative of the relative strengths of those three types of interaction.

Nonpolar molecules can interact only by London forces, but polar molecules interact by both London and dipole–dipole forces, and hydrogen-bonding substances interact by all three forces. Table 6.3 on the following page lists substances that interact by one, two, or all three secondary forces, along with their normal boiling points.

TABLE 6.3 Boiling Points and Structures of Substances Whose Molecules Interact by One, Two, or Three Types of Secondary Forces

Substance	Structure	London	Dipole–dipole	Hydrogen bonds	Boiling point (°C)
methane	H—C—H (with H above and H below)	yes	no	no	−164
trifluoromethane	F—C—H (with F above and F below)	yes	yes	no	−82
methyl fluoride	H—C—F (with H above and H below)	yes	yes	no	−78
formaldehyde	C=O (with two H on left)	yes	yes	no	−21
methanol	H—C—OH (with H above and H below)	yes	yes	yes	65
water	O with H and H	yes	yes	yes	100
ethylene glycol	HO—C—C—OH (with H above and H below each C)	yes	yes	yes	198

Concept checklist

✓ Hydrogen bonds are stronger than dipole–dipole forces, which are stronger than London forces.

✓ Although both hydrogen bonds and dipole–dipole forces are stronger than London forces, that weak force still contributes to the overall interactions among polar molecules.

The answers to two questions allow us to predict trends in physical properties (such as melting points or boiling points) for a series of compounds:

- Is a compound polar or nonpolar?
- Can a compound form hydrogen bonds?

Example 6.3 Using molecular structure to predict physical properties: I

Arrange the normal boiling points of the following compounds from highest to lowest:

$$H_2 \quad CH_4 \quad NH_3 \quad HF$$

Solution

The two polar compounds, NH$_3$ and HF, must have higher boiling points than the two nonpolar molecules, H$_2$ and CH$_4$. The polar compounds can also form H-bonds. Because fluorine is the element with the highest electronegativity (Section 3.9), HF is more polar than NH$_3$ and forms stronger H–bonds. The order of boiling points for these two compounds, then, from higher to lower, is HF > NH$_3$. Methane and hydrogen interact by weak London forces, but methane is heavier, contains more electrons, and therefore interacts by stronger London forces than does hydrogen and will boil at a higher temperature. The overall order of boiling points, from highest to lowest, is:

$$HF > NH_3 > CH_4 > H_2$$

Problem 6.3 Arrange the normal boiling points of the following substances from highest to lowest: (a) He; (b) CH$_4$; (c) I$_2$; (d) HF.

Surface Tension

Surface tension is an important property of liquids that depends on the type of attractive forces between molecules. Molecules behave differently when they are at a surface or at a boundary between phases, as between a liquid and its vapor. Molecules in the interior of a liquid undergo attractions uniformly in all directions around them. However, molecules at a surface are in contact with attracting molecules on one side but not on the other (Figure 6.6). They therefore undergo a net attraction inward toward the liquid's interior, reducing the number of molecules at the interface and consequently reducing the surface area. The result is a force, called the **surface tension,** that resists the expansion of the liquid's surface and acts, in effect, like an elastic skin. Figure 6.7 shows a metal paper clip floating on water despite the fact that its density is much greater than that of water. The surface tension of water is great enough and the mass of the paper clip is distributed in such a way as to prevent the paper clip from penetrating the surface and sinking. Figure 6.8 illustrates how some insects take advantage of this effect to walk on the surface of water. As long as the weight of the insect does not exceed the force exerted by the surface tension, the creature remains on the water's surface.

The surface tension of water is reduced by the presence of dissolved substances called **surface-active agents** or **surfactants.** Examples of surfactants are soaps and detergents. The reduction of the surface tension of water by soaps typically leads to the formation of stable foams (suds). If a surfactant were added to the pond pictured in Figure 6.8, our water strider would have to learn to swim. The bile salts synthesized by the liver are surfactants crucial to the digestion of dietary fats (Box 6.1 on the following page). Surfactants are also produced by the cells that line our lungs (Box 6.2 on page 159).

Figure 6.6 Molecules in a liquid are attracted in all directions, but molecules at the surface are attracted only in an inward direction. The inward pull results in surface tension.

Figure 6.7 A metal paper clip floats on water. Because of the hydrogen bonds between the water molecules, the surface of the water behaves like an elastic membrane. (Chip Clark.)

Figure 6.8 The water strider can walk on the surface of water even though it is more dense than the water. (Hermann Eisenbeiss/Photo Researchers.)

6.1   CHEMISTRY WITHIN US

Surface Tension and the Digestion of Dietary Fats

By the time that a dietary lipid (fat) has passed through the stomach and entered the small intestine, it has been separated into liquid droplets suspended in an aqueous medium. Digestive enzymes whose function is to convert lipid molecules into soluble substances for absorption by intestinal cells cannot enter these droplets and therefore can act on the lipid molecules only at the droplet surface. The rate of digestion of the droplets thus depends on the amount of lipid-droplet surface area accessible to the enzymes. If the surface area of the droplets can be increased, the rate of enzyme action will be significantly enhanced. Not surprisingly, our intestines contain a compound to do just that.

When an oil—any salad oil, for example—is shaken with water, the mechanical agitation breaks the oil into tiny droplets and disperses them throughout the water, but, as the droplets collide with one another, they coalesce

to form a continuous oil layer above a separate continuous water layer. However, if surface-active compounds are added to the oil and water mixture, they concentrate at the boundary between the oil and the water and prevent the oil droplets from coalescing. This process creates a stable oil–water mixture and is known as **emulsification.**

In the intestine, emulsification is accomplished by the action of biological surface-active compounds called bile salts. They are synthesized in the liver from cholesterol, stored in the gall bladder, and released into the small intestine in response to the arrival of dietary fat. Their presence at the oil–water boundary combined with the mechanical action of the intestine emulsifies the lipid droplets, greatly increasing their total surface area. Specifically, bile salts cause dietary fats in the intestines to be reduced to fat droplets about 0.0001 cm in diameter. For example, when a teaspoon of olive oil enters the small intestine, its surface area will increase about 10,000 times. This increase in the surface area of lipid droplets results in a similar increase in the rate of their digestion.

All surface-active substances have similar characteristics: they consist of rather long molecules containing both ionic (or polar) and nonpolar segments. The sodium salt of palmitic, or hexadecanoic, acid [$CH_3(CH_2)_{14}COO^-Na^+$], a soap, is a good example (Figure 6.9) of a surface-active substance. The molecule consists of a nonpolar hydrocarbon chain (Example 3.10) 15 carbons long with an ionic group called a carboxylate group at the end. The nonpolar hydrocarbon part of the molecule is described as **hydrophobic,** or "water hating." The ionic end is called **hydrophilic,** or "water loving." Molecules simultaneously possessing both these properties are called **amphipathic.**

As a result of having hydrophobic and hydrophilic properties within the same molecule, surfactants tend to stay at interfaces (air–water or oil–water interfaces, for example) rather than remain in the interior of the water phase. At an interface, the hydrophobic end can escape the water, whereas the hydrophilic end remains nestled in the water's surface. The concentration of

INSIGHT INTO FUNCTION

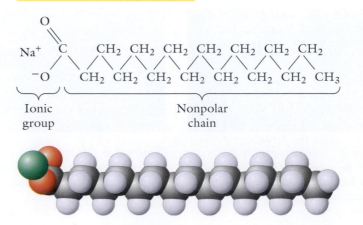

Figure 6.9 Structure and space-filling model of a surface-active agent, the sodium salt of palmitic acid. The molecule has two distinctly different ends: a nonpolar hydrocarbon chain that does not dissolve in water and, at the other end, an ionic group that does.

6.2 CHEMISTRY WITHIN US

Respiratory Distress Syndrome

Breathing requires two types of work: (1) the work of stretching the thorax and diaphragm (see Box 5.3) and (2) the work needed to "stretch" the alveoli of the lungs. Gas exchange with the blood takes place in the air space within an alveolus, whose inner surface contacting the gas phase is covered with a thin layer of water. The surface of a liquid is like an elastic membrane. Its surface tension is a measure of the work that must be done to expand its surface, and the surface tension of water is among the largest known. The surface tension of the water lining the alveoli makes the water behave like a stretched rubber band, constantly contracting and resisting expansion. Considerable work must be expended to distend and expand the bubblelike alveolar sacs lined with water. If the alveoli were lined with pure water, it is likely that no amount of muscular effort could expand the lungs, and in fact they would tend to collapse. Fortunately, this does not happen in healthy people, because alveolar cells synthesize a surface-active agent called **lung surfactant,** which markedly reduces the surface tension of the fluid lining the alveoli.

A condition called **respiratory distress syndrome** of the newborn, also called **hyaline membrane disease,** is the best-known consequence of the absence of the lung's ability to synthesize lung surfactant. This condition is often present in premature newborns whose surfactant synthesizing cells have not had time to fully develop.

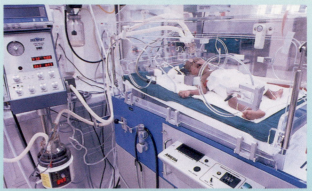

Life-support equipment surrounds an incubator holding a premature baby born after a term of 24 weeks. The infant is suffering from respiratory distress syndrome, also called hyaline membrane disease. (Pete Ryan/Science Photo Library/ Photo Researchers.)

Taking a breath requires very strenuous effort on the part of these infants and may cause exhaustion, inability to breathe, lung collapse, and death. The symptoms can be relieved by aspirating a synthetic lung surfactant directly into the lungs. As these infants mature and begin to produce their own surfactant, the problem is mitigated. Lung surfactant is dipalmitoylphosphatidylcholine, also called lecithin.

water at the interface is reduced. Therefore, the attractive forces between molecules at the interface are reduced and, as a consequence, the surface tension is reduced.

Molecular Mixtures

It is well known that oil and water do not mix. Other substances—ethyl alcohol and water, table sugar and water, or salt and water—combine so readily that their individual components cannot be easily distinguished. A mixture of substances with different kinds of molecules (or ions) uniformly distributed visually throughout is called a **solution.** The key to the formation of this intimate mixture is that the secondary forces of each substance are similar. Solutions will not form between substances that each give rise to different kinds of secondary forces. An old rule of thumb expressive of that idea is, "Like dissolves like."

To explore this idea, let's represent the secondary forces between the molecules of liquid A by the symbol A---A, and those of liquid B by the symbol B---B. (The dashed lines represent the secondary forces between molecules). For liquids A and B to mix thoroughly (form a solution), the molecules must be able to change partners to form A---B and A---B. If the forces represented by A---A and B---B are about the same, then molecules of A will be just as likely to interact with molecules of B as with each other, and a solution will probably form. Written as an equation,

$$A\text{---}A + B\text{---}B \longrightarrow A\text{---}B + A\text{---}B$$

INSIGHT INTO PROPERTIES

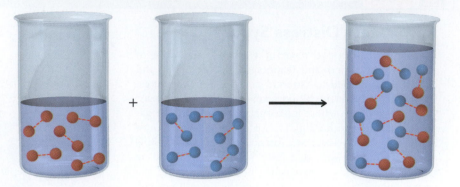

Figure 6.10 Schematic representation of the formation of a solution. A volume of liquid A (molecules colored red) is added to an equal volume of liquid B (molecules colored blue). In this case, the secondary forces indicated by the dashed lines are similar; therefore, A---B secondary attractions form and a solution forms. We will see this principle in action throughout our study of organic and biochemistry.

Figure 6.10 shows the process in diagrammatic form. In this sketch, a beakerful of A molecules (colored red) is added to a beakerful of B molecules (colored blue), and the two substances mix to form a solution (the forces between A molecules being about the same as those between B molecules) whose volume is equal to the sum of the volumes of the two individual components.

An understanding of secondary forces makes it possible to predict whether two molecular substances will form molecular mixtures, or solutions. You have only to examine the structure of the molecules in question and determine whether they are polar or nonpolar or whether H-bonding is possible between them.

Example 6.4   **Using molecular structures to predict whether solutions will form: I**

Can water and ethanol form a molecular mixture? The structural formulas for ethanol and water are

$$\begin{matrix} & H & H & & & & \\ & | & | & & & & \\ H-&C-&C-&O-H & & H-O-H \\ & | & | & & & & \\ & H & H & & & & \end{matrix}$$

Ethanol   Water

Solution

Both molecules are polar, and the presence of the hydroxyl (OH) group in both molecules indicates that both can form H-bonds not only between their own molecules but also between each other's molecules. These observations suggest that the secondary forces within each individual substance and those that would form within a mixture of the substances are about the same. Therefore, a solution should form.

Problem 6.4   Can water and carbon tetrachloride

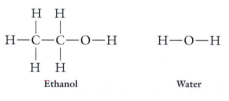

form a solution? Explain your answer.

Example 6.5 **Using molecular structures to predict whether solutions will form: II**

Will the hydrocarbon hexane, C_6H_{14}, and water, H_2O, form a solution? The structures for hexane and water are

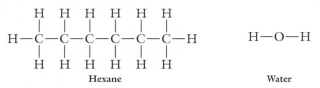

Hexane   Water

Solution

The structural formulas must be examined to see how the secondary forces within each liquid compare with those of the other. Hexane is a hydrocarbon with no permanent dipole; so its liquid state results from molecular attraction by London forces. Water molecules interact with each other through hydrogen bonds. Therefore the attractions between hexane and water molecules would be significantly different from those that exist between water molecules and between hexane molecules. The result is that water will be attracted to water and hexane to hexane. Hexane will not mix with or dissolve in water.

Problem 6.5 Explain why the following pairs of substances form or do not form solutions.
(a) CCl_4 and H_2O; (b) C_5H_{12} and CCl_4.

✓   When the attractive forces between different molecules are similar, solutions will form.

✓   When the attractive forces between different molecules are different, solutions will not form.

Concept checklist

6.5   THE VAPORIZATION OF LIQUIDS

As already stated, vaporization is the process that transforms a liquid into a gas. From a molecular point of view, vaporization is the "escape" of a molecule from a liquid. The energy of a particular molecule in a liquid continuously changes as it collides with others and exchanges energy. It can in time acquire a kinetic energy greater than the average and, should it be at the liquid's surface, it will be able to free itself of its neighbors' attractive forces and enter the gas phase.

When a substance in a sample is present as a liquid and a gas simultaneously, the gaseous part of the sample is called a **vapor.** Oxygen is never called a vapor, because, at room temperature, it does not exist in the liquid state. However, at the temperature of liquid oxygen, about −200°C, when both liquid and gas are present, oxygen in the gaseous state can be called oxygen vapor. The familiar odor of gasoline at the filling station is a vapor because both liquid and gaseous states are present simultaneously. A liquid's vapor is a gas and exerts pressure just as any gas does.

6.6   VAPOR PRESSURE AND DYNAMIC EQUILIBRIUM

Vaporization of a liquid depends on its temperature. A molecule in a liquid must have sufficient kinetic energy (be moving fast enough) to escape the attractive forces of the other molecules and enter the gas phase. **Vapor pressure** therefore increases with increases in temperature. On the other hand, its

Rate of = Rate of
vaporization condensation

Figure 6.11 After all the gas has been removed by a vacuum pump from a bell jar containing a beaker of liquid, the vapor pressure above the liquid eventually reaches a constant value. That value depends on the specific liquid and the temperature.

condensation depends on the number of vapor molecules colliding with the liquid's surface per unit time. The more vapor molecules striking the liquid surface per unit time, the greater the rate of condensation.

Consider the following experiment, illustrated in Figure 6.11. A beaker of a liquid—say, water—is placed under a bell jar. The experimental setup illustrated is called a **closed system,** a system in which mass is held constant; that is, it is physically impossible for mass to escape or enter the container. The experiment begins when the gaseous atmosphere under the bell jar is removed by a vacuum pump. At first, only vaporization takes place because, in the absence of vapor, condensation cannot take place. Shortly, however, condensation also begins to take place. In time, the amount of liquid transformed into the gaseous state per unit time (its rate of vaporization) and the amount of gas condensing into the liquid state per unit time (its rate of condensation) will be the same, and the amounts of liquid and vapor in the bell jar will remain unchanged. This kind of balance is a condition known in chemistry as an **equilibrium state.** The vapor pressure of a liquid in such a closed system is called its **equilibrium vapor pressure,** and only in a closed system can an equilibrium vapor pressure of a liquid be established.

When an equilibrium is the result of two opposing but active processes, it is called a **dynamic equilibrium.** Heat is required for the liquid-to-vapor-phase transition. That same quantity of heat is evolved when vapor condenses into a liquid.

$$\text{Liquid + heat} \xrightarrow{\text{vaporization}} \text{vapor}$$

$$\text{Vapor} \xrightarrow{\text{condensation}} \text{liquid + heat}$$

A dynamic equilibrium is represented by separating two states—such as liquid and vapor—by two arrows facing in opposite directions:

$$\text{Liquid + heat} \underset{\text{condensation}}{\overset{\text{vaporization}}{\rightleftarrows}} \text{vapor}$$

The names of the processes are sometimes inserted above and below the arrows, as shown here. At other times, labels are placed on the arrows to indicate any special conditions (heat, pressure, the presence of a certain kind of chemical) without which the process will not take place.

If the bell jar in the experimental setup illustrated in Figure 6.11 is removed, mass can then escape or enter the container. That kind of situation is described as an **open system,** and vaporization from an open system is called **evaporation.** Evaporation is an imbalance in the vaporization–condensation equilibrium. Because vaporized molecules continue to diffuse away from the liquid's surface, little or no condensation can take place. Vaporization is the dominant process, and the liquid will be continuously transformed into the gas.

To evaporate, the liquid must continuously absorb heat from its surroundings. If the flow of heat from the surroundings is somehow restricted, the liquid's temperature will fall. That is why evaporation is called a cooling process and is used to reduce body temperature. This process is illustrated in Box 6.3 and considered in Section 6.8.

6.7 THE INFLUENCE OF SECONDARY FORCES ON VAPOR PRESSURE

Vapor pressure is a measure of the **escaping tendency** of a substance's molecules. The smaller the attractive forces within a liquid sample, the greater the tendency of molecules to escape from that liquid and, consequently, the greater the vapor pressure. The larger the attractive forces between molecules,

6.3 CHEMISTRY AROUND US

Topical Anesthesia

The crowd has finally settled down for the start of the fourth game of the World Series, and the first ball has been thrown. There is no way that the unfortunate batter can prevent the errant 92 mph fast ball from making contact with his hand. The trainer rushes to the plate with an aerosol can and sprays the afflicted area. As the batter's contorted face relaxes, he aims a grateful smile at the trainer and then takes his painfully earned position at first base.

The substance sprayed on the traumatized zone is ethyl chloride (C_2H_5Cl), whose boiling point is 12.3°C. At temperatures above that limit, vapor forms throughout the liquid, which proceeds to evaporate at a speed dependant on the rate at which heat is transferred into it. When ethyl chloride is sprayed on human skin (temperature from approximately 30°C to 35°C), the rate at which heat transfers into the thin film of liquid is very rapid, and the liquid evaporates instantly. This extremely rapid evaporation of ethyl chloride occurs much faster than the rate at which the skin's heat can be replaced by the flow of blood; therefore there is a large temperature drop in the sprayed area. The cooling, in turn, significantly lowers the metabolic activity of pain receptors in the skin, and the sensation of pain diminishes as if by magic.

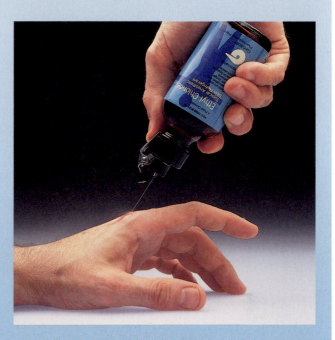

Ethyl chloride, a topical anesthetic, is sprayed on an injured joint. (Richard Megna/Fundamental Photographs.)

the smaller the vapor pressure. Liquids that are called volatile have high vapor pressures because of their relatively small secondary forces. Examples are ethyl ether, formerly used as an anesthetic, or ethyl chloride, a topical anesthetic.

Boiling is a special case of vaporization in which bubbles of vapor form throughout a liquid in an open container. Atmospheric pressure acts like a piston pressing against the surface of a liquid open to the atmosphere. A liquid's molecules must overcome this opposing force to enter the gas space over the liquid. Bubbles of vapor cannot form within the bulk of the liquid when its vapor pressure is below atmospheric pressure, because the greater atmospheric pressure will cause any bubbles to collapse. Therefore, when the temperature of a liquid is such that its vapor pressure is less than atmospheric pressure, vapor will leave the liquid only at its surface. However, when the vapor pressure is equal to atmospheric pressure, stable bubbles of vapor can form throughout the bulk of the liquid, rather than only at its surface, and the liquid is said to **boil.**

There is no reason why boiling cannot occur at any temperature. All that is required is that the liquid's vapor pressure be equal to the external pressure. Atmospheric pressure decreases with altitude; so the higher one is above sea level, the lower the temperature of boiling. The boiling point of water in New York City is 100°C, but, at the top of Mt. Everest in the Himalaya Mountains, it is 70°C. Boxed cake mixes purchased in Denver, Colorado, will be culinary disasters when baked in Pensacola, Florida, because the recipe used in Denver must be designed for cooking at high altitude—in other words, at atmospheric pressures lower than that at sea level.

Because of the variations in atmospheric pressure and their effects on boiling points, chemists have found it helpful to define the **normal boiling point.**

Figure 6.12 Vapor pressure of chloroform (CHCl$_3$), hexane (C$_6$H$_{14}$), and ethanol (C$_2$H$_5$OH) as a function of temperature. The horizontal dashed line represents atmospheric pressure of 1 atm. When the vapor pressure is equal to the atmospheric pressure, the liquids boil.

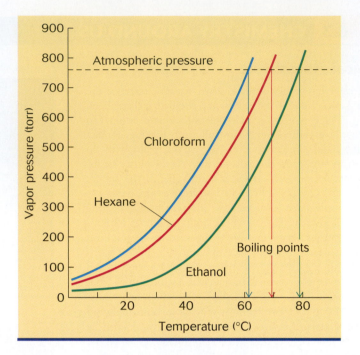

This point is the temperature at which boiling occurs under an external pressure of exactly 1 atm. Figure 6.12 shows changes in vapor pressure with changes in temperature for three compounds of varying volatility: chloroform (CHCl$_3$), hexane (C$_6$H$_{14}$), and ethanol (C$_2$H$_5$OH). As the temperature increases, the vapor pressures increase to a different extent for each compound. Of the three liquids, chloroform has the highest vapor pressure (is the most volatile), then hexane, and then ethanol. The dashed horizontal line represents atmospheric pressure. Boiling points are inversely related to vapor pressure: the more volatile a substance, the lower its boiling point.

The normal boiling point is a good qualitative indication of the secondary forces within a liquid. Differences in normal boiling points among liquids are due to differences in those forces within the liquids. The data in Table 6.3 illustrate this effect. The boiling points can be seen to increase with increasing strength of secondary forces.

Example 6.6   Reading a vapor-pressure-versus-temperature curve

Estimate the normal boiling points of the three liquids whose pressure-versus-temperature relations appear in Figure 6.12.

Solution
The normal boiling point is the temperature at which the vapor pressure of a liquid reaches atmospheric pressure. The temperatures in Figure 6.12 that correspond to that pressure for each liquid are: chloroform, approximately 62°C; hexane, approximately 69°C; and ethanol, approximately 78°C.

Problem 6.6   Estimate the vapor pressures at 50°C of the three liquids whose pressure-versus-temperature relations appear in Figure 6.12.

Concept check   ✓   The normal boiling point is the temperature at which vapor forms throughout a liquid at an external pressure of exactly 1 atm.

Example 6.7 — Relating vapor pressure to melting and boiling points

The following table lists the normal boiling points and melting points of two substances.

	Melting point	Boiling point
Substance A	−15°C	60°C
Substance B	−60°C	−10°C

(a) Compare the vapor pressure of substance A at 60°C with that of substance B at −10°C. (b) Then compare the vapor pressures of substances A and B when both substances are at −12°C.

Solution

(a) Because 60°C and −10°C are the respective normal boiling points of the two liquids, their vapor pressures must be equal to atmospheric pressure, and therefore equal to each other, at those temperatures.

(b) At −12°C, substance B is in the liquid state and close to its normal boiling point. Substance A also is in the liquid state but barely above its melting point and far from its normal boiling point. Therefore, at −12°C, the vapor pressure of liquid B must be greater than the vapor pressure of liquid A.

Problem 6.7 Methyl alcohol (CH_3OH) and methyl iodide (CH_3I) are both polar molecules, but methyl iodide's vapor pressure at room temperature is almost four times as great as that of methyl alcohol. What is the molecular basis for the difference?

Example 6.8 — Using secondary forces to predict boiling points

Referring to the Lewis structures of H_2O (water) and H_2S (hydrogen sulfide), explain the much higher normal boiling point of water, 100°C, compared with −61°C for H_2S.

Solution

Hydrogen bonds can form between molecules when hydrogen is covalently bonded within the molecule to oxygen, nitrogen, or fluorine. Of these two compounds, only water can form H-bonds, whereas H_2S is limited to interacting through weaker secondary forces.

Problem 6.8 The structures for ammonia and phosphine are:

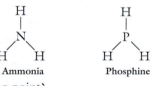

Which has the higher boiling point?

6.8 VAPORIZATION AND THE REGULATION OF BODY TEMPERATURE

The regulation of body temperature is of vital importance to humans. It is achieved by a balance between the body's heat production and its ability to lose heat. Vaporization requires heat and therefore provides an important mechanism for heat loss. Water vaporizes from the body at the skin's surface

A PICTURE OF HEALTH
Temperature Regulation in the Body

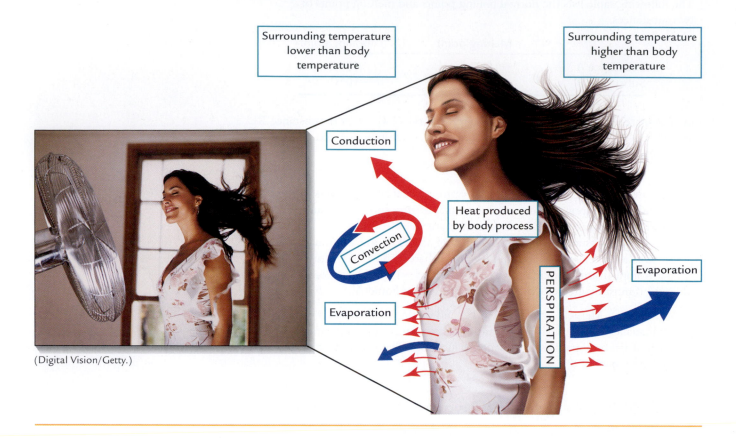

Surrounding temperature lower than body temperature

Surrounding temperature higher than body temperature

Conduction

Convection

Heat produced by body process

PERSPIRATION

Evaporation

Evaporation

(Digital Vision/Getty.)

Molar Heats of Vaporization

H_2O	9.73 kcal/mol
NH_3	5.59 kcal/mol
CH_4	1.96 kcal/mol

when we perspire, as well as from the lungs and mucous membranes of the mouth and nasal passages, removing heat from the body in the process.

The heat loss from the body through the vaporization of water is very efficient because of water's high heat of vaporization. That in turn is the result of hydrogen bonding among water molecules. When the molar heats of vaporization of water, ammonia, and methane are compared (see margin), the amount of heat required for the vaporization of water is about twice as much as that for another hydrogen-bonding molecule, ammonia, and five times that for nonpolar methane. At comfortable room temperatures, water loss from vaporization of a sedentary person might be only 600 mL/day, or about 350 kcal/day. Water loss due to sweating as a result of heavy work or running can be as high as 4 L/h, equivalent to as much as 2000 kcal. However, the rate of vaporization of body water can be seriously affected by the relative humidity, a factor discussed next.

Water vapor is one of the most important gases in Earth's atmosphere, but its concentration varies greatly from place to place. Water's atmospheric partial pressure may be as low as 0.5 torr in certain desert regions and as high as 55 torr in the tropics. That last figure is close to the equilibrium vapor pressure of water (that is, the vapor pressure of water in a closed system). Because the atmosphere is not a closed system, water's vapor pressure in the atmosphere must be expressed in relative terms. Thus we have the term **relative humidity,** defined as the ratio of the partial pressure of water vapor in the atmosphere to the equilibrium vapor pressure of water at the same temperature times 100. Mathematically:

$$\text{Relative humidity} = \left(\frac{\text{partial pressure of water}}{\text{equilibrium vapor pressure of water}} \right) \times 100\%$$

The equilibrium vapor pressure of water is 23.8 torr at 25°C. If the partial pressure of atmospheric water vapor is 12.8 torr, the relative humidity is

$$\text{Relative humidity} = \left(\frac{12.8}{23.8} \right) \times 100\% = 53.8\%$$

If the temperature goes down to 15°C, where water's equilibrium vapor pressure is 12.8 torr and if the partial pressure of atmospheric water vapor remains the same, the relative humidity is

$$\text{Relative humidity} = \left(\frac{12.8}{12.8} \right) \times 100\% = 100\%$$

(Julian Baum/Bruce Coleman.)

At that temperature, water's partial pressure is equal to its equilibrium pressure. The equilibrium pressure represents the maximum concentration of water vapor at that temperature, and the air is said to be saturated with water vapor. The important point is that, under these conditions, the rate of condensation is equal to the rate of vaporization.

At environmental temperatures below body temperature, heat is transferred from our bodies to the environment in the same way that heat is transferred when a cup of coffee cools. Three mechanisms are responsible:

1. **Conduction.** Heat flows from a warm body to a cold one. If you are in contact with any substance (air, water, or a solid) that is at a lower temperature than your body, you will lose heat to that substance.
2. **Convection.** Air warmed by the body becomes less dense, rises, and is replaced by a flow of cooler, denser air.
3. **Evaporation.** Water from the body is vaporized and lost to the environment.

Conduction and convection are largely responsible for body-temperature control when the environmental temperature is below body temperature. However, when the environmental temperature approaches body temperature, convection and conduction begin to fail, and we depend almost exclusively on the evaporation of water to control our body temperature.

We have all heard or used the phrase "It's not the heat, it's the humidity." This phrase is an expression of the fact that, when environmental temperatures are near body temperature and water vapor partial pressures are near the equilibrium vapor pressure, we feel quite uncomfortable. For some people, these conditions are not only uncomfortable, but also dangerous. A short-lived heat wave in the Midwest caused the deaths of more than 100 elderly men and women. When the same temperatures are reached in the Southwest, very few are overcome by the heat. Why, then, did so many die in the Midwest?

The answer is that the death-causing heat wave was accompanied by very high relative humidity, with the partial pressure of water vapor almost equal to the equilibrium vapor pressure of water. Under those conditions, the rate of condensation of water is close or equal to that for evaporation. The consequence is that any heat lost by evaporation is immediately gained back by condensation. Therefore no heat loss can be effected, and body temperature can rise to dangerous levels.

Air conditioning and electric fans would have allowed heat loss by conduction and convection, respectively. Not possessing either of these modern conveniences, the elderly poor succumbed rapidly to rising body temperatures.

Figure 6.13 Crystalline solids have well-defined faces and edges. Each face is the exposed end of an orderly internal arrangement of atoms, ions, or molecules. The solid shown here is the lead ore galena, PbS. (Chip Clark.)

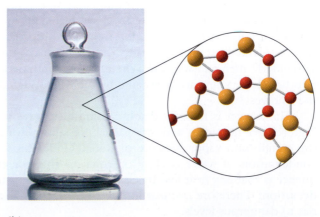

Crystal faces

6.9 ATTRACTIVE FORCES AND THE STRUCTURE OF SOLIDS

In contrast with the disorder of the gaseous state, solids are characterized by order. The atoms of a solid are arranged in regular patterns. Solids form crystals that have sharply defined melting points and can exist in geometrically well defined shapes, such as prisms, octahedrons, or cubes. The reason that **crystalline solids** have flat faces and well-defined edges is that each face is the exposed end of an orderly internal arrangement of atoms, ions, or molecules, as shown in Figure 6.13. Not all solids have an ordered crystalline form; those lacking it are called **amorphous.** For example, the noncrystalline form of SiO_2, called glass, is an amorphous solid. Figure 6.14a contains a photograph of the crystalline form of silica, SiO_2, or quartz, and a two-dimensional sketch showing how the atoms are arranged in an orderly network; Figure 6.14b shows a glass vessel formed when molten quartz is allowed to solidify, along with a sketch showing how the atoms now form a disorderly array.

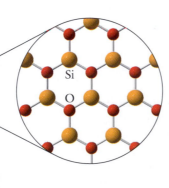

Si

O

(a)

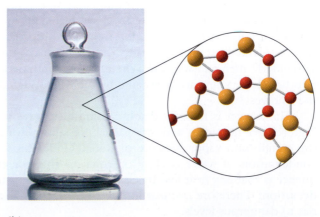

(b)

Figure 6.14 (a) The crystalline form of silica (quartz), and the orderly way in which its atoms are arranged. (b) The amorphous form of silica, called glass, and the disorderly array of its atoms. (Photo (a), Steven Smale; photo (b), W.H. Freeman photo by Ken Karp.)

TABLE 6.4	Melting Points of Covalent, Ionic, and Molecular Crystals	
Substance	Solid state	Melting point (°C)
diamond (carbon)	covalent crystal	3550
quartz (SiO_2)	covalent crystal	1610
sodium chloride	ionic crystal	800
calcium chloride	ionic crystal	782
citric acid	molecular crystal	153
cocaine (crack)	molecular crystal	98
chloroform	molecular crystal	−64
ethanol	molecular crystal	−117

The melting point is one of the most useful clues available for identifying the types of attractive forces operating in a solid. (Remember that the melting point and freezing point are identical; they are merely approached from opposite directions.) Table 6.4 lists several compounds along with their melting (freezing) points. The highest, by far, belongs to diamond, a form of elemental carbon in which the atoms are held in place in the crystal by covalent bonds. Quartz is a naturally occurring form of silicon dioxide also held together by covalent bonds. Diamond and quartz are therefore called **covalent crystals.** Next in order are NaCl and $CaCl_2$, whose crystals are formed by ionic chemical bonds. Substances such as NaCl and $CaCl_2$ are therefore called **ionic crystals.**

Molecules such as ethanol, citric acid, cocaine, or chloroform are held together in the solid state by the much weaker secondary forces—that is, hydrogen bonds, dipole–dipole forces, and London forces—and, in the solid state, are called **molecular crystals.** The melting points in Table 6.4 illustrate the relative strengths of the attractive forces within these three different kinds of crystalline solids.

Example 6.9 **Using molecular structure to predict physical properties: II**

Arrange the order of melting points for the following solids, highest to lowest:

$$CH_4 \quad C_6H_{14} \quad C_2H_6 \quad C_4H_{10}$$

Solution
All these compounds are hydrocarbons. They do not have permanent dipoles, nor can they form hydrogen bonds. They therefore can interact only by London forces. The greater the molecular mass, the more electrons, the larger the temporary dipoles, and the larger the secondary forces. The order of melting points from highest to lowest is therefore

$$C_6H_{14} > C_4H_{10} > C_2H_6 > CH_4$$

Problem 6.9 Arrange the order of melting points for the following solids, highest to lowest: (a) Br_2; (b) I_2; (c) F_2; (d) Cl_2.

✓ Crystalline solids have an orderly internal structure.

✓ The internal structure of amorphous solids is disordered.

Concept checklist

Under ordinary environmental conditions—for example, normal atmospheric pressure and room temperature—some solids can vaporize directly to the gaseous state without melting to an intermediate liquid state. Examples are solid CO_2 (dry ice); solid I_2, a solid that is always present in its container along with some purple vapor; and *p*-dichlorobenzene, a solid whose distinctive odor, as it vaporizes, is used to keep moths away from clothing. The direct transition from solid to gas is called **sublimation.** The energy required to transform the solid into a gas is about the same as it would be if the solid had gone through two phase transitions, melting and vaporization, instead of just one.

In preceding chapters, chemical bonds and the nature of molecules were the subjects of interest. In this chapter, the objective has been to broaden the view to include interactions between molecules. These interactions are far weaker than chemical bonds but are of central importance in defining the physical characteristics of substances. Of more interest to those who intend to work in the life sciences, these weak interactions are of paramount importance in all aspects of biochemical structure and function.

Summary

Molecules and the States of Matter

• Molecules retain their sizes and shapes whether in the solid, liquid, or gaseous state.

• Molecules in the solid state are highly organized and in close contact.

• Molecules in the liquid state also are in contact but are less well organized.

• In the gaseous state, the amount of space between molecules and the degree of disorder among them are at a maximum.

Attractive Forces Between Molecules

• Attractive forces between molecules are called secondary forces.

• The secondary forces that lead to the formation of liquids and solids are London forces, dipole–dipole forces, and hydrogen bonds.

• The relative strengths of these forces are in the order: hydrogen bonding > dipole–dipole attractions > London force.

• The relative strength of the secondary forces between molecules controls differences between melting and boiling points of different substances

Changes in the States of Matter

• Phase transitions take place when the balance between kinetic energy and secondary forces is changed by the addition or removal of heat.

• Melting, the conversion of solid into liquid, happens when only a part of the secondary forces is overcome.

• Vaporization, the conversion of liquid into gas, occurs when all secondary forces are overcome.

Vapor Pressure and Dynamic Equilibrium

• A gas in contact with its liquid form is called a vapor and exerts a pressure called the vapor pressure.

• Vapor pressure increases with increase in temperature.

• Vaporization and condensation of a liquid take place simultaneously. When the two processes proceed at the same rate, the amounts of liquid and gas (vapor) remain constant and the system is said to be in dynamic equilibrium.

Vaporization of Water and the Regulation of Body Temperature

• The body's ability to maintain a constant temperature chiefly depends on the large value of water's heat of vaporization.

Formation of Solutions

• If the secondary forces in one substance are of the same type as in another substance, two substances will probably mix to form a solution.

The Structure of Solids

• In contrast with the structural disorder found in gases and liquids, the atoms or molecules of solids are arranged in distinct and ordered patterns.

• The regular patterns of atoms within solids lead to their geometrically well defined shapes and fixed melting points.

Key Words

boiling point, p. 150
condensation, p. 150
crystalline solid, p. 168
dipole–dipole force, p. 151
dynamic equilibrium, p. 162
equilibrium state, p. 162
freezing point, p. 150
hydrogen bonding, p. 153

kinetic energy, p. 150
liquid, p. 149
London force, p. 151
melting point, p. 150
molar heat of fusion, p. 150
molar heat of vaporization, p. 151
phase transition, p. 150
relative humidity, p. 166

secondary force, p. 151
solid, p. 149
solution, p. 159
surface tension, p. 157
vaporization, p. 150
vapor pressure, p. 161

Exercises

Solids, Liquids, and Gases

6.1 Describe the condition of matter if there were no secondary forces operating between molecules.

6.2 Describe the condition of matter if the secondary forces operating between molecules were as strong as covalent or ionic bonds.

6.3 Propose a molecular mechanism for the melting of a solid.

6.4 Propose a molecular mechanism for the evaporation of a liquid.

6.5 True or False: The vapor pressure of a liquid increases with temperature. Explain your answer.

6.6 True or False: The vapor pressure of a boiling liquid in an open beaker increases with temperature. Explain your answer.

6.7 Why are solids and liquids virtually incompressible?

6.8 Why are gases so much more compressible than liquids or solids?

6.9 What is the relation of the molar heat of fusion to the molar heat of vaporization?

6.10 What is the molecular explanation for your answer to Exercise 6.9?

6.11 True or False: The melting point of a solid is the same as its freezing point. Explain your answer.

6.12 True or False: The normal boiling point of a liquid is the temperature at which its vapor pressure is equal to the atmospheric pressure. Explain your answer.

6.13 If it takes 9.7 kcal of heat to vaporize 1.0 mol of water, how much heat will be released when 1.0 mol of water vapor condenses into liquid water?

6.14 If it takes 1.44 kcal of heat to melt 1.0 mol of solid water (ice), how much heat will be released when 1.0 mol of liquid water freezes and forms ice?

Attractive Forces Between Molecules

6.15 How can molecules in the gaseous state condense into a liquid if their molecules are not polar?

6.16 Compare the boiling points of liquids whose molecules interact by hydrogen bonds with those whose molecules interact by London forces.

6.17 Between which of the following pairs of molecules can hydrogen bonds be formed?

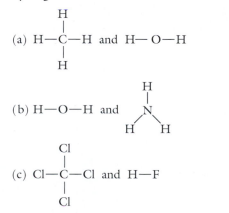

6.18 Between which of the following pairs of molecules can hydrogen bonds be formed?

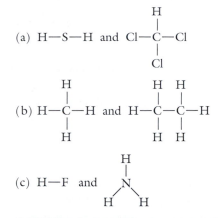

6.19 What is the relation between molecular mass and the strength of the London forces exerted between molecules?

6.20 Arrange the boiling points of the following compounds in the order from highest to lowest.

$$C_8H_{18} \qquad C_6H_{14} \qquad C_3H_8 \qquad C_4H_{10}$$

6.21 Why is it that carbon tetrachloride, CCl_4, a molecule with four polar covalent bonds, has no dipole moment?

6.22 Why is it that carbon dioxide, CO_2, a molecule with two polar covalent bonds, has no dipole moment?

6.23 Fill in the following table with a yes or no answer to the question, "Does this secondary force operate between the molecules of the given compound?"

	London force	Dipole-dipole	Hydrogen bond
CH_4	_____	_____	_____
$CHCl_3$	_____	_____	_____
NH_3	_____	_____	_____

6.24 Fill in the following table with a yes or no answer to the question, "Does this secondary force operate between the molecules of the given compound?"

	London force	Dipole-dipole	Hydrogen bond
CCl_4	_____	_____	_____
CHF_3	_____	_____	_____
H_2O	_____	_____	_____

6.25 Which of the following compounds has the greater heat of vaporization?

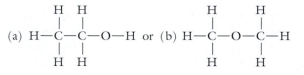

6.26 Which of the following compounds has the greater heat of vaporization?

(a) $H-\underset{\underset{H}{|}}{\overset{\overset{H}{|}}{C}}-H$ or (b) $H-O-H$

Liquids

6.27 Explain why a liquid can flow and assume the shape of its container.

6.28 Explain why a solid retains its shape and cannot assume the shape of its container.

6.29 Explain why water will continuously evaporate from an open container at room temperature.

6.30 If water continuously evaporates from an open container at room temperature, does the evaporation require energy? If it does, where does the energy come from?

6.31 From a molecular point of view, describe the relation between the volume of a solid and the volume of the resultant liquid.

6.32 From a molecular point of view, describe the relation between the volume of a liquid and its volume as a gas.

6.33 Why do we not use the word vapor to characterize oxygen or nitrogen?

6.34 Why do we use the word vapor to characterize water or ethanol?

6.35 Arrange the following compounds in order of increasing vapor pressure, lowest to highest: (a) H_2O; (b) KCl; (c) $CHCl_3$.

6.36 Arrange the following compounds in order of increasing vapor pressure, lowest to highest: (a) NaCl; (b) CH_4 (methane); (c) CH_3CH_2OH (ethanol).

6.37 Is it possible to measure the vapor pressure of a liquid in an open container? Explain your answer.

6.38 Is it possible to measure the vapor pressure of a liquid in a closed container? Explain your answer.

6.39 What is meant by equilibrium?

6.40 What is meant by a dynamic equilibrium?

6.41 What is the normal boiling point of a liquid?

6.42 Can a liquid boil at any pressure? Explain your answer.

6.43 Compare the surface tensions of water and chloroform, and explain any differences. The structures of water and chloroform are

Water

Chloroform

6.44 Compare the surface tensions of water and the hydrocarbon hexane, and explain any differences. The structures of water and hexane are

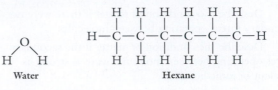
Water   Hexane

6.45 Will either (a) acetic acid or (b) pentane form molecular mixtures with (dissolve in) water? Explain your answer. Their structures are

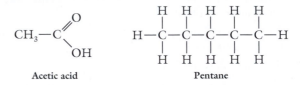
Acetic acid   Pentane

Solids

6.46 What are the molecular characteristics and an important physical property of a crystalline solid?

6.47 What are the molecular characteristics and an important physical property of an amorphous solid?

6.48 Place the melting points of the following substances in the order from lowest to highest: KCl; H_2O; H_2S.

6.49 Place the melting points of the following substances in the order from lowest to highest: C_2H_5OH (ethanol); NaCl, CH_4 (methane).

6.50 The equilibrium vapor pressure of water at 75°F is 23.6 torr. The partial pressure of water in the atmosphere is 12.3 torr. Calculate the relative humidity.

6.51 The equilibrium vapor pressure of water at 86°F is 31.8 torr. The partial pressure of water in the atmosphere is 21.6 torr. Calculate the relative humidity.

Unclassified

6.52 Give three examples of molecules that interact exclusively by London forces.

6.53 Give three examples of molecules that have permanent dipole moments.

6.54 Why does it take so much longer to boil an egg in Denver than in San Francisco?

Expand Your Knowledge

Note: The icons ▲ ▲ ▲ denote exercises based on material in boxes.

6.55 The boiling points of some hydrides of the first period are

CH_4	NH_3	H_2O	HF
$-161°C$	$-33°C$	$100°C$	$-19.5°C$

Explain why water, not HF, has the highest boiling point, even though oxygen is less electronegative than fluorine.

6.56 Explain the use of alcohol rubs to lower body temperature.

6.57 The relation of strength of secondary force to vapor pressure is inverse. That is, the greater the secondary forces, the smaller the vapor pressure. Explain this phenomenon.

6.58 Why can you detect the presence of mothballs, composed of solid para-dichlorobenzene, without any sophisticated analytical instrumentation?

6.59 Explain why dimethyl ether, $CH_3—O—CH_3$, an apparently symmetrical molecule, has a dipole moment.

6.60 Frank is backpacking with friends in the Rockies. They all complain of shortness of breath and insist that the altitude must be more than 12,000 feet. Frank knows from previous trips that the altitude cannot be more than 10,000 feet and that they're just not in shape. He declares that they're wrong and he can prove it with only a thermometer. They scoff, but he puts a pot of water on the grill and, when it boils, he announces that they are at 8000 feet. Use the accompanying data to determine the temperature at which the water boiled. (Hint: Graph the data.)

Altitude in feet	Temperature (°C)
0	100
6,400	94
8,600	92
10,800	90
13,000	88

6.61 Why is hydrogen sulfide, H_2S, a gas at $-10°C$, whereas water, H_2O, is a solid at the same temperature?

6.62 Explain why silicon carbide, SiC, and boron nitride, BN, are about as hard as diamond.

6.63 Moisture will often form on the outside of glass containing a mixture of ice and water. Can you explain this phenomenon?

6.64 Explain why the vapor pressure of ethanol is greater than the vapor pressure of hexane at the same temperature.

6.65 Explain why there is a difference in the boiling points for each of the following pairs of substances:
   (a) HF, 20°C; HCl, −85°C
   (b) $TiCl_4$, 136°C; LiCl, 1360°C
   (c) HCl, −85°C; LiCl, 1360°C
   (d) CH_3OCH_3 (dimethyl ether), 35°C; CH_3CH_2OH (ethanol), 79°C

6.66 How do the following physical properties depend on the strength of intermolecular forces? (a) Boiling point; (b) melting point; (c) vapor pressure; (d) surface tension.

6.67 Explain why the heat required for vaporization for a mole of a substance is significantly greater than the heat required for melting the same amount of the same substance.

6.68 In what way is a liquid similar to a solid?

6.69 In what way is a liquid similar to a gas?

6.70 Wet laundry hung on a clothesline on a cold winter day will freeze but eventually dry. Explain.

6.71 Explain why burns of the skin are much more severe from steam than from boiling water.

▲ **6.72** What is an emulsion? (See Box 6.1.)
6.1

▲ **6.73** Describe the function of lung surfactant. (See Box 6.2.)
6.2

CHAPTER 7

SOLUTIONS

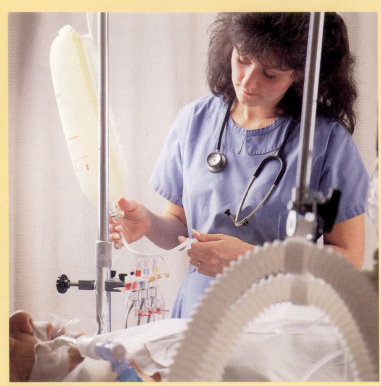

(Steve Weinrebe/Picture Cube.)

Chemistry in Your Future

When a head-injury patient arrives at the hospital, a key concern of the health-care team is to prevent or reduce excess fluid collection around the brain. Such swelling, or edema, is a natural response to injury but, when the injured organ is the brain, the added fluid pressure can cause severe damage. One way to deal with this problem is to administer an intravenous solution of mannitol, a water-soluble compound having no biological activity. Its only physiological effect is to increase the osmotic pressure of the filtrate in the kidneys' tubules, thus increasing the amount of fluid disposed of in the urine. What is osmotic pressure, and how does it affect body fluids? After reading this chapter on solutions, you will know.

For more information on this topic and others in this chapter, go to www.whfreeman.com/bleiodian2e

Learning Objectives

- Describe how the formation of a solution depends on the molecular properties of the solute and the solvent.

- Give the quantitative definitions of concentration and use them as conversion factors in calculations.

- Specify reasons and methods for preparing dilute solutions from concentrated solutions.

- Describe diffusion and the characteristics of semipermeable membranes from a molecular point of view.

- Describe the origin of osmotic pressure and how it is measured and used in calculations.

- Describe the properties of macromolecules and colloidal solutions.

Chapter 6 dealt with the properties of pure substances, but, in practice, most chemical reactions take place in solutions. A solution is a special kind of mixture in which substances are mingled so thoroughly that the separate components are no longer visible, even under the microscope. The composition of a solution is variable, and the components retain their chemical identities. However, the properties of a solution can be markedly different from the properties of the solution's components. This difference warrants special consideration of the properties of solutions.

It is difficult to find a biological process that does not take place in solution. The living cell itself consists of an aqueous medium containing a complex mixture of structures and dissolved substances. To the chemist, then, solutions offer a means of studying processes that could never be observed in the solid or gaseous states. But, to the biochemist, solutions can serve as experimental models that closely simulate conditions in the cell.

In this chapter, we will explore the qualitative and quantitative aspects of solution formation and some of the ways in which they are of central importance in biological systems.

7.1 GENERAL ASPECTS OF SOLUTION FORMATION

The simplest kind of solution is a mixture of two substances: the **solute,** the substance in smaller amount, dissolved in the **solvent,** the substance in larger amount. We have said that mixtures have no fixed recipe, so the components can be present in varying proportions. Nevertheless, there is generally a limit to the amount of solid solute that can be dissolved in a solvent. Gas mixtures are an exception to this general rule. Solvent and solute can be mixed in any proportions in all gas mixtures, as well as in certain liquid mixtures such as ethanol and water. When two or more substances can form solutions in all proportions, they are said to be **miscible.**

All sorts of substances can mix with one another, but not all can form solutions. If we categorize substances by physical state—gas, liquid, or solid—and mix the states two at a time, nine kinds of mixtures but only seven kinds of solutions are possible. The combinations are described in Table 7.1 as G/G (gas in gas), L/L (liquid in liquid), S/S (solid in solid), and so forth.

The pairs of states that cannot be combined to form true solutions are solids in gases (S/G) and liquids in gases (L/G). The S/G pairing describes smoke, and the L/G pairing describes mists or fogs; both these mixtures are called suspensions. A **suspension** consists of visible aggregates or particles

TABLE 7.1	The Nine Types of Mixtures Made from the Three States of Matter, with Examples
Mixture*	**Example**
G/G	air
G/L	carbonated water
G/S	hydrogen in steel
L/G	mist or fog
L/L	alcohol in water
L/S	dental amalgam, Hg in Ag
S/G	smoke
S/L	sugar in water
S/S	TV screen phosphors

*The letters G, L, and S stand for gas, liquid, and solid, respectively. The first letter denotes the solute, and the second is the solvent. For example, G/L means gas in liquid, and L/G means liquid in gas.

suspended in a continuous medium. Sometimes the particles, although visible, are so small that gravity has a negligible effect on them and they remain suspended indefinitely. Particles that small are described as colloidal and will be considered separately later in this chapter.

7.2 MOLECULAR PROPERTIES AND SOLUTION FORMATION

Although many types of solutions are possible, liquid solutions—that is, solids in liquids or liquids in liquids—will be the primary objects of our interest. The physical properties of liquids and solids discussed in Chapter 6 were shown to depend on the secondary forces operating between molecules—that is, London forces, dipole–dipole forces, and hydrogen bonding. Solution formation also was shown to depend on the same secondary forces. The molecules of solvent and solute must interact through secondary forces to form a solution. A solution will form when solute–solvent interactions are due to secondary forces closely similar to the solute–solute and solvent–solvent interactions. This idea is often stated simply as "like dissolves like." The interaction between any solute and solvent is described as **solvation.**

Water is a solvent in which solutions form because of strong, rather than weak, interaction between solvent and solute. Its polarity and its ability to form hydrogen bonds allow water to interact with, or solvate and dissolve, electrically charged substances, polar substances, and substances that can form hydrogen bonds (Section 6.2). The special case of solvation by water is called **hydration.** Solutions in which water is the solvent are called **aqueous solutions.** The hydration of ions or molecules is expressed by the notation aq, as in $Na^+(aq)$ and $Cl^-(aq)$. The hydration of Na^+ and Cl^- ions is illustrated schematically in Figure 7.1.

From a molecular point of view, what is the difference between a pure liquid such as water and a solution of some substance such as glucose in that same liquid? To answer this question, we must examine and compare the structures of glucose and water. The interactions between glucose molecules consist of hydrogen bonds between OH groups, the same type of bond as those within pure water. Therefore, glucose will dissolve in water because similar attractive forces act within a mixture of the two substances.

The difference between pure water and its glucose solution is that some of the water molecules have been replaced by glucose molecules. The glucose

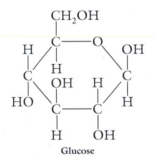

Glucose

INSIGHT INTO PROPERTIES

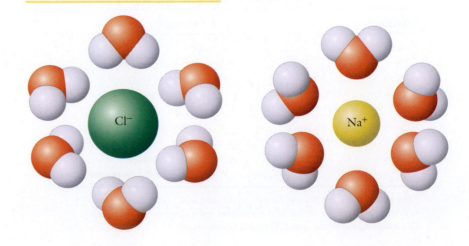

Figure 7.1 The solvation of sodium and chloride ions in water. The ions are surrounded by a shell of water molecules bound to the ions by secondary forces, which helps to explain water's ability to dissolve so many ionic substances.

molecules are said to be hydrated—that is, to have water molecules closely attracted to them by secondary forces. But that situation is not significantly different from that in pure water, in which water molecules have water molecules attracted to them. Thus, in dilute aqueous solutions, polar solute molecules dissolve because some portion of the water originally present has been replaced by some other substance similar in structure or polarity or both.

7.3 SOLUBILITY

Solubility is a quantitative term often confused with the qualitative term soluble. The designation soluble is more properly reserved to describe substances able to dissolve in significant amounts in a given solvent, perhaps a few grams in 100 mL of solvent. For example, a tablespoon of sugar can be seen to dissolve rapidly in a beaker of water and so is considered to be water soluble.

For most substances, there is a limit to the amount that can be dissolved in a given amount of solvent. When less than that limit is dissolved, the solution is said to be **unsaturated.** When the limit is reached, the solution is said to be **saturated. Solubility** is a quantitative measure of the limiting amount of solute that will dissolve in a fixed amount of solvent at a specific temperature.

It is difficult to draw a line between soluble and insoluble, but the data in Table 7.2 impart a sense of the meaning and use of these descriptors. All the compounds listed in Table 7.2 dissolve in water, but they clearly fall into two distinct categories in regard to the extent to which they can dissolve.

Chemists developed rules based on experience that are useful for making qualitative predictions of whether an ionic substance will be soluble in water (Table 7.3). The general rule of solubility or insolubility for a given class of

TABLE 7.2 The Quantitative Meaning of the Words Soluble and Insoluble

	Soluble salts		Insoluble salts
Name of salt	Solubility (g/100 mL at 20°C)	Name of salt	Solubility (g/100 mL at 20°C)
$BaCl_2$	35.7	$BaSO_4$	2.4×10^{-4}
BaI_2	203.1	$Ba(IO_3)_2$	2.2×10^{-2}
CaI_2	67.6	CaF_2	1.6×10^{-3}
$FeSO_4$	26.5	FeS	6.2×10^{-4}
$HgCl_2$	6.1	Hg_2Cl_2	2.0×10^{-4}

TABLE 7.3 Qualitative Solubility Rules for Ionic Solids in Water

Solubility	Important exceptions
SOLUBLE	
Na, K, and NH_4 salts	
nitrates and acetates	
sulfates	Ca, Sr, and Ba sulfates
chlorides and bromides	Ag, Pb, and Hg(I) chlorides and bromides
INSOLUBLE	
hydroxides	Li, Na, K, Rb, Ca, Sr, and Ba hydrides
phosphates, carbonates, and sulfides	Li, Na, K, Rb, and NH_4^+ phosphates, carbonates, and sulfides

Figure 7.2 Visualizing
concentration as a kind of molecular
density. The concentrations and
corresponding molecular densities
of these three solutions are in the
order highest to lowest: A > B > C.

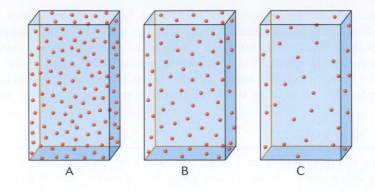

compounds is stated in the left-hand column, and important exceptions to that
general rule are presented in the right-hand column.

7.4 CONCENTRATION

Concentration is a measure of the amount of solute dissolved in a given
amount of solution. Figure 7.2 demonstrates that concentration is a kind of
molecular density, a measure of mass of solute per unit volume of solution.
Figure 7.3 shows that, no matter how small or large an amount of a solution is
observed, the amount of dissolved solute found in any fixed volume of the
solution—the concentration—remains the same. It is defined mathematically as

$$\text{Concentration} = \frac{\text{amount of solute}}{\text{amount of solution}}$$

The amount of solute can be expressed either in physical mass units, such
as grams, or in chemical mass units, such as moles. The amount of solvent can
be expressed in units of either volume or mass—that is, liters or kilograms, respectively. For example,

$$\text{Concentration} = \frac{\text{grams of solute}}{\text{liter of solution}}$$

The effectiveness of drugs depends on their concentrations in the blood.
If the concentration is too high, there will be toxic side effects; if it is too low,
then the drug will be ineffective. The fluid balance of the body depends on
the concentrations of blood components. In fact, literally no aspect of human
physiology can be adequately described without referring to concentration.
In the next sections, we will more fully develop the quantitative concept of
concentration.

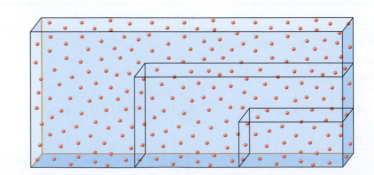

Figure 7.3 The concentration of a solution remains constant regardless of the volume
of solution observed.

7.5 PERCENT COMPOSITION

The concept of percent composition of chemical compounds was introduced in Section 2.1. The concept can also be applied to the concentration of solutions. For that use, **percent concentration** is defined as 100% × concentration, or

$$\text{Percent concentration} = \frac{\text{amount of solute}}{\text{amount of solution}} \times 100\%$$

In this designation of concentration, the amounts may be expressed either in mass or in volume or in both. It is important to note that the amount of solution in the definition consists of the sum of the amount of solute plus that of the solvent.

w/w Percent Composition

When the compositions of both the solute and the solvent are given in mass units (grams), the percent composition is designated as % w/w, meaning that the quantities of both the solute and the solvent were determined by weighing on a balance. The concentration provides us with the unit-conversion factor for calculations aimed at the determination of solution composition.

Example 7.1 Preparing % w/w solutions

How would you prepare 75.0 g of a 7.20% w/w aqueous solution of glucose?

Solution
The problem is to calculate the mass of glucose required to add to water to make up 75.0 g of a 7.20% glucose solution. Mathematically, the percent composition of the solution is

$$7.20\% \text{ w/w} = \left(\frac{7.20 \text{ g of glucose}}{100 \text{ g of glucose solution}}\right) \times 100\%$$

The concentration of the glucose solution provides us with the unit-conversion factor in the form required to cancel out the glucose-solution units. To solve the problem, we multiply the given mass of solution by the conversion factor (the concentration):

$$75.0 \text{ g of glucose solution} \left(\frac{7.20 \text{ g of glucose}}{100 \text{ g of glucose solution}}\right) = 5.40 \text{ g of glucose}$$

The aqueous solution is prepared by adding 5.40 g of glucose to 69.6 g of water. Seventy-five grams of a 7.20% w/w aqueous glucose solution would therefore have the composition

$$\frac{5.40 \text{ g of glucose}}{5.40 \text{ g of glucose} + 69.6 \text{ g of H}_2\text{O}}$$

Problem 7.1 How would you prepare 200 g of a 5.50% w/w aqueous solution of gelatin?

The w/w designation tells us precisely how much solute and how much solvent are present by mass. It is therefore useful as a conversion factor for determining the mass of solute in a given mass of solution, as illustrated in the following example.

Example 7.2 Determining mass of solute in a % w/w solution

Calculate the number of grams of glucose present in 1.00 g of a 3.00% w/w aqueous solution of glucose.

Solution

First, write the 3.00% w/w concentration as a conversion factor:

$$\text{Concentration} = \frac{3.00 \text{ g of glucose}}{100 \text{ g of solution}}$$

The next step is to use the conversion factor to calculate the number of grams of glucose in 1.00 g of solution:

$$1.00 \text{ g of solution} \left(\frac{3.00 \text{ g of glucose}}{100 \text{ g of solution}} \right) = 0.0300 \text{ g of glucose}$$

Problem 7.2   Phenol is a strong germicide. Calculate the number of grams of phenol present in 1.0 g of a 0.20% w/w aqueous solution.

Suppose you need 5.0 g of NaCl for the preparation of a saline solution, and all that you have available is several liters of a 10% w/w NaCl solution. The next example illustrates how to determine what mass of a given % w/w solution contains a desired mass of solute.

Example 7.3   **Determining the volume of solution containing a desired mass of solute**

Calculate how many grams of a 4.50% w/w aqueous lactate solution are needed to provide 9.00 g of lactate, a product of glucose metabolism that is used to prepare isotonic saline solutions. (Isotonic solutions are discussed in Section 7.15.)

Solution

There are 4.50 g of lactate in 100 g of aqueous solution. Using that relation as a unit-conversion factor gives

$$9.00 \text{ g of lactate} \left(\frac{100 \text{ g of solution}}{4.50 \text{ g of lactate}} \right) = 200 \text{ g of solution}$$

Problem 7.3   How many grams of a 6.30% w/w aqueous solution of $CaCl_2$ must be used to obtain 27.6 g of $CaCl_2$?

v/v Percent Composition

The v/v designation means that the parts in the definition of percent composition are measured in units of volume. The v/v designation is the most practical way to characterize a solution of liquids in liquids. The following example describes the preparation of a solution of ethanol in water.

Example 7.4   **Preparing % v/v solutions**

How would you prepare an aqueous 8.0% v/v solution of ethanol?

Solution

The v/v designation means that the amounts of both solvent and solute are expressed in volume units—for example, in milliliters:

$$\% \text{ v/v} = \frac{\text{mL of liquid solute}}{100 \text{ mL of solution}} \times 100\%$$

An 8.0% v/v ethanol solution in water is therefore

$$8.0\% \text{ v/v} = \frac{8.0 \text{ mL of ethanol}}{100 \text{ mL of solution}} \times 100\%$$

The solution is prepared by adding 8.0 mL of ethanol to a small amount of water and then adding enough additional water to produce 100.0 mL of solution.

Problem 7.4   How would you prepare an aqueous 4.8% v/v solution of acetone?

It is important to note that the amount of water used in preparing the ethanol solution of Example 7.4 will not be 92 mL (even though 100 mL of aqueous solution was prepared by using 8 mL of ethanol). When liquids are mixed together, their final combined volume does not always equal the sum of the individual volumes. In regard to ethanol and water, 50 mL of ethanol added to 50 mL of water results in a solution of about 95-mL volume. A change in the volume of a mixture is caused by the nature of the interaction between the molecules of solvent and solute. In a solution of ethanol and water, the strong interaction causes a reduction in the volume of the mixture compared with the sum of the volumes of the separate components.

✓ The v/v designation allows one to know the volume of solute but not the volume of solvent added to it.

Concept check

w/v Percent Composition

The most common method of preparing a solid in liquid solution in the laboratory is to weigh a sample of a solid, place it in a container calibrated to contain a fixed volume (say, 100 mL), add just enough solvent to dissolve the solid, and then add sufficient additional solvent to fill the container to the calibration mark. This procedure is illustrated in Figure 7.4.

The w/v percent designation means grams of solid per 100 mL of solution. Calculations can be handled in the same way as % w/w or % v/v problems, but the units do not cancel and the % w/v is not a true fraction, as are % w/w and % v/v. It is more useful to think of a 5% w/v solution of KCl as 5 g of KCl per 100 mL of solution, and we will express the conversion factor in those terms.

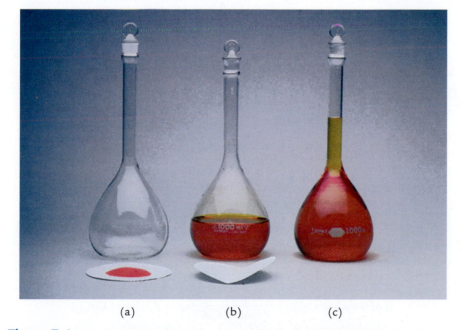

(a)   (b)   (c)

Figure 7.4 The preparation of 1.00 L of an aqueous solution containing 0.100 mol of $K_2Cr_2O_7$. The 0.100 mol of $K_2Cr_2O_7$ (29.4 g) is (a) weighed out and (b) added to a 1.00-L volumetric flask partly filled with water and dissolved in it. (c) More water is added to bring the total volume up to the 1.00-L mark on the neck of the flask. The solution is continually swirled to ensure complete mixing. (Chip Clark.)

Example 7.5 Determining the mass of solute in a given volume of % w/v solution

Calculate the number of grams of glucose contained in 45 mL of a 6.0% w/v aqueous glucose solution.

Solution

The w/v designation means that the amount of solute is expressed in grams and the amount of solution (not solvent) is expressed in milliliters. Therefore, the conversion factor is

$$\frac{6.0 \text{ g of glucose}}{100 \text{ mL of glucose solution}}$$

and the answer is

$$45 \text{ mL of glucose solution} \left(\frac{6.0 \text{ g of glucose}}{100 \text{ mL of glucose solution}} \right) = 2.7 \text{ g of glucose}$$

Problem 7.5 Calculate the number of grams of NaCl contained in 55 mL of a 12% w/v aqueous solution of NaCl.

7.6 MOLARITY

Molarity is a measure of concentration that is based on chemical mass units. It is defined as moles of solute per liter of solution. Concentrations described in this way are called **molar** and are denoted by the symbol **M.** For example, a 2.0 M solution is a 2.0 molar solution, a solution whose molarity is 2.0 mol/L. Three important contexts for the use of molarity as a measure of concentration are:

- Preparation of a solution of a specific molarity
- Calculation of how many moles of solute are contained in a certain volume of solution
- Calculation of the molarity of a certain volume of solution containing a specific mass of a solute in grams

The following examples explore each of these cases.

Example 7.6 Preparing molar solutions

How would you prepare 1.00 L of a 0.650 M aqueous solution of $CaCl_2$, and how many grams of $CaCl_2$ would be required?

Solution

To prepare 1.00 L of a 0.650 M solution, 0.650 mol of $CaCl_2$ must be dissolved in sufficient water to form 1.00 L of solution. The mass of $CaCl_2$ in grams equivalent to 0.650 mol can be calculated by using the formula mass of $CaCl_2$ as a conversion factor (Section 4.1):

$$0.650 \text{ mol of } CaCl_2 \left(\frac{111 \text{ g of } CaCl_2}{1 \text{ mol of } CaCl_2} \right) = 72.2 \text{ g of } CaCl_2$$

Weigh out 72.2 g of $CaCl_2$, and dissolve it in sufficient water to make 1.00 L of solution.

Problem 7.6 How would you prepare 500 mL of a 0.275 M aqueous solution of KH_2PO_4, and how many grams of KH_2PO_4 would be needed?

Example 7.7   Using molarity as a conversion factor: I

How many moles of solute are contained in 1.75 L of a 0.950 M solution?

Solution
In this case, rewrite the concentration as a conversion factor, 0.950 M = 0.950 mol/L, and multiply the given volume by the conversion factor in the form that will convert liters into moles:

$$1.75 \text{ L} \left(\frac{0.950 \text{ mol}}{\text{L}} \right) = 1.66 \text{ mol}$$

Problem 7.7   How many moles of solute are contained in 0.680 L of a 1.35 M solution?

Example 7.8   Using molarity as a conversion factor: II

What volume of a 1.68 M solution will contain 0.250 mol of solute?

Solution
Write the concentration term as a unit-conversion factor: 1.68 M = 1.68 mol/L. Invert the conversion factor to obtain the form that will convert moles into liters:

$$0.250 \text{ mol} \left(\frac{1.00 \text{ L}}{1.68 \text{ mol}} \right) = 0.149 \text{ L}$$

Problem 7.8   What volume of a 0.750 M solution will contain 1.25 mol of solute?

Example 7.9   Calculating the molarity of a solution

What is the molarity of a $CaCl_2$ solution in which 72.15 g of $CaCl_2$ is dissolved in 200.0 mL of solution?

Solution
Because molarity is defined as moles per liter, first calculate the number of moles in 72.15 g of $CaCl_2$ (Section 4.2), then convert 200.0 mL into liters, and finally calculate M.

Step 1. $72.15 \text{ g of } CaCl_2 \left(\dfrac{1 \text{ mol}}{111.0 \text{ g of } CaCl_2} \right) = 0.6500 \text{ mol of } CaCl_2$

Step 2. $200.0 \text{ mL} \left(\dfrac{1 \text{ L}}{1000 \text{ mL}} \right) = 0.2000 \text{ L}$

Step 3. $\dfrac{0.6500 \text{ mol}}{0.2000 \text{ L}} = \dfrac{3.250 \text{ mol}}{1.000 \text{ L}} = 3.250 \ M$

Problem 7.9   What is the molarity of a KCl solution in which 20.6 g of KCl is dissolved in 920 mL of solution?

  Solution concentration is often expressed in terms of **millimoles per milliliter**, abbreviated **mmol/mL**. Just as 1 mL is 1/1000th of a liter, 1 mmol is 1/1000th of a mole. Therefore, 1000 mmol = 1 mol and

$$\frac{1 \text{ mmol}}{\text{mL}} = \frac{1 \text{ mmol} \left(\dfrac{1 \text{ mol}}{1000 \text{ mmol}} \right)}{1 \text{ mL} \left(\dfrac{1 \text{ L}}{1000 \text{ mL}} \right)} = \frac{1 \text{ mol}}{\text{L}}$$

For example, the concentration of a 4.0 M solution of NaCl expressed in millimoles per milliliter is

$$4.0 \ M = \frac{4.0 \ \text{mol}}{\text{L}} = \frac{4.0 \ \text{mmol}}{\text{mL}}$$

and the concentration of a solution of KCl containing 7.4 mmol in 2 mL of solution is

$$\frac{7.4 \ \text{mmol}}{2.0 \ \text{mL}} = \frac{3.7 \ \text{mmol}}{\text{mL}} = \frac{3.7 \ \text{mol}}{\text{L}} = 3.7 \ M$$

The formula masses of compounds are often expressed in milligrams and millimoles. From the preceding discussion, you can see that the formula mass in grams per mole can also be expressed as milligrams per millimole:

$$\frac{1 \ \text{g}}{\text{mol}} = \frac{1 \ \text{mg}}{\text{mmol}}$$

Example 7.10   Calculating concentrations in millimoles per milliliter

Express the concentration in millimoles per milliliter of a solution in which 0.0746 g of KCl is dissolved in 100 mL of solution.

Solution
First calculate the mass of KCl in milligrams, then convert into millimoles, and finally calculate the molarity in millimoles per milliliter.

Step 1. $\qquad$ 0.0746 g of KCl $\left(\dfrac{1000 \ \text{mg}}{\text{g}} \right) = 74.6$ mg of KCl

Step 2. $\qquad$ 74.6 mg of KCl $\left(\dfrac{1 \ \text{mmol of KCl}}{74.6 \ \text{mg of KCl}} \right) = 1.00$ mmol of KCl

Step 3. $\qquad \dfrac{1.00 \ \text{mmol of KCl}}{100 \ \text{mL}} = 0.0100 \ \dfrac{\text{mmol}}{\text{mL}} = 0.0100 \ M$

Problem 7.10   Express the concentration in millimoles per milliliter of a solution in which 0.259 g of $Ca(OH)_2$ is dissolved in 175 mL of solution (Section 4.2).

7.7   DILUTION

Suppose an analytical procedure requires a $1 \times 10^{-6} \ M$ solution of a reagent. You might attempt to prepare it by weighing out 1×10^{-6} mol of the reagent, with the intention of dissolving in sufficient water to make a liter of solution, but a mass that tiny, perhaps from 0.2 to 0.3 mg, is difficult to weigh out accurately. A more practical approach is to first prepare a more concentrated solution and then dilute it to the desired concentration.

Suppose we wish to prepare a 1% w/w aqueous solution from a 2% w/w solution. Because the desired concentration is less than the solution's current concentration, we do not need to do a calculation to know that the desired concentration must be obtained by dilution. That is, water must be added to some quantity of the concentrated solution to reduce its concentration. The key point to remember is that the mass of solute that is present in the concentrated solution will still be present in the diluted solution, but it will occupy a larger volume. For example, after an 8-ounce can of frozen orange juice is made up to 1 quart, the mass of orange juice in the mixture remains the same; it is just distributed over a larger volume.

Let's say that the original mass was measured in grams or moles. Because there is no loss in mass of solute on dilution, we can write

$$\text{Grams}_{\text{initial}} = \text{grams}_{\text{final}}$$

or

$$\text{Moles}_{\text{initial}} = \text{moles}_{\text{final}}$$

If we replace the mass in moles with an equivalent expression that is the result of multiplying molar concentration by volume—namely,

$$\text{mol} = \text{L}\left(\frac{\text{mol}}{\text{L}}\right)$$

it follows that

$$\text{L}_{\text{initial}}\left(\frac{\text{mol}}{\text{L}}\right)_{\text{initial}} = \text{L}_{\text{final}}\left(\frac{\text{mol}}{\text{L}}\right)_{\text{final}}$$

or

$$\text{Volume}_{\text{initial}} \times \text{concentration}_{\text{initial}} = \text{volume}_{\text{final}} \times \text{concentration}_{\text{final}}$$

This expression can be simplified to

$$V_I \times M_I = V_F \times M_F$$

The volumes and masses can be in any units, as long as they are the same on both sides of the equation. Returning to the dilution of the 2.0% solution to a concentration of 1.0% and choosing 10 mL as the starting volume, we can now write

$$10 \text{ mL} \times 2.0\% = x \text{ mL} \times 1.0\%$$

Solving the equation for x mL gives

$$x \text{ mL} = \frac{10 \text{ mL} \times 2.0\%}{1.0\%} = 20 \text{ mL}$$

Notice how the concentration units cancel; the final volume must be 20 mL. We obtain this final volume by adding sufficient water to the original 10-mL solution to obtain a final volume of 20 mL. The result is the desired 1.0% solution.

Example 7.11   Determining the concentration of a diluted solution

What is the final volume of a 0.0015 M solution prepared by dilution of 100 mL of a 0.090 M solution?

Solution
Use the expression $V_I \times M_I = V_F \times M_F$, where V_I = 100 mL, M_I = 0.090 M, M_F = 0.0015 M, and V_F = unknown.

$$100 \text{ mL} \times 0.090 \text{ } M = V_F \times 0.0015 \text{ } M$$
$$V_F = 6.0 \times 10^3 \text{ mL} = 6.0 \text{ L}$$

Problem 7.11   What is the final volume of a 0.00750 M solution prepared by dilution of 75.0 mL of a 0.180 M solution?

Example 7.12   Preparing a dilute solution

What initial volume of a 1.25 M solution must be used to prepare a final volume of 2.00 L of a 0.500 M solution?

Solution
Use the expression $V_I \times M_I = V_F \times M_F$, where V_I = unknown, M_I = 1.25 M, V_F = 2.00 L, and M_F = 0.500 M.

$$V_I \times 1.25 \text{ } M = 2.00 \text{ L} \times 0.500 \text{ } M$$
$$V_I = 0.800 \text{ L}$$

Problem 7.12   What volume of a 0.730 M solution must be used to prepare 1.36 L of a 0.270 M solution?

7.8   CONCENTRATION EXPRESSIONS FOR VERY DILUTE SOLUTIONS

An expression of concentration often used in the clinical laboratory is mg %, which is a % w/v designation. It is defined as

$$mg\ \% = \frac{mg\ of\ solute}{100\ mL\ of\ solution} \times 100\%$$

For example, the concentration of glucose in a person's blood from 3 to 4 h after a meal ranges from 70 to 110 mg %, and normal concentrations of calcium in the blood range from 8.5 to 10.5 mg %. The SI equivalent of mg % is mg/dL, which is the preferred concentration notation and is becoming more commonly used today.

Example 7.13   Calculating the concentration of solute in a mg % solution

Creatine is produced as a result of muscle metabolism. Normal creatine concentrations in blood are reported as 1.5 mg %. Calculate the concentration of blood creatine in units of milligrams per milliliter and mg/dL.

Solution
The conversion factor derived from the 1.5 mg % concentration is

$$\frac{1.5\ mg\ of\ creatine}{100\ mL\ of\ blood} = \frac{0.015\ mg\ of\ creatine}{1.0\ mL\ of\ blood}$$

Creatine concentration = 0.015 mg/mL of blood = 1.5 mg/dL.

Problem 7.13   Assuming that the normal serum calcium concentration is 9.0 mg %, calculate its concentration in grams per milliliter and mg/dL.

Another convenient expression for describing very dilute aqueous solutions is the designation **parts per million, or ppm.** This is a weight-per-volume designation for parts of solute per million parts of solution. For example, a 0.0001% w/v solution of some salt may be written as 0.0001 g of salt per 100 mL of solution. This ratio can be converted into whole numbers by multiplying both numerator and denominator by 1×10^4. The results will be clearer if we rewrite numerator and denominator in scientific notation:

$$\frac{0.0001\ g}{100\ mL} = \frac{1 \times 10^{-4}\ g}{1 \times 10^2\ mL}$$

and then multiply numerator and denominator by 1×10^4:

$$\left(\frac{1 \times 10^{-4}\ g}{1 \times 10^2\ mL}\right)\left(\frac{1 \times 10^4}{1 \times 10^4}\right) = \frac{1\ g}{1,000,000\ mL}$$

The solution can then be seen to contain 1 gram of solute per 1 million milliliters of solution—that is, to have a concentration of one part per million, or 1 ppm.

It is more common to express parts per million as the number of milligrams of solute contained in a liter of solution. For example, 1 mg per liter can be expressed as 0.001 g per 1000 mL. Then,

$$\frac{1\ mg}{L} = \frac{0.001\ g}{1000\ mL}$$

Multiplying numerator and denominator by 1×10^3 gives

$$\left(\frac{0.001 \text{ g}}{1000 \text{ mL}}\right)\left(\frac{1 \times 10^3}{1 \times 10^3}\right) = \frac{1 \text{ g}}{1,000,000 \text{ mL}}$$

So,

$$\frac{1 \text{ mg}}{\text{L}} = 1 \text{ ppm}$$

Similarly, one part per billion is expressed as 1 ppb and, in terms of mass per liter, is

$$1 \text{ ppb} = \left(\frac{1 \text{ g}}{1 \times 10^9 \text{ mL}}\right)\left(\frac{1 \times 10^{-6}}{1 \times 10^{-6}}\right) = \frac{1 \times 10^{-6} \text{ g}}{1 \times 10^3 \text{ mL}} = \frac{1 \text{ μg}}{\text{L}}$$

Concept checklist

✓ One part per million = 1 ppm = $\dfrac{1 \text{ mg}}{\text{L}}$

✓ One part per billion = 1 ppb = $\dfrac{1 \text{ μg}}{\text{L}}$

7.9   THE SOLUBILITY OF SOLIDS IN LIQUIDS

The solubility of a solid describes the maximum amount, or mass, of it that will dissolve in a fixed volume of solvent at a particular temperature. For example, no more than 24.99 g of the amino acid glycine will dissolve in 100 g of water at 25°C. Solubility must be determined by experiment, as was done for the compounds plotted in Figure 7.5. A solution containing the maximum amount of a dissolved solid is said to be saturated at that temperature.

A saturated solution is a system at equilibrium, a condition defined in Section 6.6 as being the result of two opposing processes whose rates are identical. In a saturated solution, the two opposing processes are (1) the dissolving of the solid, in which ions or molecules leave the solid and enter the liquid phase, and (2) the crystallization of previously dissolved ions or molecules onto the surface of the solid. Let's think about what happens when enough

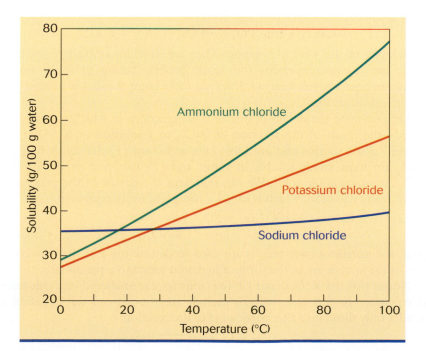

Figure 7.5 The solubilities of ammonium chloride, potassium chloride, and sodium chloride all increase with increase in temperature—however, each to a different extent.

Figure 7.6 A solution of sodium chloride (NaCl) added to a solution of silver nitrate ($AgNO_3$) produces the white precipitate silver chloride (AgCl). (Chip Clark.)

of an ionic solid, such as potassium chloride, is mixed with water to form a saturated solution. The opposing processes can be represented by the following equation:

$$KCl(s) \underset{\text{crystallization}}{\overset{\text{solution}}{\rightleftharpoons}} K^+(aq) + Cl^-(aq)$$

where $KCl(s)$ represents potassium chloride in the solid state and $K^+(aq)$ and $Cl^-(aq)$ represent hydrated ions.

At first, the rate of dissolving is greater than the rate of crystallization, but, as the concentration of dissolved ions increases, the rate of crystallization increases. When the two rates become equal, the solution is saturated. The maximum amount of dissolved solid (for that particular temperature) has been reached.

In the body, most nitrogen-containing compounds are transformed into urea, which becomes the principal solid component of urine. However, a certain group of nitrogen-containing compounds, the nucleotides (from which the genetic molecules RNA and DNA are synthesized, undergo metabolic degradation to form uric acid. Under normal conditions, uric acid and its sodium salt are excreted in the urine. A small group of people suffer from a disorder in nucleotide metabolism that results in an overproduction of uric acid and leads to the clinical condition known as gout. Uric acid and its sodium salt are thousands of times less soluble than urea, and their overproduction rapidly saturates body fluids. In consequence, needlelike uric acid crystals are deposited in the cartilaginous structures of the joints, causing inflammation and great pain.

7.10 INSOLUBILITY CAN RESULT IN A CHEMICAL REACTION

When two ionic compounds are added to water, one of the reasons that new compounds can form is that one of the products is insoluble and forms a precipitate. Reactions of this type are shown in Figures 7.6 and 7.7. Reactions between $BaCl_2$ and K_2SO_4 or K_3PO_4 and $CaCl_2$, represented in Chapter 4 by

$$BaCl_2 + K_2SO_4 \longrightarrow BaSO_4 + 2 KCl$$

and

$$3 CaCl_2 + 2 K_3PO_4 \longrightarrow Ca_3(PO_4)_2 + 6 KCl,$$

are examples of this kind of reaction. They are stoichiometrically correct (balanced) but do not represent the actual process taking place in solution, because, in aqueous solution, all dissolved ionic compounds are present in the form of individual ions.

Let's examine the details of one of these reactions in which a precipitate is formed—the result of the addition of $BaCl_2$ to a solution of K_2SO_4. Tables 7.2 and 7.3 list soluble and insoluble salts and show (1) that both $BaCl_2$ and K_2SO_4 are ionic solids soluble in water and (2) that one of the products, $BaSO_4$, is an insoluble solid. We can represent the details of the process with the following equation, which is called a **complete ionic equation.**

$$Ba^{2+}(aq) + 2 Cl^-(aq) + 2 K^+(aq) + SO_4{}^{2-}(aq) \rightleftharpoons$$
$$BaSO_4(s) + 2 K^+(aq) + 2 Cl^-(aq)$$

Again, the notations (aq) and (s) are used to denote the fact that the components of the system are in the form of hydrated ions and a precipitate.

Notice that the $K^+(aq)$ and $Cl^-(aq)$ ions appear unaltered on both sides of the equation. In this case, we can exclude such ions from the equation and consider only those ions that participate in the reaction:

$$Ba^{2+}(aq) + SO_4{}^{2-}(aq) \rightleftharpoons BaSO_4(s)$$

Figure 7.7 A yellow precipitate of lead(II) iodide, PbI_2, forms immediately when colorless aqueous solutions of lead(II) nitrate, $Pb(NO_3)_2$, and potassium iodide, KI, are mixed. (Chip Clark.)

This is called the **net ionic equation.** The K$^+$(aq) and Cl$^-$(aq) ions omitted from the net ionic equation are called **spectator ions** because they do not take part in the reaction and would appear on both sides of the equation if included.

Example 7.14 **Finding the net ionic equation**

Write the net ionic equation that describes the result of adding silver nitrate, AgNO$_3$, to a solution of NaCl, a reaction in which a precipitate is formed. Refer to Table 7.3 for solubility data.

Solution

Table 7.3 shows that silver chloride is an insoluble salt. The reaction equation is

Ag$^+$(aq) + NO$_3^-$(aq) + Na$^+$(aq) + Cl$^-$(aq) $\rightleftharpoons$
$$AgCl(s) + Na^+(aq) + NO_3^-(aq)$$

The net ionic equation is

$$Ag^+(aq) + Cl^-(aq) \rightleftharpoons AgCl(s)$$

Problem 7.14 Write (a) the complete ionic equation and (b) the net ionic equation for the reaction that takes place when calcium chloride, CaCl$_2$, is added to sodium phosphate, Na$_3$PO$_4$. (Refer to Table 7.3.)

7.11 DIFFUSION

Molecules free to move about a space will tend, in time, to distribute themselves uniformly throughout that space. To achieve such a uniform distribution, molecules move from regions of higher concentration to regions of lower concentration. This movement is called **diffusion.**

Diffusion happens very rapidly in gases but much more slowly in liquids. In gases, there are comparatively few collisions between molecules, and progress is rapid. However, in liquids, diffusing molecules encounter and collide with large numbers of solvent molecules, and such collisions impede their progress. After adding sugar to your coffee or tea, you would never consider drinking it without first stirring. From experience, you know that it would take a long time for the sugar to become uniformly distributed throughout the drink. Would you be surprised to learn that it actually takes several months?

Figure 7.8 depicts events at the molecular level when a solute diffuses in a liquid. At first, the concentration of solute particles is greater on the left-hand side (in this representation) than on the right-hand side. The arrows are meant

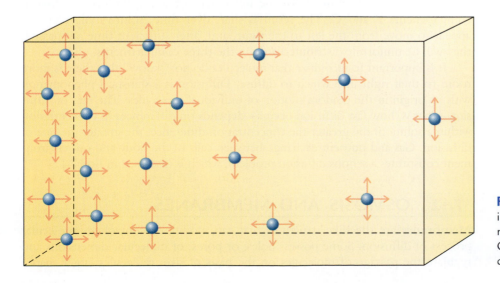

Figure 7.8 A molecular model illustrating random motion of solute molecules diffusing from left to right. Only back-and-forth and up-and-down motions are indicated.

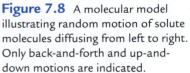

Figure 7.9 Random motion of solute molecules in a solution. The solute molecules are distributed in a nonuniform manner, and only the left-to-right random motions are shown.

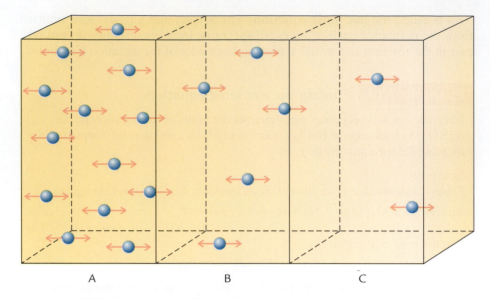

A   B   C

to indicate that the motion of individual solute particles is random; each particle is as likely to move in one direction as another. In this representation, only solute molecules are seen; the solvent is not shown.

In Figure 7.9, the molecular picture has been modified in two ways. First, the container of the solution has been divided into three imaginary compartments, labeled A, B, and C. Second, the arrows are redrawn so as to emphasize what, in this case, are the most relevant components of the particles' random motion—those movements that affect diffusion from left to right and cause particles to approach the compartments' imaginary boundaries directly from either side. There is no need to consider random motion in other directions.

Figure 7.9 represents our thought experiment at the start: more solute molecules are in compartment A than in compartment B, which in turn contains more solute than compartment C. Because motion is entirely random, any solute molecule is just as likely to move to the right as to the left. Let's assume some convenient number of solute molecules in each compartment—say, 15 in A, 10 in B, and 5 in C. Some solute molecules in compartment A may leave A and enter B. The rate at which they leave is proportional to their number, and so we may assign a rate at which they enter B as 15 per unit time. The rate at which solute leaves compartment B and enters A is proportional to their number, or 10 per unit time. If 10 solute molecules leave B but 15 molecules enter B, then, over time, solute will accumulate in compartment B. The same argument can be made for the movement of solute molecules between compartments B and C. The result is that, over time, solute molecules will appear to move from the left to the right, and, after sufficient time has passed, they will be uniformly distributed among the three compartments.

It is important to recognize that molecules do not "know" that there is more room to their right into which to diffuse. Diffusion is a statistical phenomenon with its origin in the random motion of individual molecules. The rate of diffusion—that is, how fast diffusion occurs—depends on differences in concentration within a solution: the greater the concentration difference, the greater the rate of diffusion. Gas and nutrient exchange between cells of the body and our environment constitutes a serious diffusion problem, which is discussed in Box 7.1.

7.12 OSMOSIS AND MEMBRANES

Membranes are sheetlike structures that can modify or control the molecular process of diffusion. Some possess holes, or pores, of molecular dimensions and regulate the passage of molecules on the basis of size: molecules larger than a

7.1 CHEMISTRY WITHIN US

Diffusion and the Cardiovascular System

As oxygen diffuses into cells, it is consumed by the many different metabolic processes taking place there. If all the oxygen is used up before some of it can diffuse to the most remote regions of the cell, certain aspects of metabolism will fail, and the cell will die. This puts a limit on the size to which a cell can safely grow, and in fact no cell of an oxygen-consuming organism, whether single cellular or multicellular, ever exceeds a thickness of about 0.5 mm. Some multicellular organisms, such as kelp and flatworms, transcend this limitation by growing in thin sheets, and so all their cells are in contact with their oxygen-containing environment. In multicellular organisms such as humans,

oxygen has a long way to go before it can even begin to diffuse into cells. This severe transport problem has been solved by the creation of a strong pump (the heart) that sends an oxygen-rich fluid (the blood) along a superhighway of "pipes," or vessels (the vascular system).

In the lungs, large quantities of oxygen diffuse rapidly into the blood across a very thin external membrane (lung alveoli). The blood is then pumped quickly through the vascular system, whose vessels constantly decrease in diameter from aorta to artery to arteriole and finally to capillary, a vessel approximately equal in diameter to the size of cells. At this point, the oxygen has only a short distance to diffuse to reach all parts of the metabolizing cells. Thanks to this elegant solution to the diffusion problem, oxygen travels from the outside world to all the cells of the body in a matter of seconds.

---

certain diameter cannot pass, but those smaller than the limiting pore size will penetrate. Other membranes control diffusion by means of electrical charge: charge repulsions allow only negatively charged ions to pass through a positively charged membrane, and vice versa. Membranes that allow some substances but not others to pass through them are called **semipermeable.** They are essentially molecular filters. Biological membranes use these properties and others, achieving an extraordinary array of specificities with regard to the ions and molecules that they will allow into and out of cells and subcellular compartments.

The solute in a solution normally diffuses under a concentration difference, but it cannot do so when it must diffuse across a semipermeable membrane. The solvent, however, is not hindered by the membrane and is free to diffuse under the influence of its concentration difference across the membrane. In aqueous systems, because the concentration of water increases as the concentration of solute decreases, water will always flow to the high-solute side of a membrane impermeable to the solute.

The flow of water through a semipermeable membrane under these circumstances is called **osmosis.** Figure 7.10 illustrates the process, with the use

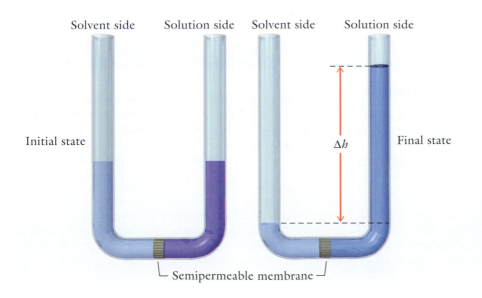

Solvent side   Solution side   Solvent side   Solution side

Initial state Δh Final state

Semipermeable membrane

Figure 7.10 A U-tube manometer measures osmotic pressure. The right-hand compartment contains an aqueous 1.0% glucose solution, and the left-hand side contains only water (pure solvent). Osmosis causes water to cross the membrane into the solution compartment, which makes the liquid in the solution compartment rise to a higher level than the level in the solvent compartment. The difference in liquid levels results, in turn, in the generation of a force that causes water to flow back into the solvent compartment. When the liquid levels remain constant, the difference in the levels is a measure of the osmotic pressure.

of a U-tube divided into two compartments by a semipermeable membrane. The right-hand compartment contains an aqueous 1.0% glucose solution, and the left-hand compartment contains only water (pure solvent).

Osmosis causes water to cross the membrane into the solution compartment, a movement of solvent that makes the liquid level in the solution compartment rise higher than the level in the solvent compartment. The difference in liquid levels results, in turn, in the generation of a force (the weight of the column of liquid) that causes water to flow back into the solvent compartment. Eventually, an equilibrium is established in which the inflow of solvent caused by the concentration difference becomes balanced by the outflow of solvent caused by the weight of the higher liquid level. At that point, the liquid levels no longer change.

The weight of the higher liquid column is measured as a pressure—the weight, or the force, per unit area. It is called a **hydrostatic pressure** because it is the result of the difference in water levels. In many cases, water flows from faucets at home because taps are at lower levels than the water at the reservoir. The difference in liquid levels leads to a pressure difference, hence flow.

7.13   OSMOTIC PRESSURE

The hydrostatic pressure is measured by the difference in height of the two liquid levels (Δh in Figure 7.10). Because the liquid levels do not change over time, the final pressures in both compartments must be equal. Therefore, the difference in liquid levels is also a measure of the opposing pressure generated by the concentration difference. That pressure is called the **osmotic pressure.**

The U-tube experimental system is a useful reminder that water can flow through a semipermeable membrane in two ways:

- Osmotic pressure, a pressure generated by a difference in concentration of dissolved substances separated by a semipermeable membrane

- Hydrostatic pressure, a pressure generated within a liquid by mechanical means

A physiological example of hydrostatic pressure is the pressure generated by the muscular contraction of the heart, forcing blood into arterial circulation. The production of urine furnishes another example: water can be squeezed out of a protein solution that is forced against a semipermeable membrane by hydrostatic pressure. The emerging protein-free solution is called an **ultrafiltrate.** The first step in urine formation is the formation, by similar means, of an ultrafiltrate from blood. This aspect of urine formation is described in Box 7.2.

7.2   CHEMISTRY WITHIN US

Semipermeability and Urine Formation

The blood supply to the kidneys, delivered by a major branch of the aorta, flows into those organs under very high pressure, about 100 torr. After blood enters the kidney, it is conducted by water-impermeable arteries to microscopic specialized tubular structures called glomeruli. The glomeruli contain capillaries whose walls are semipermeable. In consequence, under sufficient pressure, all molecules that are smaller than the membrane's smallest pore size will be forced through the capillary wall into the glomerulus tubule. Molecules larger than that minimum pore size will not pass through and will remain in the plasma.

In short, when whole blood enters the kidney glomerulus, hydrostatic pressure generated by the heart literally squeezes a protein-free solution through the semipermeable kidney glomerulus. This protein-free filtrate—called an ultrafiltrate—is the first step in urine formation. Not surprisingly, one of the earliest signs of kidney disease is the presence of protein in the urine, indicating some kind of pathology of the glomerular membranes. Unless the problem is remedied, the loss of protein from the blood plasma will cause a drop in plasma osmotic pressure, and generalized edema or swelling of the tissues will result.

Figure 7.11 A commercial reverse-osmosis unit. The unit shown removes more than 95% of the salt initially present. (Courtesy of Ionics.)

An industrial example is a process, called **reverse osmosis,** used to desalinate water on a large scale in some of the Persian Gulf states and at a number of U.S. military bases. Figure 7.11 shows a commercial reverse-osmosis unit that uses hydrostatic pressure to force the water from saline (salt water) solutions through special semipermeable membranes that are able to retain salts, thus producing pure water. It is called reverse osmosis because the water is made to flow from a region of low water concentration to one of high water concentration.

A PICTURE OF HEALTH
Examples of Solutions in the Body

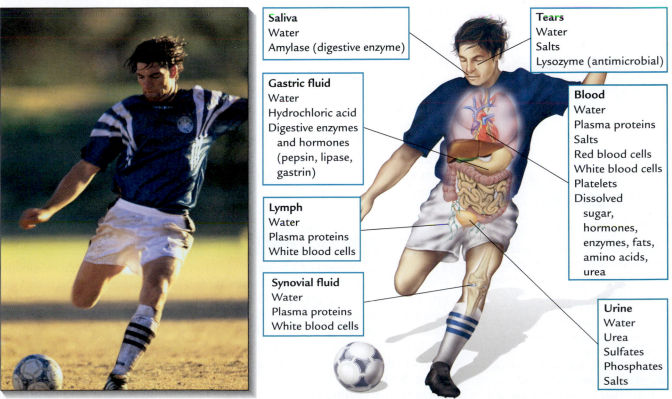

Saliva
Water
Amylase (digestive enzyme)

Gastric fluid
Water
Hydrochloric acid
Digestive enzymes
   and hormones
   (pepsin, lipase,
   gastrin)

Lymph
Water
Plasma proteins
White blood cells

Synovial fluid
Water
Plasma proteins
White blood cells

Tears
Water
Salts
Lysozyme (antimicrobial)

Blood
Water
Plasma proteins
Salts
Red blood cells
White blood cells
Platelets
Dissolved
   sugar,
   hormones,
   enzymes, fats,
   amino acids,
   urea

Urine
Water
Urea
Sulfates
Phosphates
Salts

(Tim Pannell/Corbis.)

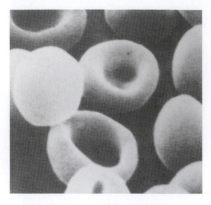

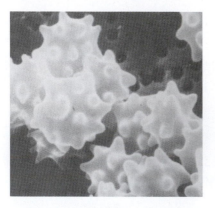

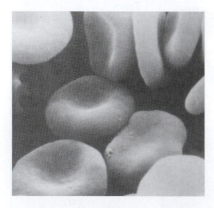

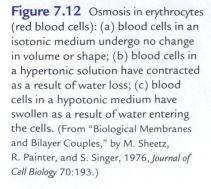

Figure 7.12 Osmosis in erythrocytes (red blood cells): (a) blood cells in an isotonic medium undergo no change in volume or shape; (b) blood cells in a hypertonic solution have contracted as a result of water loss; (c) blood cells in a hypotonic medium have swollen as a result of water entering the cells. (From "Biological Membranes and Bilayer Couples," by M. Sheetz, R. Painter, and S. Singer, 1976, *Journal of Cell Biology* 70:193.)

7.14   OSMOLARITY

The osmotic pressure of a solution depends on the sum of molar concentrations of all independent particles in that solution, without regard to the particles' specific identities. One mole of chloride ions in solution exerts the same osmotic pressure as one mole of dissolved glucose molecules (which cannot ionize or dissociate). In certain ways, the behavior of a dilute solution is similar to that of an ideal gas, with the solute behaving like gas particles and the solvent representing empty space (just more viscous). Because of this similarity, the osmotic pressure can be calculated by using a law that is virtually identical with the ideal gas law:

$$P = \frac{nRT}{V} = \left(\frac{n}{V}\right)RT$$

However, in regard to osmotic pressure, an additional factor must be added to the equation to explicitly specify the number of particles contained in 1 mol of the solute. The reason for adding this factor is that ionic substances in solution provide more than one mole of particles per mole of solute, and n in the gas law, $PV = nRT$, must represent the actual number of moles present. Therefore, the osmotic pressure relation is

$$\Pi = i\left(\frac{n}{V}\right)RT$$

Π is the osmotic pressure in atmospheres
R is the gas constant, 0.0821 L·atm·K^{-1}·mol^{-1}
T is the temperature in kelvins
V is the volume in liters
n is the number of moles present; (n/V) is the molar concentration, mol/L
i is the factor that corrects for the number of moles of particles per mole of solute

The quantity n/V is the molarity of the solution, but the quantity $[i \times (n/V)]$ is called the **osmolarity.** It is a measure of the total particle concentration of a solution. One osmole (1 osmol) is equal to one mole (1 mol) of an ideal nonionizing solute such as glucose. If n is the number of moles of particles that 1 mol of a compound will add to a solution when dissolved, then $[i \times n]$ is the number of osmoles.

Osmolarity recognizes that ionic solids such as NaCl dissolved in solution yield more than one mole of particles per mole of solute. A 1 molar (1 M) solution of NaCl contains 2 mol of dissolved particles per liter (1 mol of Na$^+$ ions and 1 mol of Cl$^-$ ions), or 2 osmol/L. A solution that is 1 M in glucose and at the same time also 1 M in NaCl has an osmolarity of 3 osmol/L. The factor i accounts for dissociation. For glucose, $i = 1$ because glucose does not form ions in solution; for NaCl, $i = 2$, for CaCl$_2$, $i = 3$; for AlCl$_3$, $i = 4$; and so forth.

7.15   OSMOSIS AND THE LIVING CELL

The osmotic condition of a living cell bounded by a semipermeable membrane is the result of a balance between the rate of water entering the cell and the rate of water leaving the cell. The relations between a cell's inner and outer concentrations lead to three different osmotic conditions: hypotonic, hypertonic, and isotonic, illustrated in Figure 7.12.

When a solution external to a cell such as a red blood cell has a lower solute concentration than that of the internal solution, the outer solution is called **hypotonic.** Under these circumstances, water will diffuse into the cell, thereby causing an increase in the hydrostatic pressure within the cell and consequent swelling. The flow may continue until the cell bursts. When this occurs in hypotonic solutions of red blood cells, the rupturing is called **hemolysis.**

When a solution external to a cell has a higher solute concentration than that of the solution inside the membrane, the outer solution is called **hypertonic.** Cells placed in a hypertonic solution begin to lose water and consequently shrink. Red blood cells in hypertonic solution shrink and appear crumpled and are described as **crenate.**

When the solute concentrations external and internal to a cell are identical, no net flow will occur. Therefore, no change in cell volume will be observed, and the external solution is said to be **isotonic.** In medical practice, any fluids added to the blood must be isotonic. If they are not isotonic, serious imbalances of fluid or electrolytes (ions in solution) will ensue, potentially leading to malfunction of all the major organs. Box 7.3 shows how different concentrations of sodium chloride (saline) and glucose (dextrose) solutions can be isotonically identical.

Concept checklist

✓ Under hypotonic conditions, the rate of water entering the cell is greater than the rate of water leaving.
Result: swelling and possible bursting.

✓ Under hypertonic conditions, the rate of water leaving the cell is greater than the rate of water entering.
Result: shrinkage and crumpling.

✓ Under isotonic conditions, the rates of water entering and leaving are the same.
Result: no change in volume or shape.

7.3 CHEMISTRY WITHIN US

The Osmotic Pressure of Isotonic Solutions

In an isotonic medium, the rates of flow of water into and out of erythrocytes are the same. Why are physiological saline (0.90% w/v NaCl) and 5.4% w/v dextrose (glucose) solutions both called isotonic solutions, when their concentrations are not the same? The answer is that the osmotic equilibrium of a blood cell is regulated by the surrounding solution's osmolarity, not by its molarity. To understand why physiological saline and dextrose solutions are both isotonic, we will calculate their osmolarities.

The osmotic pressure of an aqueous solution is calculated with the following relation:

$$\Pi = i\left(\frac{n}{V}\right)RT$$

where $\left[i \times \left(\frac{n}{V}\right)\right]$ is the osmolarity.

To compare the osmotic characteristics of the two solutions, it is sufficient to calculate their osmolarities. The first step is to calculate their molarities.

For glucose:

$$5.4\% \text{ w/v} = \frac{5.4 \text{ g}}{100 \text{ mL}} = \frac{54 \text{ g}}{\text{L}}$$

$$\frac{\text{mol glucose}}{\text{L}} = 54 \text{ g}\left(\frac{1 \text{ mol}}{180 \text{ g}}\right) = \frac{0.30 \text{ mol}}{\text{L}}$$

Because $i = 1$ for an aqueous glucose solution,

$$\left[i \times \left(\frac{n}{V}\right)\right] = \frac{0.30 \text{ osmol}}{\text{L}}$$

For a 0.90% w/v solution of NaCl (physiological saline),

$$0.90\% \text{ w/v NaCl} = \frac{0.90 \text{ g}}{100 \text{ mL}}$$

$$\left(\frac{0.90 \text{ g}}{100 \text{ mL}}\right)\left(\frac{1000 \text{ mL}}{\text{L}}\right)\left(\frac{1 \text{ mol}}{58.45 \text{ g}}\right) = 0.15 \text{ mol/L}$$

Second, recognizing that each mole of NaCl provides 2 mol of ions in solution, we assign i a value of 2. The osmolarity of 0.90% w/v NaCl is

$$\left[i \times \left(\frac{n}{V}\right)\right] = \frac{0.30 \text{ osmol}}{\text{L}}$$

The osmolarities of the glucose and saline solutions are the same. Therefore, they will have identical effects on osmotic equilibria.

TABLE 7.4   Dimensions and Masses of Various Molecules, Macromolecules, and Biological Entities

Molecule	Largest dimension (cm)	Molecular mass Atomic mass units	Molecular mass Grams
Na$^+$ (sodium ion)	0.4×10^{-7}	23	
H$_2$O	0.5×10^{-7}	18	
C$_6$H$_{12}$O$_6$ (glucose)	0.7×10^{-7}	180	
glutamic acid	0.7×10^{-7}	147	
oxytocin	1.5×10^{-7}	1,040	
myoglobin	3.5×10^{-7}	17,000	
serum albumin	5.5×10^{-7}	68,000	
γ-globulin	8.5×10^{-7}	250,000	
bacterial virus	25×10^{-7}	6,200,000	1.0×10^{-17}
plant virus	300×10^{-7}	40,000,000	6.0×10^{-17}
E. coli cell	$2,000 \times 10^{-7}$		2×10^{-12}
human liver cell	$20,000 \times 10^{-7}$		$2,000 \times 10^{-12}$

7.16   MACROMOLECULES AND OSMOTIC PRESSURE IN CELLS

Proteins, polysaccharides, and nucleic acids are members of a class of high-molecular-mass molecules called **macromolecules** or **polymers.** They are formed by covalently linking together many small molecules. Proteins are formed from amino acids such as glutamic acid. Polysaccharides are formed from sugars such as glucose. Table 7.4 lists a number of biologically significant molecules and macromolecules along with their dimensions and masses. The volume of the plasma protein molecule γ-globulin is more than 150 times as great as that of the glucose molecule. Several other biological entities are included in Table 7.4 to provide additional perspective.

Biological macromolecules are usually too large to penetrate cell membranes, and this characteristic is at the heart of a number of interesting biological phenomena. Some of these phenomena are caused by the fact that solvent but not solute will penetrate cell membranes. Consequently, an osmotic pressure will develop across such a barrier. Fluid balance in the body—that is, water in versus water out—depends on osmotic phenomena in the vascular system. Urine formation and the need for the digestive system are discussed in Box 7.2 (on page 192) and Box 7.4.

Particles ranging in size from about 1×10^{-7} cm to 1×10^{-4} cm are called **colloidal.** Although too small to be seen with the naked eye, colloidal particles can be identified by the fact that a beam of light passing through a colloidal solution can be seen at right angles to the beam's direction. (When a beam of light is passed through a solution containing molecules the size of glucose, the beam remains invisible). This quality of colloidal solutions is called the **Tyndall effect** and is illustrated in Figure 7.13.

Figure 7.13 Colloidal particles are larger than molecules such as glucose but are too small to be seen even with a microscope. However, their size is great enough to intercept the passage of a light beam, scatter it, and reveal their presence. (E.R. Degginer, Convent Station, New Jersey.)

7.4 CHEMISTRY WITHIN US

Semipermeability and the Digestive System

Aside from their specific roles in functions such as circulatory water balance and urine formation, the semipermeable membranes of the body serve a crucial general purpose. That is, they prevent key proteins and other macromolecules from leaking out of cells.

Cells contain a wide variety of macromolecules that are central for sustaining life. Reproduction and metabolism could not take place without them. If the membranes that surround the cells were permeable to such large molecules, the molecules would continually escape. Cells would have to synthesize new macromolecules at an impossibly rapid rate to keep up with the loss.

The evolution of semipermeable membranes was an excellent solution to two problems: (1) how to retain critical macromolecular components within the cells and (2) how to enable low-molecular-mass nutrients to enter the cell so that new macromolecules can be synthesized when necessary. However, this solution led to another physiological dilemma.

We obtain nutrition by ingesting the cells and tissues of other organisms, and these cells and tissues are composed primarily of biological macromolecules. How can our cells obtain the benefit of nutrients whose molecular dimensions are too large to penetrate the cells' semipermeable walls? This dilemma was solved by the evolution of our digestive system.

The digestive system is a tube about 15 feet long running through the body from mouth to anus. The lumen, or interior, of this tube is continuous with the external environment, in analogy with the hole in a doughnut. Because it is continuous with the external environment, its contents are technically "outside" of our bodies. Within this tube, macromolecular foodstuffs are chemically converted into molecules of small dimensions and mass. Nutrient molecules such as amino acids and sugars are small enough to penetrate the membranes of intestinal-wall cells, enter the bloodstream, and finally penetrate cell membranes to fuel the metabolic engine.

Soap and detergent solutions form colloidal solutions. The physical basis underlying the formation of colloidal-sized particles (Figure 7.14) in soap and detergent solutions is discussed in Box 7.5 on the following page, explaining how globular proteins assume their characteristic shapes.

INSIGHT INTO FUNCTION

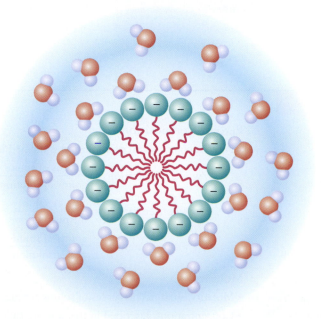

Figure 7.14 Sketch of a typical spherical micelle formed in water by an amphipathic substance such as a sodium soap. The water-soluble groups at one end of each amphipathic molecule (green circles) form the surface of the micelle and interact with the surrounding water molecules. The nonpolar hydrocarbon tails (zigzag lines) excluded from the water form the liquidlike core of the micelle.

7.5 | CHEMISTRY IN DEPTH

Association Colloids, Micelles, and Protein Structure

In many ways, the properties of amphipathic substances in solution (Section 6.4) are very much like those of nonamphipathic substances such as glucose or sodium chloride. For example, at low concentrations, solution properties such as osmotic pressure and surface tension change gradually with increase in concentration, no matter what the solute. However, at some concentration unique to each amphipathic substance, a sudden and dramatic change occurs in a solution's properties. In soaps, for example, the solutions begin to form stable foams and exhibit their cleaning powers.

The sudden change in properties seen in amphipathic solutions is caused by the formation of large molecular assemblies called **micelles,** also called **association colloids.** A typical spherical micelle is sketched in Figure 7.14. The concentration at which micelles are created is the **critical micelle concentration,** or **CMC.** The formula masses of micelles vary, but they usually exceed 50,000 and may be much higher, depending on molecular type. Each amphipathic substance has its own CMC.

The CMC of an amphipathic substance is the point at which an aqueous solution of the substance becomes "saturated" with respect to the solute molecules' hydrophobic tails. However, the molecules' hydrophilic head groups are still quite water soluble. Thus, one part of the molecule tends to leave the aqueous solution while the other part of the molecule is impelled to enter into solution. A compromise is reached in which groups of molecules assemble into a colloid-sized aggregate with the hydrophobic tails on the inside and the hydrophilic head groups on the outside. In this way, the hydrophobic groups escape from the aqueous environment, but, at the

same time, the colloidal particle remains soluble by virtue of the large number of water-soluble groups on its surface. Micelles are called association colloids because they are of colloidal dimensions but are held together by secondary forces rather than by covalent bonds (Section 6.2).

Soluble proteins, such as those in blood plasma, are large enough so that we can also think of them as having an inside and an outside. The inside can, in fact, provide an environment that is less polar than water and can solvate molecules not normally soluble in water. For example, the protein serum albumin, an important component of blood plasma, transports water-insoluble nutrients such as palmitic acid. The acid is able to dissolve in the interior of the protein, which, because of its water-soluble surface, is then able to carry the acid through the aqueous environment of the bloodstream.

The properties of micelles allow us to explain these structural characteristics of proteins, as well as the fact that each type of protein possesses a unique topography on its polar outside surface. A protein consists of a long covalently bonded chain of amino acids, and significant numbers of these amino acids contain structural parts that do not interact readily with the macromolecule's aqueous environment. Consequently, a number of amino acids seek an interior position in the protein, forcing the covalently bonded chain to organize itself in the way that a soap micelle does, with water-soluble groups on the outside and water-incompatible groups on the inside. Because each protein possesses a unique sequence of amino acids, the folding of each protein is unique—and so is the topography of its surface. These unique surface properties are at the heart of the ability of proteins such as enzymes and antibodies to recognize a given type of molecule out of the thousands of kinds of molecules within a cell.

Summary

General Aspects of Solution Formation

• Solutions contain a solute dissolved in a solvent, the component present in larger amounts.

• Concentration is a measure of the amount of solute in physical or chemical mass units dissolved in an amount of solution expressed in units of volume or mass.

Solubility

• Solutions form when the secondary forces between solvent and solute molecules are similar to those within a pure solvent and pure solute.

• Solubility quantitatively describes the largest amount of solute that will dissolve in a fixed amount of solvent at a specific temperature.

Percent Composition

• Percent composition of a solution is defined in three ways:

$$\text{w/w \% concentration} = \frac{\text{grams of solute}}{\text{grams of solution}} \times 100\%$$

$$\text{v/v \% concentration} = \frac{\text{mL of solute}}{\text{mL of solution}} \times 100\%$$

$$\text{w/v \% concentration} = \frac{\text{grams of solute}}{\text{mL of solution}} \times 100\%$$

Molarity

• Molarity is defined as moles of solute per liter of solution.

• Concentrations described in this way are called molar, a property identified by the symbol M.

Diffusion and Membranes

• Diffusion is a statistical phenomenon resulting from the random motion of molecules or ions.

• Membranes prevent or limit the process of diffusion.

• Membranes possessing the ability to select among penetrating ions and molecules are called semipermeable.

• The flow of water to the high-solute side of a semipermeable membrane is called osmosis, and the driving force for the flow is called the osmotic pressure.

• The osmotic pressure depends on the sum of molar concentrations of all particles in solution without regard to the particles' identity.

Macromolecules

• Macromolecules can consist of many small molecules linked to one another by covalent chemical bonds or by weak secondary forces.

• Important classes of biological macromolecules are proteins, carbohydrates, and nucleic acids.

Key Words

Exercises

Percent Composition

7.1   How many grams of each solute are present in the following solutions? (a) 25.2 g of a 4.25% w/w solution of glucose; (b) 125 g of a 6.55% w/w solution of sodium sulfate.

7.2   Calculate the number of grams of each compound contained in (a) 60 g of a 4.5% w/w aqueous $CaCl_2$ solution; (b) 26.0 g of a 0.450% w/w solution of insulin.

7.3   How many grams of $KHCO_3$ must you add to 250 mL of water to prepare a 0.600% w/v solution?

7.4   What volume of 7.50% w/v KCl solution can be prepared from 30.0 g of KCl?

7.5   What volume of a 6.20% w/v NH_4Cl solution contains 5.40 g of salt?

7.6   What mass of glucose is required to prepare 210 g of a 8.50% w/w solution of glucose?

7.7   What weight of a 5.00% w/w solution of glucose must you take to obtain 6.25 g of glucose?

7.8   What volume of a 5.50% v/v solution of ethanol in water can be prepared with 22.6 mL of ethanol?

7.9   What volume of a 7.50% v/v solution of propylene glycol in water can be prepared with 33.6 mL of propylene glycol?

7.10 What is the concentration in mg % of a glucose solution that contains 0.049 g of glucose in 70 mL of solution?

7.11 What is the concentration in mg % of a uric acid solution that contains 0.230 mg of uric acid in 19.0 mL of solution?

7.12 What is the concentration in ppm of a solution that contains 0.147 g of glucose in 105 mL?

7.13 What is the concentration in % w/v of a NaCl solution whose concentration is 42 ppm?

Molarity

7.14 What is the volume of a 2.8 M solution that contains 0.70 mol of solute?

7.15 How many moles of solute are there in 2.5 L of a 0.40 M solution?

7.16 What are the molarities of solutions containing (a) 0.60 mol in 0.80 L; (b) 0.75 mol in 1.5 L?

7.17 What are the molarities of solutions containing (a) 2.6 mol in 1.3 L; (b) 0.810 mol in 0.240 L?

7.18 Calculate the number of moles contained in (a) 0.200 L of a 0.700 M solution; (b) 2.60 L of a 0.750 M solution.

7.19 Calculate the number of moles contained in (a) 1.05 L of a 2.65 M solution; (b) 452 mL of a 0.850 M solution.

7.20 Calculate the volumes containing the specified number of moles from the following solutions: (a) 0.750 mol of a 0.850 M solution; (b) 0.50 mol of a 2.5 M solution.

7.21 What are the volumes containing the specified number of moles from the following solutions? (a) 1.35 mol of a 0.900 M solution; (b) 2.80 mol of a 0.731 M solution.

7.22 What is the molarity of a solution containing 21.55 g of KCl in 643 mL?

7.23 What is the molarity of 275 mL of a solution containing 23.83 mg of $MgCl_2$?

7.24 How many grams of $MgCl_2$ are contained in 225 mL of a 0.650 M solution?

7.25 How many milligrams of $CaCl_2$ are contained in 75.0 mL of a 0.450 M solution?

7.26 What is the volume of a 0.750 M $MgCl_2$ solution containing 45.0 g of $MgCl_2$?

7.27 What is the volume of a 0.620 M KCl solution containing 25.0 g of KCl?

Dilution

7.28 Given 20.0 mL of a solution of 2.50 M HCl, to what volume must we dilute it to obtain a concentration of 0.200 M?

7.29 The concentration of sulfuric acid, H_2SO_4, is 36 M. What volume is required to prepare 6.0 L of 0.12 M acid?

7.30 What volumes of the following concentrated solutions are required to prepare solutions of the final concentrations indicated? (a) 12 M H_2SO_4 to prepare 2.0 L of 1.5 M H_2SO_4; (b) 2.0 M KCl to prepare 200 mL of 0.50 M KCl.

7.31 What volumes of the following concentrated solutions are required to prepare solutions of the final concentrations indicated? (a) 0.750 M glucose to prepare 4.50 L of 0.250 M glucose; (b) 0.360 M lactose to prepare 650 mL of 0.18 M lactose.

7.32 What volume of 0.900% w/v saline solution can be prepared from 0.300 L of a 3.00% w/v saline solution available in stock?

7.33 From 0.80 L of a glucose stock solution, 4.0 L of a 1.8% w/v solution of glucose was prepared. What was the concentration of the stock solution?

7.34 What is the molarity of a solution prepared by diluting 0.15 L of 0.80 M HCl to a final volume of 0.48 L?

7.35 Calculate the molarity of a solution prepared by diluting 0.100 L of 2.10 M KOH to a final volume of 0.420 L.

Solubility

7.36 An unknown amount of KCl was added to 250 mL of water at 35°C and formed a clear solution. The temperature was lowered to 25°C, and the solution remained clear. Was the solution at 35°C saturated? (See Figure 7.5.)

7.37 The solubility of strontium acetate is 43 g/100 mL at 10°C and 37.4 g/100 mL at 50°C. A clear solution of strontium acetate at 10°C was heated to 50°C and remained clear. Was the solution at 10°C saturated?

7.38 The solubility of aluminum sulfate in water at 20°C is 26.7 g/100 mL. Calculate its molarity.

7.39 The solubility of calcium chloride in water at 20°C is 74.5 g/100 mL. Calculate its molarity.

Diffusion and Membranes

7.40 Calculate the osmolarity of the following aqueous solutions: (a) 0.40 M glucose; (b) 0.10 M $CaCl_2$.

7.41 Calculate the osmolarity of the following aqueous solutions: (a) 0.10 M $AlCl_3$; (b) 0.01 M KCl.

7.42 What happens to erythrocytes (red blood cells) placed in a 0.05% w/v saline solution?

7.43 What happens to erythrocytes (red blood cells) placed in a 8.0% w/v dextrose (glucose) solution?

7.44 What happens to erythrocytes (red blood cells) placed in a 0.90% w/v sodium chloride solution?

Unclassified Exercises

7.45 Calculate the percent by weight of the following solutions: (a) 13.0 g of $CaCl_2$ in 133 g of H_2O; (b) 12.5 g of ethyl alcohol in 165 g of H_2O; (c) 2.5 g of procaine hydrochloride in 60 g of propylene glycol.

7.46 Given an aqueous stock solution that is 5.0% by weight in iron(II) chloride, how many milliliters of that solution must you take to obtain 7.5 g of $FeCl_2$?

7.47 How much of each compound is present in the following mixtures? (a) 55.0 g of a 4.00% w/w solution of glucose; (b) 175 g of a 7.50% w/w solution of ammonium sulfate.

7.48 Calculate the volume percent of the following mixtures: (a) 15.0 mL of ethyl alcohol made up to a final volume of 200 mL of solution with water; (b) 8.40 mL of chloral hydrate made up to 35.0 mL with ethyl alcohol; (c) 10.6 mL of propylene glycol made up to 50.0 mL with water.

7.49 Calculate the weight/volume percent of the following mixtures made up to final volume with water: (a) 6.40 g of LiCl in 66.0 mL; (b) 1.30 g of $NaHCO_3$ in 125 mL; (c) 3.60 g of $CaCl_2$ in 90.0 mL.

7.50 Calculate the molarity of the following aqueous solutions: (a) 0.320 mol in 1.25 L; (b) 0.220 mol of NaCl in 0.471 L; (c) 1.70 mol of $NaHCO_3$ in 2.26 L; (d) 0.325 mol HCl in 116 mL.

7.51 Calculate the number of moles in each of the following aqueous solutions: (a) 46 mL of 0.066 M $CaCl_2$ solution; (b) 1.2 L of 0.065 M KNO_3 solution; (c) 495 mL of 0.250 M LiCl solution; (d) 0.625 L of 1.75 M $NaHCO_3$ solution.

7.52 Calculate the number of grams of solute in each of the following aqueous solutions: (a) 75 mL of 1.75 M KCl; (b) 0.075 L of 0.82 M LiCl; (c) 2.75 L of 0.415 M NaSCN; (d) 62.5 mL of 18.0 M H_2SO_4.

7.53 How many moles of a 0.375 M solution of glucose will be contained in the following volumes of that solution? (a) 1.49 L; (b) 6.72 L; (c) 2.56 L; (d) 1.81 L.

7.54 How many grams of $KHCO_3$ must you add to 250 mL of water to prepare a 0.650 M solution?

7.55 How would you prepare 500.0 mL of a 0.160 M aqueous solution of KI?

7.56 What weight of a 5.00% w/w solution of glucose must you take to obtain 6.25 g of glucose?

7.57 Calculate the weight percent of the following solutions: (a) 15.0 g $CaCl_2$ in 250 g H_2O; (b) 16.5 g NaCl in 325 g H_2O; (c) 2.20 g C_2H_5OH in 122 g H_2O.

7.58 What volume of physiological saline (0.900% w/v NaCl) can be prepared from 90.0 mL of 2.75% w/v NaCl solution?

7.59 Calculate the grams of solute present in each of the following solutions: (a) 0.950 L of 1.25 M NaCl; (b) 625 mL of 0.750 M $Mg(NO_3)_2$; (c) 325 mL of 2.20 M K_2CO_3.

7.60 Calculate the molarity of the following aqueous solutions: (a) 62.0 g $K_2C_2O_4$ in 747 mL of solution; (b) 126 g $C_6H_{12}O_6$ in 350 mL of solution; (c) 26.0 g $BaCl_2$ in 0.825 L of solution.

7.61 What volume of 0.180 M HCl can be prepared from 1.50 L of 1.80 M HCl?

7.62 What volume of 0.1250 M KOH is required to prepare 750.0 mL of 0.06500 M KOH?

7.63 A 150-mL portion of a concentrated stock saline solution was used to prepare 1.50 L of 0.90% w/v saline solution. What was the concentration of the stock solution?

7.64 Calculate the molarity of a solution prepared by diluting 0.100 L of 2.10 M KOH to a final volume of 0.420 L.

When the chemical reactions of metabolism are undertaken in test tubes, extraordinary conditions of temperature, concentration, and acidity are required to induce the reactions to take place. However, when these same reactions take place in cells, they proceed rapidly and efficiently under quite mild conditions. Unique protein molecules called enzymes make these intracellular reactions possible. When you reach the end of this chapter, you should have a clearer understanding of the reasons for the chemical ingenuity of cells. Our goals are to enable you to learn how chemical reactions take place, the molecular mechanisms of how they take place, and how to characterize their outcome quantitatively.

8.1 REACTION RATES

In a chemical reaction, the concentrations of the starting components, the **reactants,** and those of the emerging chemical species, the **products,** change with the passage of time. The **rate** of a chemical reaction is the speed with which these changes take place. The study of the factors affecting the rates of chemical reactions is called **chemical kinetics.**

Consider reactions between substances such as nitrogen and hydrogen to form ammonia or between nitric oxide and bromine to form nitrosyl bromide. The balanced reaction equations for those processes are

$$3 H_2 + N_2 \longrightarrow 2 NH_3 \tag{1}$$

$$2 NO + Br_2 \longrightarrow 2 NOBr \tag{2}$$

The equations imply that, in reaction 1, the concentrations of hydrogen and nitrogen decrease and that of ammonia increases and, in reaction 2, the concentrations of nitric oxide and bromine decrease and that of nitrosyl bromide increases. Theoretical data representing these kinds of reactions are charted in Figure 8.1. As time passes, the molar concentrations of reactants are seen to decrease while the molar concentrations of products simultaneously increase.

The **reaction rate** is defined as the change in the number of moles of a reactant or a product per unit time. That is,

$$\text{Rate} = \frac{\text{moles of product appearing}}{\text{elapsed time}}, \text{ or } \left(\frac{\text{moles of reactant disappearing}}{\text{elapsed time}}\right)$$

The rate is positive for the appearance of product and for the disappearance of reactant. For example, if 2.5 mol of a reactant is used up in a 10-min period, the average rate of reaction is 0.25 mol/min, and, if 2.5 mol of a product is formed in a 10-min period, the average rate of reaction also is 0.25 mol/min.

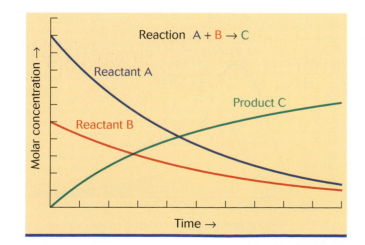

Figure 8.1 In a chemical reaction, product appears and reactant disappears with the passage of time.

7.65 Aqueous solutions of the following compounds are added together. Consult Table 7.3 to predict whether a reaction will take place. If a reaction will take place, write the net ionic equation that describes it. (a) NaCl + AgNO$_3$; (b) KOH + MgCl$_2$; (c) Pb(NO$_3$)$_2$ + KCl; (d) K$_3$PO$_4$ + CaCl$_2$.

7.66 Aqueous solutions of the following compounds are added together. Consult Table 7.3 to predict whether a reaction will take place. If a reaction will take place, write the net ionic equation that describes it. (a) K$_2$SO$_4$ + BaCl$_2$; (b) AlCl$_3$ + KOH; (c) NaBr + AgNO$_3$; (d) KCl + Na$_2$CO$_3$.

Expand Your Knowledge

Note: The icons ▲ ▲ ▲ denote exercises based on material in boxes.

7.67 Seventy-five milliliters of propylene glycol, a pure liquid, is dissolved in 55 mL of water. Which component is the solute and which is the solvent?

7.68 A chemist has 75 mL of 0.25 *M* HCl available and requires 125 mL of 0.175 *M* HCl. Will she be able to obtain the required quantity? Explain your answer.

7.69 The hydrostatic pressure at the renal (kidney) artery is about 100 torr. The osmotic pressure of the blood is about 15 torr. Does water move into or out of the arterial blood?

7.70 How would you prepare an isotonic saline solution described by a medical dictionary as a 0.90% NaCl w/v solution?

7.71 What is the osmolarity of a solution that is a mixture of 0.45% w/v NaCl and 2.7% w/v glucose?

7.72 Seawater is approximately 0.600 *M*. What is the minimum pressure that must be applied to the seawater side of a reverse-osmosis membrane at 298 K to purify seawater by reverse osmosis?

7.73 Wine contains 12.5% ethanol by volume. The density of pure ethanol is 0.790 g/cm^3. Calculate the mass percent of ethanol in wine.

7.74 How would you prepare 500 mL of an aqueous glucose (C$_6$H$_{12}$O$_6$) solution that would have an osmotic pressure of 12.0 atm?

7.75 If the fluid inside a tree is 0.600 *M* and that of the fluid bathing the roots is 0.500 *M*, how high will the fluid rise in the tree, assuming the internal fluid's density is 1.00 g/cm^3? (The density of mercury is 13.6 g/cm^3.)

7.76 A 0.10 g sample of human serum albumin is dissolved in 3.0 mL of water and produces an osmotic pressure of 9.7 torr at 298 K. Calculate the protein's molar mass. (Hint: Number of mols = grams/molar mass. See Box 7.5.)

7.77 Concentrated sulfuric acid H$_2$SO$_4$ is sold as a solution that is 98.0% sulfuric acid by mass, and 2.00% water by mass. The density of this solution is 1.84 g/cm^3. Determine the solution's molarity.

7.78 Two solutions are separated by a membrane permeable only to water. For each of the following solutions, determine whether water will flow from one compartment to the other and the direction of the water flow.

(a) 0.1 *M* NaCl(*aq*) and 0.1 *M* KBr(*aq*)
(b) 0.1 *M* Al(NO$_3$)$_3$ (*aq*) and 0.2 *M* NaCl(*aq*)
(c) 0.2 *M* CaCl$_2$(*aq*) and 0.5 *M* KCl(*aq*)

7.79 Both of the paired solutions:

(a) Na$_2$SO$_4$(*aq*) + BaCl$_2$(*aq*) $\longrightarrow$
(b) NaCl(*aq*) + AgNO$_3$(*aq*) $\longrightarrow$

are mixed. In each case, if a precipitate forms, write both the complete reaction equation and the net ionic equation. Use Table 7.3 to determine whether reaction takes place, and write "no reaction" if that is the case. Assume that, before mixing, all solutions are 0.25 *M* and that equal volumes are mixed.

7.80 Both of the paired solutions:

(a) Pb(NO$_3$)$_2$(*aq*) + KCl(*aq*) $\longrightarrow$
(b) CaCl$_2$(*aq*) + Na$_2$CO$_3$(*aq*) $\longrightarrow$

are mixed. In each case, if a precipitate forms, write both the complete reaction equation and the net ionic equation. Use Table 7.3 to determine whether reaction takes place, and write "no reaction" if that is the case. Assume that, before mixing, all solution are 0.25 *M* and that equal volumes are mixed.

7.81 What takes place at the critical micelle concentration? (See Box 7.5.)

7.82 What medical problem is indicated by the presence of protein in the urine? (See Box 7.2.)

CHAPTER 8

CHEMICAL REACTIONS

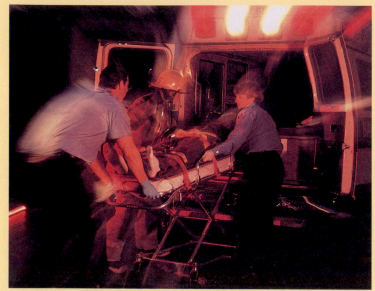

(Pete Saloutos/The Stock Market.)

Chemistry in Your Future

A house painter is brought to the emergency room in a state of collapse after complaining of visual problems. You and the other firefighter paramedics bring all his paints and solvents along to help the physicians find clues to his problem. One solvent is "mineral spirits" (methyl alcohol, also called wood alcohol), a toxic substance known to affect vision. The physicians immediately begin feeding small doses of ethyl alcohol (a nontoxic alcohol) to the patient, and overnight he completely recovers. You overhear one physician say that many biochemical processes are reversible. Chapter 8 explains this statement and how reactions can be reversed.

For more information on this topic and others in this chapter, go to www.whfreeman.com/bleiodian2e

Learning Objectives

- Describe how the rates of chemical reactions are affected by the concentrations of reactants, the temperature, and the presence of catalysts.

- Describe how chemical reactions are the result of collisions that lead to the formation and decay of an activated complex.

- Explain how a chemical system reaches a state of dynamic equilibrium.

- Describe Le Chatelier's principle.

- Describe how an equilibrium constant provides a quantitative description of chemical equilibrium.

Chemical reactions require that collisions occur between the reacting molecules. The rate of collisions (and hence of reaction) depends on four variables: (1) the concentration of reactants, (2) their spatial orientation when they collide, (3) the temperature at which the reaction takes place, and (4) the presence of catalysts.

8.2   REACTIVE COLLISIONS

For a chemical reaction to take place, a collision must result in the breaking or the rearrangement of bonds. This result can occur only if (1) the molecules collide with sufficient energy and (2) they collide in a favorable spatial orientation. A collision having both characteristics is called a **reactive collision.**

The kinetic energy (energy of motion) of molecules increases as the temperature increases. Molecules move about much faster in hot systems than in cold ones. It is important to note that, in any assembly of molecules, only a small fraction are moving about with velocities great enough to produce reactive collisions. This small fraction of fast-moving molecules can increase dramatically with an increase in temperature. Figure 8.2 plots the number of collisions between different gas molecules as a function of their kinetic energies at two different temperatures, T_1 and T_2. The minimum energy needed for a reactive collision—the **activation energy,** E_a—is noted on the x-axis. The graph shows that the number of collisions having kinetic energies above the activation energy increases rapidly with increase in temperature. The increase in the number of collisions in the high-kinetic-energy range is so rapid that, for many reactions, the reaction rate can be doubled by a mere 10-kelvin increase in temperature. An interesting application of this effect used by biological systems is discussed in Box 8.1 on page 206.

Figure 8.3 is a schematic representation of the reaction between a molecule of fluorine and a molecule of nitrogen dioxide to form nitrogen oxyfluoride. Three scenarios depict the collisions of the reactants. Figure 8.3a shows a nonreactive collision in which the kinetic energies of the molecules are not sufficient to cause bonds to break and reform. The molecules merely bounce off each other. Figure 8.3b shows a reactive collision in which both the kinetic energies and the spatial orientation are optimized for correct bond breakage and new bond formation. The new transient structure formed on collision is called an **activated complex.** The new structure is also known as the **transition state.** It is unstable and rapidly decays either into products or back into the original reactants. Figure 8.3c shows a nonreactive collision in which the kinetic energies were sufficient but the orientation was not correct. The

Figure 8.2 The number of collisions between different gas molecules is plotted versus their kinetic energies at two different temperatures (T_1 and T_2). The minimum energy needed for a reactive collision, the activation energy, is indicated on the x-axis.

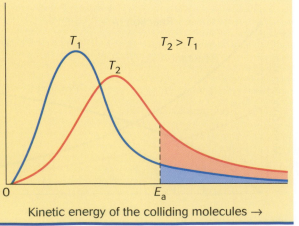

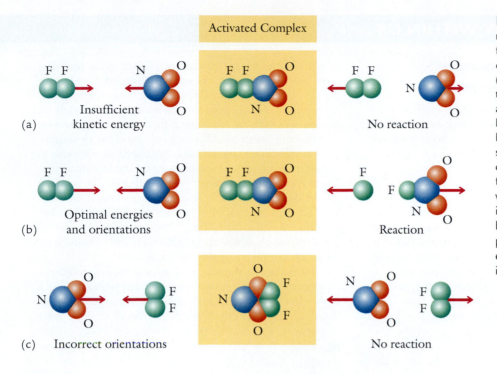

Activated Complex

(a) Insufficient kinetic energy   No reaction

(b) Optimal energies and orientations   Reaction

(c) Incorrect orientations   No reaction

Figure 8.3 Three scenarios for the reaction between a molecule of fluorine and a molecule of nitrogen dioxide to form nitrogen oxyfluoride. (a) A nonreactive collision in which the kinetic energies of the molecules are insufficient to cause bonds to break. (b) A reactive collision in which both the kinetic energies and the spatial orientations are optimal for correct bond breakage and new bond formation. (c) A nonreactive collision with sufficient kinetic energy but incorrect orientations for bond breakage. The shorter arrows in part (a) indicate that the kinetic energies of the reactants are insufficient for bond breakage.

fluorine molecule must strike the nitrogen atom of nitrogen dioxide for the complex to form and the reaction to take place. (We use the word complex to describe any combination of molecules that has a short lifetime.)

The activation energy, then, is the minimum energy required to form an activated complex. The energy of motion of the colliding molecules is incorporated into the overall energy of the activated complex and is expressed by greater vibrations of the constituent atoms. In consequence, the bond lengths (the distances between atoms) become greater than normal and therefore unstable.

Figure 8.4, called an activation-energy diagram, is a plot of the changes in energy (the y-axis) of a pair of colliding molecules as they form an activated complex; that is,

$$A + B \longrightarrow \text{activated complex} \longrightarrow C + D$$

The x-axis in Figure 8.4 represents the progress of the reaction in which A approaches B to form the complex, which proceeds to form products C and D. The difference between the energy of the activated complex and that of the reactants is the activation energy. In Figure 8.4, the difference in energy between the A + B level and the C + D level signifies that energy is lost when

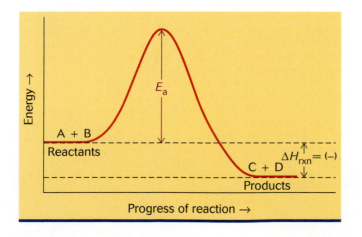

Energy →

E_a

A + B
Reactants

C + D

$\Delta H_{rxn} = (-)$

Products

Progress of reaction →

Figure 8.4 An activation-energy diagram plots the changes in the energy of an activated complex as an exothermic chemical reaction progresses from reactants to products. The heat of reaction for the process is designated by ΔH_{rxn}.

8.1 CHEMISTRY WITHIN US

The Influence of Temperature on Physiological Processes

One very important aspect of temperature regulation in warm-blooded organisms is the **febrile response,** the physiological changes that lead to an increase in body temperature in response to infection. Fever has been primarily considered a medical problem and thus a problem pertaining to humans. As such, it has not been the subject of experimental studies, because of the difficulties of conducting studies that use human subjects. More important, humans and other mammals are designed to maintain their body temperatures within a very narrow range, a fact that severely limits their potential usefulness in any study of how temperature can cause changes in biochemical processes.

Fortunately, there are organisms that do not possess physiological mechanisms for controlling their body temperatures. These organisms are the **ectothermic,** or "cold-blooded," animals, such as lizards, snakes, and insects. Ectothermic animals regulate their body temperatures behaviorally. For example, they warm their bodies in the sun to raise their metabolic rates to be able to pursue food

or mates or to escape predation. If their body temperatures become too high because of extended exposure to the sun, they seek shade. Because their body temperatures are not fixed, ectothermic animals are excellent models for studies of the effects of temperature on selected processes.

In both warm-blooded and cold-blooded organisms, body temperatures are primarily regulated in response to signals from a region of the brain called the hypothalamus. There appears to be a "thermostatic setpoint" about which regulation occurs. If the body temperature drops below the setpoint, physiological or behavioral mechanisms that raise it to the desired value begin. If body temperature exceeds the setpoint, physiological or behavioral mechanisms leading to a lower body temperature are invoked.

In the febrile response in humans, the hypothalamus behaves as if the thermostatic setpoint has been raised to a higher level by approximately 2 to 3 Celsius degrees. The chills that humans experience at the onset of fever are a response to this new internal environment in which the hypothalamus is convinced that the body temperature should be much higher than it is. The person then shivers, experiencing violent muscle contractions, which are

the reactants are chemically transformed into products. This difference in energy levels represents a loss of heat, and the reaction is characterized as **exothermic.** Such reactions give off heat. The reaction as written, with reactants on the left and products on the right, is called a **forward reaction.**

There is no reason to suppose that the products of our hypothetical reaction (C + D) cannot react to form A + B; that is, the reaction is reversible (see Section 8.5). That chemical reaction,

$$C + D \longrightarrow \text{activated complex} \longrightarrow A + B,$$

called a **back reaction,** is illustrated in Figure 8.5. It is the opposite of the exothermic forward reaction and is therefore accompanied by an absorption of heat in an amount corresponding to the increased energy of the products (A + B). A reaction characterized by the absorption of heat is called **endothermic.**

The energy difference between reactants and products is called the **heat of reaction,** ΔH_{rxn} (the subscript rxn stands for reaction), and it is experimentally determined as the heat absorbed or evolved (produced) as a result of reaction.

Figure 8.5 An activation-energy diagram plots the changes in the energy of an activated complex as an endothermic chemical reaction progresses from reactants to products.

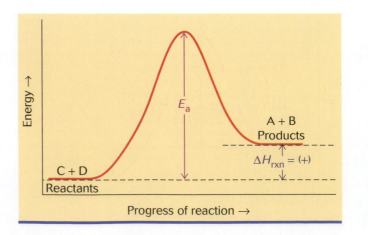

exothermic and therefore raise the body temperature. When body temperature reaches the new higher setpoint, shivering ceases and the person feels comfortable. When the fever breaks, the original setpoint is restored, and the person feels very warm because the body temperature is now a few degrees above the setpoint. The physiological response is to perspire copiously and to feel an urge to throw off the covers.

Research in the past 30 or 40 years has established that the febrile response occurs in nearly all animal phyla: fishes, amphibia, reptiles, and birds, as well as mammals. The phenomenon is so widespread that it seems to be a fundamental, adaptive biological response of multicellular organisms to infection.

The most interesting research probing this possibility was performed with lizards. First, they are prone to suffer from a particular bacterial infection. Second, they can regulate their body temperatures behaviorally, moving about to find a location having the temperature that best suits their current metabolic needs. When researchers placed healthy lizards in an aquarium outfitted with a floor whose temperature varied along its length, low at one end, high

at the other, the lizards regularly moved to the part of the aquarium floor where the temperature was 38°C. The researchers injected several groups of lizards with pathogenic bacteria and placed them in other aquaria whose temperatures also varied along the floor length. However, the high-end temperature of each aquarium was different and varied from 42°C to below 30°C. All infected lizards experienced a febrile response and tried to raise their body temperatures above the normal setpoint of 38°C, but only those in the aquaria with high-end temperatures above 38°C succeeded. The result was nearly 100% survival among those lizards who could raise their body temperatures to above 40°C. There was significant mortality among those who could not raise their body temperatures above the normal setpoint. The specific mechanisms leading to survival are not known.

These results indicate that the febrile state is an adaptive and useful response to infection. Although too high a fever for too long can be hazardous, researchers now question the generally held view that, at the first sign of fever, antipyretic, or fever-reducing, drugs such as aspirin should be taken to reduce the body temperature.

It can also be calculated as the difference between the activation energies of the forward and back reactions:

$$\Delta H_{rxn} = E_{forward} - E_{back}$$

When the energy of the products is greater than the energy of the reactants, heat must have been absorbed, and ΔH_{rxn} is positive. When the energy of the products is less than the energy of the reactants, heat must have been lost, and ΔH_{rxn} is negative.

✓ In an exothermic reaction, heat is evolved (lost) and the heat of reaction is negative $[\Delta H_{rxn} = (-)]$.

✓ In an endothermic reaction, heat is absorbed (gained) and the heat of reaction is positive $[\Delta H_{rxn} = (+)]$.

Concept checklist

Example 8.1   Using activation energy to calculate the heat of reaction

The activation energies characteristic of the reversible chemical reaction A + B $\rightleftharpoons$ C are $E_{forward} = 53$ kJ and $E_{back} = 82$ kJ. Calculate ΔH_{rxn}, the heat of the forward reaction. Is the forward reaction exothermic or endothermic?

Solution

$$\Delta H_{rxn} = E_{forward} - E_{back}$$
$$= 53 \text{ kJ} - 82 \text{ kJ}$$
$$= -29 \text{ kJ}$$

The heat of reaction is negative, which means that heat has been lost and the forward reaction is exothermic.

Problem 8.1   The activation energies characteristic of the reversible chemical reaction A + B $\rightleftharpoons$ C are $E_{forward} = 74$ kJ and $E_{back} = 68$ kJ. Calculate ΔH_{rxn}, the heat of the forward reaction. Is the forward reaction exothermic or endothermic?

8.3 CATALYSTS

A **catalyst** is a substance that increases the rate of a reaction but emerges unchanged after the reaction has ended. Catalysts in cells are called enzymes. For example, the enzyme carbonic anhydrase, found in red blood cells, is a catalyst that specifically increases the rate of formation of bicarbonate ion from carbon dioxide and hydroxide ion by a factor of 3.5×10^6. Without this rapid reaction, ridding the body of the carbon dioxide generated by cellular oxidative processes would be impossible.

When sulfur burns, it forms two different oxides: sulfur dioxide and sulfur trioxide. The reactions are

$$S + O_2 \longrightarrow SO_2 \tag{1}$$

and

$$2 SO_2 + O_2 \longrightarrow 2 SO_3 \tag{2}$$

Reaction 1 is much faster than reaction 2; so, when a sample of coal, which contains sulfur compounds, is burned, we might expect the bulk of the oxide produced to be SO_2. However, measurements have shown that, in the course of coal combustion, the formation of large quantities of SO_3 contributes significantly to the acid rain descending on our cities and forests. Acid rain forms when SO_3 dissolves in water to form sulfuric acid:

$$SO_3 + H_2O \longrightarrow H_2SO_4$$

The reason that SO_3 is formed in unexpected excess is that coal contains nitrogen compounds as well as sulfur; so its combustion also produces nitric oxide (NO). Nitric oxide, in turn, reacts very rapidly with SO_2 in a series of steps to form SO_3:

$$2 NO + O_2 \longrightarrow 2 NO_2$$
$$2 NO_2 + 2 SO_2 \longrightarrow 2 SO_3 + 2 NO$$
$$\text{Sum:} \quad 2 SO_2 + O_2 \longrightarrow 2 SO_3$$

As you examine this reaction sequence, notice that NO enters the first reaction as a reaction component but emerges unchanged at the end of the second reaction. In other words, nitric oxide speeds up the production of SO_3 and emerges unchanged. It is a catalyst. Notice, too, the role of NO_2, which is formed in the first step and used up in the second step (as emphasized by the connecting line). The NO_2 is called a **reaction intermediate.** The two steps are connected in a sequence because the product of the first step—NO_2—becomes a reactant in the second step. The final stoichiometry of the production of SO_3 is derived from an algebraic summation of the two steps in the reaction sequence.

All catalysts provide a different pathway for a reaction that has a lower activation energy, as illustrated in Figure 8.6. Although the activation energy is lowered by a catalyst, the heats (endothermic or exothermic) of reaction and the products of a reaction are unchanged.

The serious damage to this statue reveals the presence of acid rain. (Spencer Grant/Photo Researchers.)

Concept checklist

✓ A catalyst lowers the activation energy of a reaction by providing a different pathway leading to products.

✓ It does so by becoming an active participant in the chemical process, but it emerges unchanged.

✓ It does not alter the results of a reaction; it changes only the speed at which the reaction takes place.

✓ Any catalyst for the forward reaction of a chemical equilibrium also acts as a catalyst for the reverse reaction.

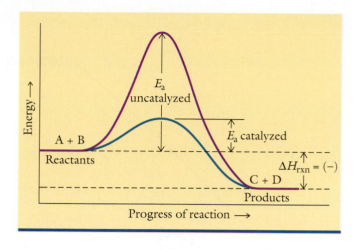

Figure 8.6 Reaction progress for a catalyzed reaction (compare Figure 8.4). The activation energy of the catalyzed reaction is significantly lowered, but the heat of reaction is unchanged.

8.4 BIOCHEMICAL CATALYSTS

The protein molecules called **enzymes** are catalysts whose structures allow them to interact selectively with particular molecules or families of molecules. The selectivity can be great enough that an enzyme can select one of thousands of different molecules in the cell for conversion into a product. Recall that carbonic anhydrase selects only CO_2 for its conversion into HCO_3^-.

The site, or location, where catalysis takes place on the enzyme—called an **active site**—is structurally related to the substance being converted—the **substrate**—as a lock is to a key. We will refer to this spatial relation as **complementarity.** The substrate is bound to the active site by secondary forces. The key to understanding enzyme-catalyzed reactions is that the catalytic properties of enzymes depend on the formation of an activated complex (see Section 8.2). In all cases, the formation of the complex of enzyme and substrate provides a reaction pathway of low overall activation energy and, consequently, a rapid reaction rate.

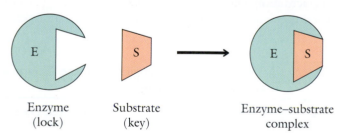

The activation energy controls the rate of a chemical reaction. The greater the activation energy, the smaller the percentage of colliding species that have the required activation energy and the slower the rate of reaction. The relation between activation energy, reaction rate, and temperature is an important feature of physiological function and adaptation. In Figure 8.7, on the following page, the relative rates of chemical reactions possessing different activation energies are plotted as a function of temperature over a small temperature range. The temperature effects on a cricket's chirping rate are included in Figure 8.7. The temperature range corresponds to temperatures encountered by most organisms on our planet. Note that the greater the activation energy, the steeper the slope. In other words, in reactions with a large activation energy, the rate of reaction changes very rapidly with a small increment in temperature. Conversely, in reactions with a small activation energy, the rates of reaction are relatively insensitive to changes in temperature.

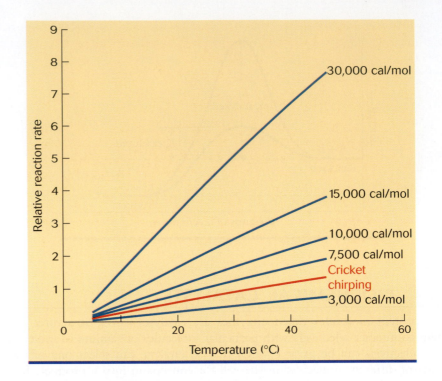

The survival of organisms depends in large part on the control of the chemical reactions that constitute metabolism. The body temperatures of warm-blooded animals are restricted to fairly constant values, and even so-called cold-blooded, or ectothermic, animals function within very narrow temperature ranges. Consequently, most animals are equipped with biochemical processes of low activation energies and therefore are not significantly dependent on changes in temperature to modify the rates of their metabolic reactions. Chemical reactions in cells are controlled by changes in enzyme activity and concentration.

Data on activation energies of uncatalyzed chemical reactions, enzyme-catalyzed reactions, and some typical biological processes are listed in Table 8.1.

TABLE 8.1	The Activation Energies of Some Chemical and Biological Processes
Process	**Activation energy (cal)**
DECOMPOSITIONS	
HI	45,600
N_2O	53,000
C_2H_5Br	54,800
sucrose	25,600
ENZYME-CATALYZED REACTIONS	
sucrose hydrolysis by invertase	9,000
tributanoyl glycerate hydrolysis by lipase	7,600
BIOLOGICAL PROCESSES	
cricket chirping	5,300
bacterial decomposition of urea	8,700
α-rhythm of human brain	8,000

They show that enzyme-catalyzed reactions and biological processes have significantly smaller activation energies than those of uncatalyzed chemical reactions.

8.5 CHEMICAL EQUILIBRIUM

The time course of a reaction such as that of hydrogen with nitrogen to produce ammonia is plotted in Figure 8.1. Figure 8.8 plots a similar reaction over a long period of time. After enough time, the concentrations of product and reactants no longer change. At that point in time, the system has reached a state of **chemical equilibrium.** The amounts of reactants left can vary from large quantities to none.

The idea of a dynamic equilibrium, a state or condition in which two opposing processes take place at the same rate, was introduced in Section 6.6. In the equation describing the liquid–vapor equilibrium, these processes were represented by arrows pointing in opposite directions:

$$\text{Liquid} \underset{\text{condensation}}{\overset{\text{vaporization}}{\rightleftharpoons}} \text{vapor}$$

As stated in Section 8.2, chemical reactions also consist of simultaneous opposing processes—the forward and back reactions. Therefore, chemists have adopted the convention of representing equilibrium in chemical reactions by two arrows facing in opposite directions:

$$A + B \rightleftharpoons C + D$$

Whenever you see this representation, you may assume that

- The chemical system is either at or is capable of reaching an equilibrium state.
- The equilibrium state may be reached from either direction; that is, by adding either A to B or C to D.

In other words, reactions are **reversible.** In theory, all reactions are reversible. However, there are reactions in which the amounts of reactants left at the point of equilibrium are much too small to be of practical significance. In other reactions, such as the combustion of a hydrocarbon, reaction of the products will not produce the original reactants. Such reactions are called **irreversible.** Often but not always, equations are written with a single arrow pointing to products to denote this condition.

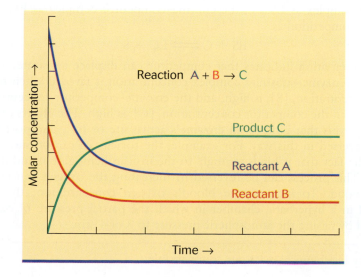

Figure 8.8 When the concentrations of products and reactants no longer change with the passage of time, a chemical reaction has reached equilibrium.

A PICTURE OF HEALTH
Examples of Enzymes in the Body

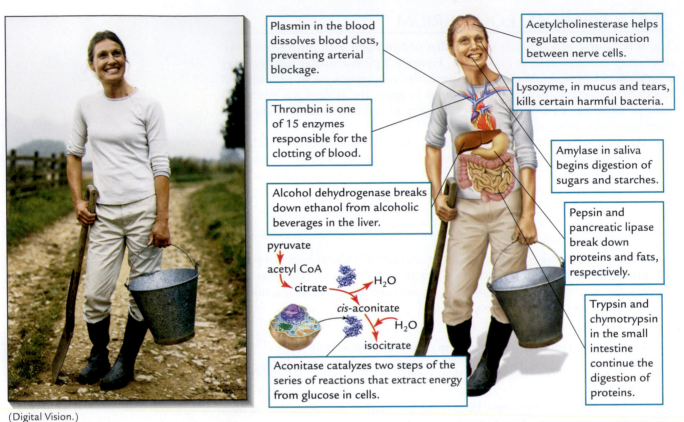

Plasmin in the blood dissolves blood clots, preventing arterial blockage.

Thrombin is one of 15 enzymes responsible for the clotting of blood.

Alcohol dehydrogenase breaks down ethanol from alcoholic beverages in the liver.

Acetylcholinesterase helps regulate communication between nerve cells.

Lysozyme, in mucus and tears, kills certain harmful bacteria.

Amylase in saliva begins digestion of sugars and starches.

Pepsin and pancreatic lipase break down proteins and fats, respectively.

Trypsin and chymotrypsin in the small intestine continue the digestion of proteins.

pyruvate
acetyl CoA
citrate
H_2O
cis-aconitate
H_2O
isocitrate

Aconitase catalyzes two steps of the series of reactions that extract energy from glucose in cells.

(Digital Vision.)

In many enzymatic reactions, a molecule closely resembling the normal substrate can bind to the active site. If such a molecule is present simultaneously with the normal substrate, both will compete for the active site. The most strongly bound molecule will win out if it is present in high enough concentrations. This is the basis for the action of drugs such as HIV protease inhibitors. If there is little difference in the degree of binding, the equilibrium can be shifted by increases in concentration of either the substrate or the substrate mimicker—that is, the competitor C. If we represent the enzyme–substrate complex as ES and the enzyme–competitor complex as EC, we can write the competition as

$$ES + C \rightleftharpoons EC + S$$

This representation indicates that competitor can displace substrate, and vice versa. The enzyme–substrate complex concentration is favored when the substrate concentration, [S], is high, and the enzyme–competitor concentration is favored when the competitor concentration, [C], is high. The effects of certain kinds of toxins that bind to an enzyme's active site or to a transport protein can be reversed in this way. For example, the toxicity of carbon monoxide is due to the fact that it binds strongly to the iron in hemoglobin and displaces oxygen. This process can be reversed by increasing the oxygen concentration in the blood. To increase the blood's oxygen concentration, the patient is placed in a hyperbaric chamber in which the partial pressure of oxygen is increased by raising the atmospheric pressure to 3 or 4 atm.

Concept check   ✓ Many biochemical reactions are reversible.

8.6 EQUILIBRIUM CONSTANTS

The quantitative study of chemical systems at equilibrium has shown that all chemical reactions can be described by an **equilibrium constant, K_{eq},** a mathematical function of the molar concentrations of reactants and products. If we propose that a reaction equation takes the form

$$a\,A + b\,B \rightleftharpoons c\,C + d\,D$$

where lowercase letters represent balancing coefficients and capital letters represent substances, the equilibrium constant is written as

$$K_{eq} = \frac{[C]^c[D]^d}{[A]^a[B]^b}$$

To write an equilibrium constant, four rules serve as a guide:

1. The products of the reaction always appear in the numerator, with reactants in the denominator.
2. Square brackets are used to indicate molar concentrations at equilibrium.
3. The molar concentration of each component at equilibrium is raised to a power equal to the component's coefficient in the balanced equation.
4. Pure solids or liquids do not appear in the equilibrium-constant expression for a reaction in which they take part.

Rules for writing equilibrium constants

The first three of the four rules apply to equilibria of reactions in the gas phase or in solutions, defined as **homogeneous equilibria.** Many other equilibria consist of more than one phase and are called **heterogeneous equilibria.** These phases can be pure solid phases and pure liquid phases. Heterogeneous equilibria require rule 4. Because the equilibrium-constant expression must take the physical states of the reaction components into account, these states must be specified for all components in the balanced equation. This specification is done by placing the following notations next to the components in the reaction equation: (g) for the gaseous state, (l) for the liquid state, (s) for the solid state, and (aq) for aqueous solution. You have already seen this notation in Sections 7.2, 7.9, and 7.10.

The equilibrium constant changes with a change in temperature. The direction of change depends on whether the reaction is exo- or endothermic. If the reaction is exothermic—evolves heat—increasing the temperature will decrease the numerical value of the equilibrium constant, and vice versa, as described in more detail in Section 8.8. The temperatures of reactions specified in the present section in the following examples and problems are to be considered constant.

Example 8.2 | **Writing the equilibrium-constant expression for a homogeneous equilibrium**

Write the equilibrium constant for the following reaction in which all components are present as gases:

$$2\,N_2(g) + 3\,Br_2(g) \rightleftharpoons 2\,NBr_3(g)$$

Solution

The reaction equation specifies the balancing coefficients of all components, and so the reaction's K_{eq} is expressed as

$$K_{eq} = \frac{[NBr_3]^2}{[N_2]^2[Br_2]^3}$$

Problem 8.2 Write the equilibrium constant for the following balanced equation:

$$H_2(g) + I_2(g) \rightleftharpoons 2\,HI(g)$$

Example 8.3 Writing the equilibrium-constant expression for a heterogeneous equilibrium

Write the equilibrium constant for the reaction of solid carbon with gaseous oxygen to form gaseous carbon dioxide:

$$C(s) + O_2(g) \rightleftharpoons CO_2(g)$$

Solution

Because carbon is a pure solid, it will not appear in the equilibrium constant:

$$K_{eq} = \frac{[CO_2]}{[O_2]}$$

Problem 8.3 Write the equilibrium constant for the following reaction:

$$2\,C(s) + O_2(g) \rightleftharpoons 2\,CO(g)$$

The most important qualitative information that we get from the equilibrium constant is how far the reaction of the components proceeds toward products. If the equilibrium constant is much larger than 1.0—for example, 1×10^3—then the concentrations of product (components in the numerator) must be much larger than the concentrations of the reactants (components in the denominator). Such a reaction therefore yields significant amounts of product, and the equilibrium is called **favorable.** Conversely, if the equilibrium constant is much smaller than 1.0—say, 1×10^{-3}—small amounts of products are the result, and the reaction is called **unfavorable.**

Equilibrium constants tell us nothing about how fast a reaction proceeds. The equilibrium constant for the reaction between hydrogen and oxygen to form water is very large, but the reaction at room temperature takes many years to go to completion. This seems counterintuitive to the idea that a large equilibrium constant leads to large amounts of product. However, with a little help from a spark, which provides the required activation energy, the equilibrium state of large amounts of water can be rapidly attained.

Example 8.4 Estimating relative concentrations of reactants and products at equilibrium

Let's assume two possibilities for the equilibrium constant of the hypothetical reaction $A \rightleftharpoons B$: one where the equilibrium constant is much larger than 1.0, and the equilibrium concentration of B is far greater than that of A, and another where the equilibrium constant is much smaller than 1.0, and the equilibrium concentration of A is far greater than that of B.

Possibility 1. The equilibrium constant $K_{eq(1)} = \dfrac{[B]}{[A]} = 100$

Possibility 2. The equilibrium constant $K_{eq(2)} = \dfrac{[B]}{[A]} = 0.01$

For each possibility, estimate the concentration of product with respect to reactant at equilibrium.

Solution

Possibility 1. Rearrangement of the equilibrium-constant expression gives

$$[B] = 100 \times [A]$$

In other words, the equilibrium concentration of B is 100 times the equilibrium concentration of A.

Possibility 2. Rearrangement of the equilibrium-constant expression gives

$$[B] = 0.01 \times [A]$$

In other words, the equilibrium concentration of B is only 1/100th the equilibrium concentration of A.

Problem 8.4 Estimate the concentration of product with respect to reactant at equilibrium for the hypothetical reaction $A \rightleftharpoons B$, for which $K_{eq} = [B]/[A] = 1.0$.

✓ The size of the equilibrium constant tells us about the yield of the reaction.

✓ When $K_{eq} > 1$, the reaction is favorable; and, at equilibrium, the product-to-reactant ratio is greater than one.

✓ When $K_{eq} < 1$, the reaction is unfavorable; and, at equilibrium, the product-to-reactant ratio is less than one.

Concept checklist

| Example 8.5 | Calculating an equilibrium constant |

For the following reaction at 25°C,

$$CO(g) + H_2O(g) \rightleftharpoons H_2(g) + CO_2(g),$$

the equilibrium concentrations are

$$[CO] = [H_2O] = 0.09 \text{ mol/L}$$
$$[H_2] = [CO_2] = 0.91 \text{ mol/L}$$

Calculate the equilibrium constant.

Solution
The equilibrium constant expression is

$$K_{eq} = \frac{[CO_2][H_2]}{[CO][H_2O]}$$

Substitute the equilibrium concentrations:

$$K_{eq} = \frac{[0.91][0.91]}{[0.09][0.09]} = 1 \times 10^2$$

Problem 8.5 In the reaction $2\,HI(g) \rightleftharpoons H_2(g) + I_2(g)$, the equilibrium concentrations at 25°C are $[H_2] = [I_2] = 0.0011\,M$ and $[HI] = 0.0033\,M$. Calculate the value of the equilibrium constant.

✓ If we know the equilibrium concentrations of reactants and products of a reaction, we can calculate the value of the equilibrium constant.

Concept check

The equilibrium constant can be used to calculate the concentrations of components of a reaction at equilibrium if we know the starting concentrations of reactants. The general method for this calculation is shown in Box 8.2 on the following page.

8.7 BIOCHEMICAL REACTIONS ARE CONNECTED IN SEQUENCES

Cellular metabolism consists of sequences of reactions. Many of the individual reactions have unfavorable equilibrium constants. However, each sequence as a whole produces adequate quantities of the necessary end product by incorporating highly favorable reactions as intermediate steps. This idea is best illustrated by an example.

8.2 CHEMISTRY IN DEPTH

The Quantitative Description of Chemical Equilibrium

The distribution of mass among products and reactants of a reaction can be calculated by using the reaction's equilibrium constant. To illustrate the procedure, we will examine the reaction

$$CO(g) + H_2O(g) \rightleftharpoons H_2(g) + CO_2(g)$$

The equilibrium constant for this reaction is 1.00×10^2 at 25°C. Our approach will be to calculate how many moles of H_2 and CO_2 will be left at equilibrium if the reaction begins with 1.00 mol each of CO and H_2O in a 1.00-L container at 25°C. The stoichiometry of the reaction equation indicates that, for every molecule of CO and H_2O that disappears, one molecule each of CO_2 and H_2 will appear.

An accounting system to deduce the equilibrium concentrations for all components is constructed by defining the concentrations of all components of the equilibrium system for three phases of the reaction: at the start, in the course of the reaction when changes take place, and, finally, at the end, or at equilibrium. This accounting system takes the form of a table:

	$[CO]$	$[H_2O]$	$[H_2]$	$[CO_2]$
Start	1.00	1.00	0.00	0.00
Change	$-x$	$-x$	$+x$	$+x$
Final	$1.00 - x$	$1.00 - x$	x	x

The first line shows that, at the start of the reaction, 1 mol each of CO and H_2O are present and no products have yet formed. When the reaction has reached equilibrium, the

---

Example 8.6 Calculating the equilibrium constant for a reaction sequence

Consider a sequence of two separate reactions that have a common intermediate. Assume that the equilibrium constant for the first reaction is large and that for the second reaction is small. What will the overall equilibrium constant be for the total process?

Solution

To solve this problem, we must first construct a hypothetical two-step chemical process in which a reactant A leads to a product C in two steps connected by a common intermediate B. Each step written separately can be characterized by an equilibrium constant:

Reaction 1. $A \rightarrow B$, equilibrium constant $K_1 = \dfrac{[B]}{[A]}$

Reaction 2. $B \rightarrow C$, equilibrium constant $K_2 = \dfrac{[C]}{[B]}$

The overall reaction, $A \rightarrow C$, is simply the result of adding reactions 1 and 2. This overall reaction is characterized by an equilibrium constant that represents the overall process:

$$K_3 = \frac{[C]}{[A]}$$

The equilibrium constant for the reaction, K_3, is the product of the equilibrium constants for reactions 1 and 2:

$$K_3 = \frac{[B]}{[A]} \times \frac{[C]}{[B]} = \frac{[C]}{[A]} = K_1 \times K_2$$

Now assume a small value for K_1—say, 1×10^{-2}—and a large value for K_2—say, 1×10^4. The overall equilibrium constant would then be

$$K_3 = K_1 \times K_2 = 1 \times 10^2$$

The overall equilibrium constant is quite large, showing that an unfavorable reaction can still produce significant amounts of product when it is one of a sequence of more favorable reactions. Sequences of this kind are characteristic of metabolic processes.

stoichiometry tells us that, for every mole of CO that has reacted, 1 mol of H_2O has disappeared and 1 mol each of H_2 and CO_2 has appeared. Because the actual amounts are unknown, we can insert the value of x in the table for the amounts that have disappeared and appeared. The 1:1 ratio of moles disappearing and appearing is maintained.

The final concentrations, as they appear in the bottom line of the accounting table, are substituted into the equilibrium constant expression:

$$K_{eq} = \frac{[CO_2][H_2]}{[CO][H_2O]} = 1.00 \times 10^2$$

$$K_{eq} = \frac{[x][x]}{[1.00 - x][1.00 - x]} = \frac{[x]^2}{[1.00 - x]^2} = 1.00 \times 10^2$$

The relation is now in the form of a quadratic equation, which has a formally prescribed method of solution.

However, because the expressions on both sides of the equal sign are perfect squares, their square roots can be taken directly and the simplified expression then solved for x:

$$\frac{[x]}{[1.00 - x]} = 10.0$$

$$[x] = 10.0 - 10.0[x]$$

$$11.0[x] = 10.0$$

$$[x] = 0.910$$

Therefore, at equilibrium,

$$[CO] = [H_2O] = (1.00 \text{ mol/L} - 0.910 \text{ mol/L})$$
$$= 0.090 \text{ mol/L}$$
$$[H_2] = [CO_2] = 0.910 \text{ mol/L}$$

This method will be used in Chapter 9 to quantitatively characterize solutions of acids and bases.

Problem 8.6 Consider a set of three hypothetical reactions connected by common intermediates:

$$A \longrightarrow B \quad K_1 = 1 \times 10^{-3}$$
$$B \longrightarrow C \quad K_2 = 1 \times 10^2$$
$$C \longrightarrow D \quad K_3 = 1 \times 10^3$$

Calculate the overall equilibrium constant for the following reaction:

$$A \longrightarrow D$$

✓ In a reaction sequence, the products of one reaction become reactants in a second reaction; the products of that reaction may become reactants in a third, and so forth.

✓ In such cases, the equilibrium constants can be combined into one that characterizes the overall process.

Concept checklist

8.8 LE CHATELIER'S PRINCIPLE

The equilibrium state of a liquid–vapor system is defined solely by its vapor pressure, which has a unique value at a fixed temperature (Section 6.6). Decreasing the volume of the gas phase of a liquid-water–water-vapor system (increasing its pressure) at a fixed temperature causes some of the gaseous water to condense to the liquid phase. This change reduces the temporarily increased gas pressure to its original equilibrium value. Conversely, an increase in the volume of the gas phase at a fixed temperature (decreasing its pressure) causes some vaporization of liquid to the gas phase, with the same result: the vapor pressure returns to its equilibrium value.

Just as a physical system in equilibrium can be temporarily displaced from equilibrium, a system in chemical equilibrium also can be temporarily displaced from equilibrium. However, a chemical system returns to its state of equilibrium by undergoing changes in the relative amounts of reactants and products. These amounts readjust so that the value of the equilibrium constant remains unchanged. In regard to the balanced chemical equation, there is a shift in mass from left to right or right to left, depending on the nature of the imposed change, or stress.

Qualitative predictions of how a chemical system will respond to a disturbance in its equilibrium conditions are described by **Le Chatelier's principle:**

- If a system in an equilibrium state is disturbed, the system will adjust so as to counteract the disturbance and restore the system to equilibrium.

This principle applies to both chemical and physical systems. For example, consider the following reaction to be at equilibrium:

$$N_2(g) + 3 H_2(g) \rightleftharpoons 2 NH_3(g)$$

How would changes in concentration of any of the components of the system affect the distribution of all the chemical components? This question is best answered by focusing on the rates of the forward and back reactions of the dynamic equilibrium and noting that the rate of a chemical reaction increases with increases in the concentrations of the reactants.

Adding nitrogen or hydrogen or both to the reaction mixture at equilibrium will increase the rate of the forward reaction, resulting in an increase in the concentration of ammonia and a new equilibrium mixture of N_2, H_2, and NH_3. In other words, adding mass to the left-hand side of the reaction equation causes the equilibrium to shift to the right. Adding ammonia to an equilibrium mixture will increase the rate of the back reaction, resulting in an increase in the concentration of hydrogen and nitrogen and a new equilibrium mixture of H_2, N_2, and NH_3. As was the case with the forward reaction, adding mass to the right-hand side of the reaction equation causes the equilibrium to shift to the left. Le Chatelier's principle is important in many biochemical processes.

Example 8.7   Using Le Chatelier's principle

Suppose some $N_2O_4(g)$ were added to the following equilibrium system: $N_2O_4(g) \rightleftharpoons 2 NO_2(g)$. What would the effect on the equilibrium mixture be?

Solution

The stress on the disturbed equilibrium mixture would be relieved and equilibrium would be reestablished by a reduction in the N_2O_4 concentration, which would lead to an increase in the NO_2 concentration. The reaction would shift to the right. However, note that both the N_2O_4 and the NO_2 concentrations would now be greater than in the preceding equilibrium state.

Problem 8.7   Suppose some $H_2(g)$ were added to the following reaction: $2 HI(g) \rightleftharpoons H_2(g) + I_2(g)$. What would the effect on the equilibrium mixture be?

Many reactions yield not only new substances, but heat as well. In these cases, the equilibrium state of the chemical system may also be displaced by the addition or removal of heat. The equilibrium will shift to either the product or the reactant side of the reaction, depending on the direction in which heat is evolved or absorbed. Thus, changes in temperature cause quantitative changes in the equilibrium constant.

Chemical equations can be written to show the exothermic or endothermic nature of a reaction. In exothermic reactions, heat is shown as a product; in endothermic reactions, it is listed as a reactant:

Exothermic: $A + B \rightleftharpoons C + D + \text{heat}$ $\Delta H_{rxn} = (-)$
Endothermic: $A + B + \text{heat} \rightleftharpoons C + D$ $\Delta H_{rxn} = (+)$

Example 8.8   Predicting the effect of heat addition or removal on a reaction

The reaction of N_2 with H_2 not only produces NH_3, but heat as well:

$$N_2(g) + 3\,H_2(g) \rightleftharpoons 2\,NH_3(g) + 98\text{kJ}$$

What changes would the system undergo if (a) heat were added to it and (b) heat were removed?

Solution

The balanced equation indicates that the reaction evolves heat—that is, it is exothermic—and the heat of reaction is designated as a negative quantity, $\Delta H_{rxn} = -98$ kJ (the system has lost heat).

(a) Because this equilibrium system evolves heat on reaction, adding heat to it is equivalent to increasing one of the components on the right-hand, or product, side. This stress is relieved by an equilibrium shift to the left, resulting in the absorption of heat with more reactants being formed.

(b) Removal of heat is equivalent to reducing one of the components on the left-hand, or reactant, side. This stress is relieved by an equilibrium shift to the right, resulting in the evolution of heat and more ammonia formed.

Problem 8.8   Consider the following equilibrium system:

$$CaCO_3(s) + 158 \text{ kJ} \rightleftharpoons CaO(s) + CO_2(g)$$

In which direction will the equilibrium shift if heat is (a) added or (b) removed from the system?

8.3   CHEMISTRY AROUND US

Nitrogen Fixation: The Haber Process

Living organisms require nitrogen, but most of them cannot take it directly out of the air. Instead, metabolizable nitrogen is acquired in the form of nitrates, ammonia, or ammonia derivatives. Before 1915, the guano (shorebird droppings) beds of coastal Chile and Peru were the principal source of nitrogen-containing compounds used commercially for both life and death—fertilizer and explosives. During World War I, the Allies placed an embargo on these guano fields to keep Germany from acquiring the guano for the manufacture of explosives. Pressed for survival, the Germans developed a chemical process for fixing atmospheric nitrogen. The key to this synthetic process was a catalyst invented by Fritz Haber, who received a Nobel Prize in chemistry in 1918 for his discovery.

It is difficult to overemphasize the effect of this process on modern civilization. Nitrogen fixation by the Haber process is the foundation of both modern agriculture and the commercial synthesis of a wide variety of chemicals and chemical products.

The process uses the direct reaction between hydrogen and nitrogen gases to produce ammonia. This reaction is exothermic; so the percentage of ammonia in the equilibrium mixture decreases as the temperature is raised. However, the proportion of ammonia in the equilibrium mixture increases as the total pressure is increased. The way to get a good yield would therefore seem to be to raise the pressure and operate at low temperatures. Unfortunately, the rate of the reaction at low temperatures is extremely slow, and one must wait for commercially unfeasible lengths of time for a significant yield. Haber's solution to this problem was to invent a catalyst that significantly reduced the activation energy of the reaction and therefore reduced the time required to reach equilibrium. When the catalyst was employed at high pressures and moderate temperatures, it quickly led to a very favorable yield of ammonia.

Modern agriculture has developed in large measure because of the invention of the Haber process for fixing nitrogen. (Richard R. Hansen/Photo Researchers.)

Box 8.3 on the previous page describes the industrial synthesis of ammonia by the Haber process. This process is an excellent example of how chemists have learned to use concentrations, temperature, and catalysts to obtain economically useful yields of chemical products.

In the next chapter, we will apply the general principles of chemical equilibrium to systems in which substances called acids and bases react with water to form new ionic species. The normal physiological function of cells and cellular systems depends strongly on these ionic equilibria.

Summary

Chemical Reaction Rates

• The rate of a chemical reaction is measured in moles of reactant disappearing or moles of product appearing per unit time.

• Rates of reaction increase with increase in concentration of reactants and increase in temperature.

Molecular Collisions and Chemical Reactions

• Reactive collisions between properly oriented molecules result in the formation of an activated complex.

• The energy required to form an activated complex is called the activation energy.

• The greater the activation energy, the slower the reaction.

• In an exothermic reaction, heat is lost, and the energy of the products is less than that of the reactants.

• In an endothermic reaction, the reaction absorbs heat and the energy of the products is greater than that of the reactants.

Effect of Catalysts on Reaction Rate

• Catalysts are agents that increase the rate of a reaction but emerge unchanged after the reaction has ended.

• A catalyst lowers the activation energy of a reaction by providing a different pathway leading to products.

• A catalyst provides a different pathway by becoming an active participant in the chemical process.

• A catalyst does not change the final result, only the rate of reaction.

Biological Catalysts

• Chemical processes within living cells are catalyzed by enzymes that form an activated complex with cellular reactants.

• The formation of the enzyme–substrate complex results in a reaction with a low overall activation energy and a faster reaction rate.

Chemical Equilibrium

• Chemical equilibrium is reached when the concentrations of reacting substances no longer change.

• The equilibrium constant is defined by the equilibrium concentrations of reactants and products.

• The equilibrium constant allows an estimation of the amount of products relative to reactants at equilibrium.

Le Chatelier's Principle

• In response to displacement from chemical equilibrium, a chemical system will reestablish its equilibrium through compensating changes in the relative amounts of reactants and products.

Key Words

activated complex, p. 204
activation energy, p. 204
catalyst, p. 208
chemical equilibrium, p. 211
chemical kinetics, p. 203
complementarity, p. 209

endothermic reaction, p. 206
enzyme, p. 209
equilibrium constant, p. 213
exothermic reaction, p. 206
heat of reaction, p. 206
product, p. 203

reactant, p. 203
reaction rate, p. 203
substrate, p. 209
transition state, p. 204

Exercises

Reaction Rates

8.1　In the reaction $3\,O_2(g) \rightarrow 2\,O_3(g)$ (ozone), the rate of appearance of ozone was 1.50 mol/min. What was the rate of disappearance of oxygen?

8.2　In the reaction $3\,H_2(g) + N_2(g) \rightarrow 2\,NH_3(g)$, the rate of disappearance of hydrogen was 0.60 mol/min. What was the rate of disappearance of nitrogen?

8.3　In the following reaction of nitrogen and oxygen,

$$N_2(g) + O_2(g) \rightleftharpoons 2\,NO(g)$$

why does the reaction rate double when the temperature is raised only 10 kelvins.

8.4　In the following reaction of nitrogen and oxygen,

$$N_2(g) + O_2(g) \rightleftharpoons 2\,NO(g)$$

why does the reaction rate increase when the concentration of nitrogen is increased.

8.5   The activation energies for each of two reactions were found to be (a) 24 kJ and (b) 53 kJ. If the temperatures of both reactions are identical, which has the greater rate of reaction?

8.6   Hydrogen peroxide, H_2O_2, is unstable and slowly decomposes over time to form water and gaseous oxygen. However, in the presence of the enzyme catalase, its rate of decomposition was increased by a factor of more than 1×10^6, with no change in the H_2O_2 concentration. Explain.

8.7   The heat of a reaction, ΔH_{rxn}, is -21 kJ and the activation energy of the forward reaction, $E_{forward}$, is $+37$ kJ. Calculate the activation energy of the reverse reaction, E_{back}.

8.8   The heat of a reaction, ΔH_{rxn}, is $+12$ kJ, and the activation energy of the forward reaction, $E_{forward}$, is -46 kJ. Calculate the activation energy of the reverse reaction, E_{back}.

Chemical Equilibrium

8.9   Calcium carbonate is heated in a crucible open to the atmosphere. The reaction that takes place is a decomposition:

$$CaCO_3(s) \rightleftharpoons CaO(s) + CO_2(g)$$

Is the reaction reversible as performed? Explain.

8.10   The decomposition reaction

$$2\ HCl(aq) + CaCO_3(s) \longrightarrow H_2O(l) + CO_2(g) + CaCl_2(aq)$$

is used as a field test for the detection of limestone. When the reaction is conducted in the field in the open air, is it reversible? Explain.

8.11   For the reaction $CaCO_3(s) \rightleftharpoons CaO(s) + CO_2(g)$, which is the forward and which is the back reaction?

8.12   For the reaction $CaO(s) + CO_2(g) \rightleftharpoons CaCO_3(s)$, which is the forward reaction and which is the back reaction?

8.13   Write the equilibrium-constant expression for the reaction in Exercise 8.11.

8.14   Write the equilibrium-constant expression for the reaction in Exercise 8.12.

8.15   In the reaction $H_2(g) + I_2(g) \rightleftharpoons 2\ HI(g)$, the equilibrium concentrations were found to be $[H_2] = [I_2] = 0.86\ M$ and $[HI] = 0.27\ M$. Calculate the value of the equilibrium constant.

8.16   In the reaction $NH_4Cl(s) \rightleftharpoons NH_3(g) + HCl(g)$, the equilibrium concentrations were found to be $[NH_3] = [HCl] = 3.71 \times 10^{-3}\ M$. Calculate the value of the equilibrium constant.

Le Chatelier's Principle

8.17   For the reaction $N_2(g) + 3\ H_2(g) \rightleftharpoons 2\ NH_3(g)$, in what direction will the equilibrium shift if some NH_3 is removed from the equilibrium mixture?

8.18   For the reaction $N_2(g) + 3\ H_2(g) \rightleftharpoons 2\ NH_3(g) + 92$ kJ, what will the effect of increasing the temperature be on the extent of the reaction?

8.19   In each of the following aqueous equilibria, a visual change can be observed when there is a shift in the equilibrium.

Predict the visual change that occurs when the indicated stress is applied.

(a) Cool the following reaction:

$$Heat + Co^{2+}(aq)(pink) + 4\ Cl^-(aq)(colorless) \rightleftharpoons$$
$$CoCl_4{}^{2-}(aq)(blue)$$

(b) Add Cl^- to the following reaction:

$$Heat + Co^{2+}(aq)(pink) + 4\ Cl^-(aq)(colorless) \rightleftharpoons$$
$$CoCl_4{}^{2-}(aq)(blue)$$

8.20   In each of the following aqueous equilibria, a visual change can be observed when there is a shift in the equilibrium. Predict the visual change that occurs when the indicated stress is applied.

(a) Add Fe^{3+} (aq) to the following reaction:

$$Fe^{3+}(aq)(brown) + 6\ SCN^-(aq)(colorless) \rightleftharpoons$$
$$Fe(SCN)_6{}^{3-}(aq)(red)$$

(b) Heat the following reaction:

$$Pb^{2+}(aq)(colorless) + 2\ Cl^-(aq)(colorless) \rightleftharpoons$$
$$PbCl_2(white\ solid) + heat$$

Unclassified Exercises

8.21   What is the meaning of the square brackets around the chemical components in equilibrium-constant expressions?

8.22   Write the equilibrium-constant expressions for each of the following reactions:

(a) $C(s) + O_2(g) \rightleftharpoons CO_2(g)$
(b) $COCl_2(g) \rightleftharpoons CO(g) + Cl_2(g)$

8.23   Write the equilibrium-constant expressions for each of the following reactions:

(a) $CH_4(g) + H_2O(g) \rightleftharpoons CO(g) + 3\ H_2(g)$
(b) $C_8H_{16}(l) + 12\ O_2(g) \rightleftharpoons 8\ CO_2(g) + 8\ H_2O(g)$

8.24   Write the equilibrium-constant expressions for each of the following reactions:

(a) $S_2(g) + C(s) \rightleftharpoons CS_2(g)$
(b) $ZnO(s) + CO(g) \rightleftharpoons Zn(s) + CO_2(g)$

8.25   Write the equilibrium-constant expressions for each of the following reactions:

(a) $NH_2COONH_4(s) \rightleftharpoons 2\ NH_3(g) + CO_2(g)$
(b) $2\ Cu(s) + 0.5\ O_2(g) \rightleftharpoons Cu_2O(s)$

8.26   Given the equilibrium reaction,

$$131\ kJ + C(s) + H_2O(g) \rightleftharpoons CO(g) + H_2(g),$$

complete the following table:

Change	Effect on equilibrium
(a) Increase in temperature	_____
(b) Increase in concentration of $H_2O(g)$	_____
(c) Increase in concentration of products	_____
(d) Decrease in temperature	_____
(e) Adding a catalyst	_____

8.27 Write the balanced equation corresponding to each of the following equilibrium constants:

(a) $K_{eq} = \dfrac{[CO][Cl_2]}{[COCl_2]}$ (b) $K_{eq} = \dfrac{[CO][H_2]^3}{[CH_4][H_2O]}$

8.28 Write the balanced equation corresponding to each of the following equilibrium constants:

(a) $K_{eq} = \dfrac{[PCl_5]}{[Cl_2][PCl_3]}$ (b) $K_{eq} = \dfrac{[O_3]^2}{[O_2]^3}$

8.29 Write the equilibrium-constant expression for each of the following reactions:

(a) $N_2O_5(g) \rightleftharpoons NO_2(g) + NO_3(g)$
(b) $4\,NH_3(g) + 3\,O_2(g) \rightleftharpoons 2\,N_2(g) + 6\,H_2O(g)$

8.30 Write the equilibrium-constant expression for each of the following reactions:

(a) $2\,NO(g) + 2\,H_2(g) \rightleftharpoons N_2(g) + 2\,H_2O(g)$
(b) $2\,Mg(s) + O_2(g) \rightleftharpoons 2\,MgO(s)$

8.31 The following equilibrium constants refer to the reaction

$$A(g) \rightleftharpoons B(g)$$

Predict the ratios of $[A]/[B]$ at equilibrium in each of the following cases: (a) $K_{eq} = 1 \times 10^{-3}$; (b) $K_{eq} = 1 \times 10^2$; (c) $K_{eq} = 0.14$; (d) $K_{eq} = 0.0832$.

8.32 Write the equilibrium constants for each of the following reactions:

(a) $N_2O_4(g) \rightleftharpoons 2\,NO_2(g)$
(b) $2\,H_2(g) + O_2(g) \rightleftharpoons 2\,H_2O(g)$

8.33 When ammonium thiocyanate (NH_4CNS) dissolves in water, the solution becomes cold. If an aqueous solution of NH_4CNS together with additional solid NH_4CNS were heated, would some of the solid dissolve or would more solid precipitate?

8.34 When lithium iodide (LiI) is dissolved in water, the solution becomes hot. If an aqueous solution of LiI together with additional solid LiI were heated, would some of the solid dissolve or would more solid precipitate?

8.35 The ΔH_{rxn} of a reaction is $+15$ kJ, and the activation energy of the back reaction is $E_{back} = 57$ kJ. What is the activation energy of the forward reaction, $E_{forward}$?

Expand Your Knowledge

Note: The icons ▲▲▲ denote exercises based on material in boxes.

8.36 One of the ways in which ozone is destroyed in the upper atmosphere is described by the following sequence of reactions:

$$O_3(g) + NO(g) \longrightarrow O_2(g) + NO_2(g)$$
$$\underline{NO_2(g) + O(g) \longrightarrow NO(g) + O_2(g)}$$
Sum: $O_3(g) + O(g) \longrightarrow 2\,O_2(g)$

Which chemical component is a catalyst and which is a reaction intermediate?

8.37 The heat of reaction of a chemical process was found to be $+16$ kJ/mol. Which energy of activation was larger, that of the forward reaction or that of the back reaction?

8.38 Magnesium hydroxide, $Mg(OH)_2$, is slightly soluble in water. The solubility can be described by the following equilibrium:

$$Mg(OH)_2(s) \rightleftharpoons Mg^{2+}(aq) + 2\,OH^-(aq)$$

What will happen to the equilibrium if hydrogen ion, H^+, is added to the solution?

8.39 State Le Chatelier's principle.

8.40 Many substances crystallize from aqueous solution in the form of a hydrate. The formula unit of a hydrate indicates the number of water molecules associated with it. Write the equilibrium constant for the following reaction:

$$CuSO_4 \cdot 5\,H_2O(s) \rightleftharpoons CuSO_4(s) + 5\,H_2O(g)$$

Deep blue solid copper(II) sulfate pentahydrate is heated, water vapor is driven off, and white solid copper(II) sulfate is formed.

8.41 Suggest an experimental method to measure the rate of reaction between hydrogen and nitrogen to form ammonia as in the following reaction:

$$N_2(g) + 3\,H_2(g) \rightleftharpoons 2\,NH_3(g)$$

8.42 True or false: A catalyst for a forward reaction can never be a catalyst for the back reaction. Explain your answer.

8.43 The human pulse rate at $98.6°F$ is 75 beats per minute. If the energy of activation for the heartbeat is 10 kcal, what will the pulse rate be at $72°F$? (Hint: Consult Figure 8.7.)

8.44 Henry is convinced that the reaction that he is studying is catalyzed by the surface of the reactions' container. Propose an experiment to investigate his hypothesis.

8.45 Ilean thinks that Henry's experiment (Exercise 8.44) may be catalyzed by light. She proposes an experiment to test her hypothesis. Describe her experiment.

8.46 Hydrogen is produced commercially by the "water reaction":

$$CO(g) + H_2O(g) \rightleftharpoons H_2(g) + CO_2(g)$$

Decide how the equilibrium will shift in each of the following cases. (a) Gaseous carbon dioxide is removed. (b) Water vapor is added. (c) The pressure is increased by adding helium gas.

8.47 In the reaction

$$2\,SO_3(g) + heat \rightleftharpoons 2\,SO_2(g) + O_2(g),$$

what will happen to the number of moles of SO_3 in equilibrium with SO_2 and O_2 in each of the following cases? (a) Oxygen is added. (b) The pressure is increased by decreasing the volume. (c) The overall pressure is increased by adding helium. (d) The temperature is decreased. (e) A catalyst is added. (f) Gaseous sulfur dioxide is removed.

8.48 Harry thinks that reactions with large equilibrium constants are very fast. Ellen thinks that he's wrong. What do you think? Explain your answer.

8.49 Under what circumstances can you calculate an overall equilibrium constant for a sequence of reactions whose individual equilibrium constants are known?

8.50 Reactions 1 and 2 with E_a values of 4 and 10 kcal/mol, respectively, proceed at the same reaction rate at 25°C. Which reaction proceeds faster at 100°C?

8.51 The activation energy, E_a, of a catalyzed reaction = 4 kcal/mol and $\Delta H_{rxn} = -18$ kcal/mol. Is the E_a of the uncatalyzed reaction larger, smaller, or the same as that of the catalyzed reaction? Is the ΔH of the uncatalyzed reaction larger, smaller, or the same as that of the catalyzed reaction? Explain.

8.52 The activation energy, E_a, of a catalyzed reaction = 4 kcal/mol and $\Delta H_{rxn} = 18$ kcal/mol. Is the E_a of the uncatalyzed reaction larger, smaller, or the same as that of the catalyzed reaction? Is the ΔH of the uncatalyzed reaction larger, smaller, or the same as that of the catalyzed reaction? Explain.

8.53 Phosgene, a toxic gas, decomposes at 800°C according to

$$COCl_2(g) \rightleftharpoons CO(g) + Cl_2(g)$$

A sample of $COCl_2(g)$ in a reaction vessel at an initial concentration of 0.500 M is heated to 800°C. The equilibrium concentration of $CO(g)$ was found to be $[CO(g)] = 0.046$ M. Calculate the equilibrium constant for the reaction at 800°C (see Box 8.2).

8.54 The equilibrium constant for the reaction

$$ICl(g) \rightleftharpoons I_2(g) + Cl_2(g)$$

is $K = 0.11$. Calculate the equilibrium concentrations of $ICl(g)$, $I_2(g)$, and $Cl_2(g)$ when 0.433 mol/L of $I_2(g)$ and 0.22 mol/L of $Cl_2(g)$ are mixed in a 1.5-L vessel (see Box 8.2).

CHAPTER 9

ACIDS, BASES, AND BUFFERS

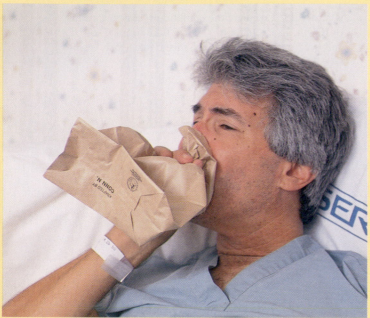

(Richard Megna/Fundamental Photographs.)

Chemistry in Your Future

Mr. Keenan, who was admitted to the medical-surgical ward at 4 this afternoon, is nervous about tomorrow's operation. His anxiety increases as the evening wears on, and he tries to breathe deeply to calm his rising fears. Instead, he begins to breathe more and more rapidly, until he cannot catch his breath. His room partner calls you (the night nurse), and you produce a paper bag and place it over Mr. Keenan's mouth and nose. His breathing becomes less labored and finally returns to normal. You give him a tranquilizer, and he rests comfortably. Apparently, you are familiar with the material presented in Chapter 9.

For more information on this topic and others in this chapter, go to www.whfreeman.com/bleiodian2e

Learning Objectives

- Describe the central role of water in acid–base chemistry.
- Explain the difference between strong and weak acids and strong and weak bases.
- Use the ion-product of water to calculate hydronium ion and hydroxide ion concentrations.
- Define pH and describe its use as a measure of acidity.
- Describe the behavior of weak acids and bases in terms of chemical equilibria.
- Use the Brønsted–Lowry theory to explain the properties of salts of weak acids and bases and of buffers.
- Describe how the concentrations of acids and bases are determined by titration.

In this chapter, we focus on an area of chemical equilibrium that is central to the life of the cell: the equilibria established when water reacts with two large groups of substances—acids and bases.

Foods, beverages, and household cleaning aids are familiar substances composed of acids and bases or containing them. Acids impart the typical tang to carbonated drinks and the tartness of citrus fruits, and bases are responsible for the bitterness of, for example, soapy water. When we eat the "wrong" foods and complain of heartburn (in reality, excess acid produced in the stomach), relief is available in the form of antacids (compounds that react with the excess acid and neutralize its effects). The health of cells critically depends on a balance between the acids produced in metabolism and the body's ability to neutralize them.

Some of the properties of acids and bases were presented in earlier chapters (Sections 2.6 and 4.6). A brief summary of some other properties of acids and bases is given in Table 9.1. One of the more important chemical properties listed is neutralization, the reaction of acids with bases to form salts, which is considered in detail in Section 9.10.

A property very useful to chemists is that acidic and basic solutions can change the colors of various dyes. Litmus is a naturally occurring dye that turns red in acidic solutions and blue in basic solutions. Figure 9.1 on the next page shows how paper strips impregnated with litmus can be used to characterize the acidity or basicity of a solution. Red litmus paper turns blue in contact with basic solution, and blue litmus paper turns red in contact with acidic solutions.

Although we encountered many of the chemical properties of acids and bases earlier, we will now enhance our practical understanding of their effects in biochemical and physiological processes by taking a quantitative look at these properties and by considering the central role of water in acid–base chemistry. Our study of acids and bases begins with a look at how water molecules react with each other to form hydronium and hydroxide ions.

9.1 WATER REACTS WITH WATER

Recall from Section 6.3 that, in liquid water, water molecules interact with each other through hydrogen bonding. The δ^+ and δ^- in the drawing in the margin indicate partial electric charges that result from the strong electronegativity of the oxygen atom. (Only ions have full electric charges.) These partial charges and the molecule's bent geometry give rise to the dipole moment of the water molecule. The hydrogen bond, represented by a dashed line, is a weak secondary force and is somewhat longer than the covalent bonds of the water molecule. A hydrogen bond in water is about 1.8 Å in length; the covalent bond in water is about 1.0 Å long (see Table 1.3).

Part of the motion of atoms within molecules consists of vibrations, or back-and-forth movements, that alternately stretch and shorten the chemical

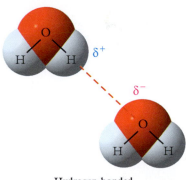

Hydrogen-bonded
water molecules

TABLE 9.1	Properties of Acids and Bases

Acids	Bases
When dissolved in water, produce hydrogen ions	When dissolved in water, produce hydroxide ions
Neutralize bases to produce water and salts	Neutralize acids to produce water and salts
Solutions turn blue litmus paper red	Solutions turn red litmus paper blue

Figure 9.1 The use of litmus paper as a quick test for acidity or basicity. Litmus turns red with acid and blue with base. In the upper row, from left to right, blue litmus paper strips have been treated with acid, base, and pure water. In the lower row, from left to right, red litmus paper has been treated with acid, base, and pure water. (Chip Clark.)

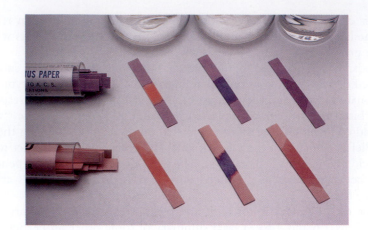

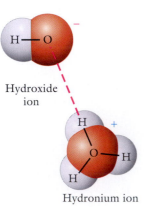

Hydroxide ion

Hydronium ion

Formation of hydronium and hydroxide ions from hydrogen-bonded water molecules

bonds. At any instant in time, the hydrogen atom participating in a hydrogen bond may find itself much closer to the oxygen atom of the adjacent water molecule than to the oxygen atom of its own water molecule. The diagram in the margin shows that the former hydrogen bond has shortened to become a covalent bond and the former covalent bond has lengthened to become a hydrogen bond. The electrons of the hydrogen atom are left behind, and the resulting proton is covalently bound to the water molecule. The process creates two new species that bear full electric charges: an H_3O^+, or **hydronium ion,** and an OH^-, or **hydroxide ion.** In subsequent discussions, it will become convenient to refer to "the protons of an acid" and to occasionally use the symbol H^+, rather than H_3O^+, in reaction equations. However, you must never forget that there can be no free protons in water, only hydronium ions.

In pure water, at any instant in time, only about one in a billion water molecules has undergone this reaction. The chemical equilibrium that is established can be represented by structural formulas:

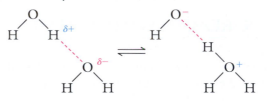

or by the more usual chemical equation:

$$H_2O(l) + H_2O(l) \longrightarrow H_3O^+(aq) + OH^-(aq)$$

This equilibrium reaction is of crucial importance to biological systems, and we need to look at it in detail.

A constant called the **ion-product of water (K_w)** is used to represent this equilibrium process. K_w is defined as the product of the molar concentrations of H_3O^+ and OH^- ions in pure water:

$$K_w = [H_3O^+][OH^-]$$

The concentrations of the H_3O^+ and OH^- ions in pure water are equal. They have been carefully measured and found to be $1.00 \times 10^{-7}\ M$ at 25°C. (Unless otherwise stated, all examples, problems, and exercises in this chapter refer to 25°C.) Therefore,

$$K_w = [H_3O^+][OH^-] = 1.00 \times 10^{-14}$$

We can obtain two pieces of information from the expression for the ion-product of water. First, very little water exists in the form of its ion products; that is, K_w is very small. Second, given that the product of hydronium ion and hydroxide ion concentrations is a constant, we know that, when the concentration of H_3O^+ in water increases, the concentration of OH^- must decrease, and vice versa. Thus three situations can exist for the relative concentrations of hydronium and hydroxide ions in water:

- In a neutral solution, $[H_3O^+]$ is equal to $[OH^-]$.
- In acidic solutions, $[H_3O^+]$ is greater than $[OH^-]$.
- In basic solutions, $[OH^-]$ is greater than $[H_3O^+]$.

These definitions of acidic and basic solutions will become clearer in the next section.

9.2 STRONG ACIDS AND STRONG BASES

Experimental results have shown that, when hydrogen chloride, HCl (g), is added to water, no HCl molecules are found in solution, only H_3O^+ ions and Cl^- ions. This observation means that each HCl molecule reacts with a water molecule to form a hydronium ion (H_3O^+) and a chloride ion. The aqueous HCl solution is called hydrochloric acid.

$$HCl(g) + H_2O(l) \longrightarrow H_3O^+(aq) + Cl^-(aq)$$

The process is called **dissociation** because it results in the formation of ions. Because HCl undergoes complete dissociation—every molecule added to water leads to the formation of a hydronium ion—it is called a **strong acid.** Hydrochloric acid is one of six strong acids that undergo complete dissociation in water (Table 9.2). It is the acid secreted by the parietal glands of your stomach as one of the first steps in the digestive process.

Similarly, when the base potassium hydroxide (KOH) is added to water, no KOH molecules are found in the solution, only hydrated K^+ and OH^- ions. The reason is that potassium hydroxide is already an ionic solid, composed of hydroxide anions and potassium cations. Its component ions simply separate when the ionic solid dissolves in water. Bases that are completely dissociated in water are called **strong bases** (see Table 9.2).

These substances are called "strong" acids and bases because there are many other acids and bases that do not fully dissociate in water. Such substances, called "weak" acids and bases, are discussed later in this chapter. Note that the terms strong and weak as used here do not denote the concentrations of acids and bases. The words strong and weak tell us only that the substance does or does not fully dissociate in solution. The listings in Table 9.2 include all the common strong acids and bases.

TABLE 9.2 | **Names and Formulas of All the Strong Acids and Bases**

Strong acids		Strong bases	
Formula	Name	Formula	Name
$HClO_4$	perchloric acid	LiOH	lithium hydroxide
HNO_3	nitric acid	NaOH	sodium hydroxide
H_2SO_4	sulfuric acid*	KOH	potassium hydroxide
HCl	hydrochloric acid	RbOH	rubidium hydroxide
HBr	hydrobromic acid	CsOH	cesium hydroxide
HI	hydroiodic acid	TlOH	thallium(1) hydroxide
		$Ca(OH)_2$	calcium hydroxide
		$Sr(OH)_2$	strontium hydroxide
		$Ba(OH)_2$	barium hydroxide

*Only the first proton in sulfuric acid is completely dissociable. The first product of dissociation, HSO_4^-, is a weak acid.

Concept checklist

✓ Strong acids are acidic compounds that undergo complete dissociation in water.

✓ Strong bases are basic compounds that undergo complete dissociation in water.

Many metals react with acids to produce gaseous hydrogen and metallic salts. Figure 9.2 shows the difference between the reactions of magnesium metal with a solution of a strong acid and with a solution of a weak acid of the same concentration. Because the concentration of hydronium ion in the strong acid solution is greater than that in the weak acid solution, the reaction of magnesium is more vigorous with a strong acid.

Because strong acids and strong bases are fully dissociated in water, we can easily calculate the hydronium or hydroxide ion concentrations in solutions of strong acids and bases as long as we know the solution concentrations. Example 9.3 will show, however, that it is also important to keep stoichiometry in mind.

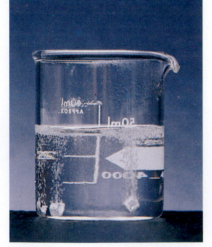

Figure 9.2 The reactions of a strong acid (top) and a weak acid (bottom) at the same solution concentrations. The reaction is between hydronium ion and magnesium. The strong acid provides much more hydronium ion than does the weak acid at the same solution concentration and therefore causes a more vigorous reaction. (W.H. Freeman photographs by Ken Karp.)

Example 9.1 Calculating ion concentrations in solutions of a strong acid

What is the concentration of hydronium ion in a 0.010 M aqueous solution of HCl? What is the total ionic concentration?

Solution

HCl, listed in Table 9.2 as a strong acid, is completely dissociated in aqueous solution; that is, 1 mol of HCl gives rise to 1 mol of H_3O^+ ion and 1 mol of Cl^- ion. In a 0.010 M solution, the concentrations of H_3O^+ and Cl^- are each 0.010 M. The total concentration of ions is therefore 0.020 M.

Problem 9.1   What is the concentration of hydronium ion in a 0.050 M aqueous solution of HCl? What is the total ionic concentration?

Example 9.2 Calculating ion concentrations in solutions of a strong base

What is the concentration of hydroxide ion in a 0.0250 M aqueous solution of NaOH? What is the total ionic concentration?

Solution

NaOH is listed in Table 9.2 as a strong base and therefore consists completely of ions in aqueous solution: 1 mol of NaOH gives rise to 1 mol of Na^+ ion and 1 mol of OH^- ion. In a 0.0250 M solution, the concentrations of OH^- and Na^+ are each 0.0250 M, and the total ionic concentration is 0.0500 M.

Problem 9.2   What is the concentration of hydroxide ion in a 0.050 M aqueous solution of NaOH? What is the total ionic concentration?

Example 9.3 Calculating ion concentrations in solutions of a strong base of a divalent cation

What is the concentration of hydroxide ion in a 0.020 M aqueous solution of $Ca(OH)_2$? What is the total ionic concentration?

Solution

Table 9.2 lists $Ca(OH)_2$ as a strong base. For this calculation, it is important to note the stoichiometry: 2 mol of OH^- ion per 1 mol of base. Thus the concentration of Ca^{2+} is 0.020 M, but the concentration of OH^- is twice that of Ca^{2+}, or 0.040 M. The total ionic concentration is 0.060 M.

Problem 9.3 What is the concentration of hydroxide ion in a 0.030 M aqueous solution of $Sr(OH)_2$? What is the total ionic concentration?

The ion-product of water at 25°C, $K_w = 1.00 \times 10^{-14}$, provides us with a method for calculating either the hydronium ion concentration of an aqueous solution, given the hydroxide ion concentration, or the hydroxide ion concentration, given the hydronium ion concentration.

Example 9.4	Using the ion-product of water to calculate the H_3O^+ concentration in a solution of known OH^- concentration

Calculate the hydronium ion concentration in an aqueous 0.00100 M NaOH solution.

Solution
Because NaOH is a strong base, the hydroxide ion concentration of a 0.00100 M solution is 0.00100 M. We can substitute the value for hydroxide ion concentration into the ion-product expression and calculate the hydronium ion concentration:

$$K_w = [H_3O^+][OH^-] = [H_3O^+] \times 0.00100 = 1.00 \times 10^{-14}$$
$$[H_3O^+] = 1.00 \times 10^{-11} \ M$$

Problem 9.4 Calculate the hydronium ion concentration in an aqueous 0.00500 M KOH solution.

Example 9.5	Using the ion-product of water to calculate the OH^- concentration in a solution of known H_3O^+ concentration

Calculate the hydroxide ion concentration in an aqueous 0.00200 M HCl solution.

Solution
Because HCl is a strong acid, the hydronium ion concentration of a 0.00200 M HCl solution is 0.00200 M. Substituting the value for hydroxide ion concentration into the ion-product expression yields

$$K_w = [H_3O^+][OH^-] = 0.00200 \times [OH^-] = 1.00 \times 10^{-14}$$
$$[OH^-] = 5.00 \times 10^{-12} \ M$$

Problem 9.5 Calculate the hydroxide ion concentration in an aqueous 0.00500 M HNO_3 solution.

9.3 A MEASURE OF ACIDITY: pH

The acidity of a solution is measured in terms of the molar concentration of hydronium ions, a concentration that typically ranges from 1.0 M to $1.0 \times 10^{-14} \ M$. Because the use of scientific notation for these often very small concentrations can become quite cumbersome, a physiologist, S. P. L. Sørenson, devised the **pH** scale defined by the following relation:

$$[H_3O^+] = 10^{-pH}$$

- This form of the definition can be used to calculate either the hydronium ion concentration, given the pH, or the pH, given the hydronium ion concentration.

Figure 9.3 A pH meter responds to the pH when a special electrode is immersed in a solution. This meter displays the measured pH digitally. The two samples are (a) orange juice and (b) lemon juice. The pH of lemon juice is lower than that of orange juice and therefore has a higher concentration of hydronium ion. (W.H. Freeman photographs by Ken Karp.)

(a)   (b)

Any number y may be expressed in the form $y = 10^x$. The number x to which 10 must be raised—that is, the exponent of 10—is called the **logarithm** or the **log** of y and is written as

$$x = \log y$$

Thus an alternative definition of pH is

$$pH = -\log [H_3O^+]$$

- This alternative form of the definition also can be used to calculate the pH, given the hydronium ion concentration, or the hydronium ion concentration, given the pH.

The pH of a solution is measured with a pH meter (Figure 9.3). The pH meter was invented in 1939 by Arnold O. Beckman. He found that a very thin glass electrode can produce an electrical voltage that is proportional to hydronium ion concentrations. By calibrating the electrode response with known hydronium ion concentrations, the pH meter can determine hydronium ion concentrations in test solutions. Since that time, glass electrodes that can measure concentrations of other ions such as sodium (Na^+) and calcium (Ca^{2+}) have been developed. The pH scale is illustrated in Figure 9.4, and the pH values of various aqueous solutions are presented in Figure 9.5.

Table 9.3 lists hydronium ion concentrations in decimal notation, scientific notation, and the corresponding values of pH. The first digit in a logarithm is a decimal place holder and is therefore insignificant. The number of significant figures in a logarithm is equal to the number of digits after the decimal point. For example, the log 7.1 contains one significant figure, the log 7.10 contains two significant figures, and so forth. Two things important

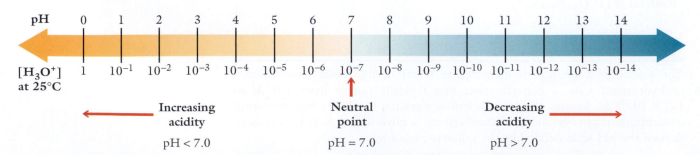

Figure 9.4 The pH scale, with pH values above and corresponding molar concentrations of hydronium ion below. Note how decreasing values of pH denote increasing acidity and how increasing values of pH denote increasing basicity.

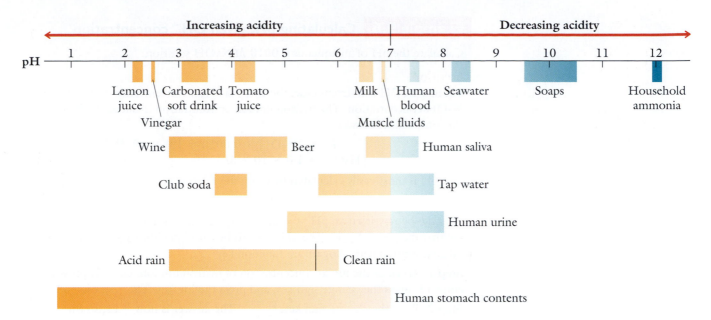

Figure 9.5 The ranges of pH found in some aqueous solutions.

to note here are that, as Figure 9.4 also shows, the pH is equal to the negative exponent of the hydronium ion concentration and that a tenfold change in hydronium ion concentration results in a change in pH of one unit. Solutions that differ in acidity by one pH unit have a tenfold difference in hydronium ion concentration.

We ended Section 9.1 by noting three possible relations between the concentrations of hydronium ions and hydroxide ions in aqueous solutions. By referring to Table 9.3, we can now describe these conditions in terms of pH:

Concept checklist

✓ In a neutral solution, pH = 7, because $[H_3O^+] = [OH^-] = 1.00 \times 10^{-7}$.

✓ In an acidic solution, pH is less than 7, because $[H_3O^+]$ is greater than $[OH^-]$.

✓ In a basic solution, pH is greater than 7, because $[H_3O^+]$ is less than $[OH^-]$.

TABLE 9.3 | **Hydronium Ion Concentrations with Corresponding pH Values**

	$[H^+]$	
Decimal notation	Scientific notation	pH
1.0	1.0×10^0	0.00
0.10	1.0×10^{-1}	1.00
0.010	1.0×10^{-2}	2.00
0.0010	1.0×10^{-3}	3.00
0.000010	1.0×10^{-5}	5.00
0.00000010	1.0×10^{-7}	7.00
0.0000000010	1.0×10^{-9}	9.00

Example 9.6 Calculating pH, given OH⁻ concentration

Calculate the pH of an aqueous 0.0010 M NaOH solution.

Solution

Because NaOH is a strong base, the hydroxide ion concentration is equal to the NaOH concentration. The hydronium ion concentration is calculated by using the ion-product of water:

$$K_w = [H_3O^+] \times 0.0010 = 1.0 \times 10^{-14}$$
$$[H_3O^+] = 1.0 \times 10^{-11} M$$

The pH is most easily calculated by using the relation

$$[H_3O^+] = 10^{-pH}$$

This can be rewritten as: pH = $-\log[H_3O^+]$. By consulting Table 9.3, we can see that the pH = 11.00. The pH can also be calculated directly with a hand calculator fitted with a logarithm function key:

Step 1. To enter the molar concentration of hydronium ion, enter 1, press EE, enter 11, press the change sign key, and press the log key. The result is −11.

Step 2. Press the change sign key, $\boxed{+/-}$. The answer is now +11, so the pH = 11.00.

Problem 9.6 Calculate the hydronium ion concentration and the pH of an aqueous 0.010 M KOH solution.

If the concentration of HCl in a solution is a value other than 0.0010 M, say, 0.0016, the $[H_3O^+]$ would be expressed as 1.6×10^{-3} M, and the pH could not be determined merely by inspection.

Example 9.7 The general case of calculating pH from hydronium ion concentration

Calculate the pH of an HCl solution whose concentration is 0.0024 M.

Solution

To solve this problem, we must use the alternative definition of pH:

$$pH = -\log [H_3O^+]$$

The H_3O^+ ion concentration in scientific notation equivalent to 0.0024 M HCl is 2.4×10^{-3} M. Substituting that value into the definition of pH gives

$$pH = -\log [2.4 \times 10^{-3}]$$

The determination of pH is now a matter of obtaining the log of 2.4×10^{-3}. In this case, we must obtain the logarithm by using a hand calculator fitted with a logarithm function key. The pH = 2.62.

Problem 9.7 Calculate the pH of an aqueous 0.0037 M HCl solution.

Chemists have extended the use of the "p" notation (as used in pH) to simplify other constants that are expressed in scientific notation. Among them are hydroxide ion concentration, expressed as **pOH,** and a variety of equilibrium constants, expressed as **pK** with an identifying subscript. (The pK notation is used for acids and bases, as we will see at the end of Section 9.4.)

We can establish a useful relation between pH and pOH. Taking the negative logarithms of both sides of the equation for the ion-product of water, $K_w = [H_3O^+][OH^-]$, we get

$$-\log K_w = -\log [H_3O^+] + (-\log [OH^-])$$

Using the "p" notation gives

$$pK_w = pH + pOH$$

Substituting $pK_w = 14$ gives

$$pH + pOH = 14$$

If the pH is known, the pOH can be calculated directly by difference, and vice versa.

Example 9.8   Calculating pH from pOH

What is the pH of an aqueous 0.0010 M NaOH solution?

Solution
NaOH is a strong base; so the concentration of OH^- ion is 0.0010 M, or 1.0×10^{-3} M.

$$pOH = -\log [OH^-] = -\log (1.0 \times 10^{-3}) = 3.00$$
$$pH = 14.00 - pOH = 14.00 - 3.00 = 11.00$$

Problem 9.8   What is the pH of a 0.00010 M aqueous solution of TlOH?

Example 9.9   Calculating pOH from pH

What is the pOH of an aqueous 0.00010 M HCl solution?

Solution
HCl is a strong acid; so the concentration of H_3O^+ ion is 0.00010 M, or 1.0×10^{-4} M.

$$pH = -\log [H_3O^+] = -\log (1.0 \times 10^{-4}) = 4.00$$
$$pOH = 14.00 - pH = 14.00 - 4.00 = 10.00$$

Problem 9.9   What is the pOH of an aqueous 0.0020 M HNO$_3$ solution?

As we have seen, for a strong acid or base, the pH of an aqueous solution is easily calculated if we know the concentration of acid or base in the solution. However, very often the pH itself is experimentally determined in the laboratory. From the measured pH, the hydronium ion concentration and thus the concentration of the strong acid or base can be calculated. This calculation uses the definition

$$[H_3O^+] = 10^{-pH}$$

and requires determination of the **antilogarithm**, or **antilog**, of the measured pH value. The antilog can be found by using a calculator, as described in Example 9.10.

Example 9.10   Calculating the concentration of a strong acid solution from the measured pH

The pH of an aqueous HCl solution was found to be 3.50. Calculate the molar concentration of the acid.

Solution
We can substitute 3.50 in the definition of pH:

$$[H_3O^+] = 10^{-pH} = 10^{-3.50}$$

Your hand calculator can provide the antilog in a number of different ways. Here are three of the ways to calculate the antilog of -3.50. If the first one does not work on your calculator, then try the next one.

- Enter 3.50, press the change-sign key, press the **10^x** key. The answer is 0.00032.

- Enter 10, press the **Y^x** key, enter 3.50, press the change-sign key, press the equal sign key. The answer is 0.00032.

- Enter 3.50, press the change-sign key, press the **INV** (inverse) key, then press the log key. The answer is 0.00032.

Problem 9.10   The pH of an aqueous HI solution was found to be 2.46. Calculate the molar concentration of the acid.

9.4   WEAK ACIDS AND WEAK BASES

Although thus far we have focused on strong acids and strong bases, the vast majority of acids and bases (hundreds of thousands of substances) are **weak acids** and **weak bases.** Aqueous solutions of these substances react only partly with water; that is, only a small percentage of the acid and base molecules dissociate in water to form hydronium and hydroxide ions, respectively. The acids and bases found in biological systems are principally weak acids and bases, and we will encounter many such molecules in our study of biochemistry. In fact, any acid or base that you encounter in your study of inorganic and organic chemistry and biochemistry that is not listed in Table 9.2 is weak.

The weak inorganic acids include HCN (hydrocyanic acid), HNO_2 (nitrous acid), and HF (hydrofluoric acid). All organic compounds containing the carboxyl group (—COOH) are weak acids. When the carboxyl group is ionized (—COO$^-$), it is called the carboxylate group or anion. (Organic acids have the general formula RCOOH, where R stands for H, CH_3, or some other group.)

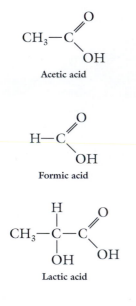

Acetic acid

Formic acid

Lactic acid

Most acid–base problems in this chapter use acetic acid as an example of a weak acid. Vinegar is an aqueous 5% acetic acid solution. The structural formula for acetic acid is shown in the margin. The hydrogen atom of the OH in the carboxyl group is the dissociable hydrogen, or proton. Other examples of organic acids are formic acid, the irritant found in ant bites, and lactic acid, an important intermediate in carbohydrate metabolism.

Ammonia (NH_3) is an example of a weak base. Many organic compounds containing nitrogen are also weak bases. These organic bases can be considered to be derivatives of ammonia, with other groups substituting for one or more of the hydrogen atoms of ammonia. Examples of organic bases are methylamine (CH_3NH_2), pyridine (C_5H_5N), and morphine ($C_{17}H_{19}O_3N$). The organic bases include a number of physiologically active compounds such as morphine.

Reaction of Weak Acids and Bases with Water

Acetic acid and other weak acids react with water, in a way similar to the way that water reacts with water (Section 9.1), to produce a hydronium ion and a carboxylate anion:

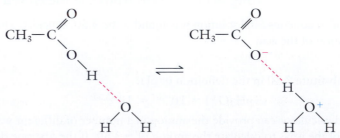

In pure water, one water molecule acts as an acid by donating a hydrogen ion, or proton, to a second water molecule, which acts as a base by accepting the proton. A weak acid simply takes the place of a water molecule, but its presence imposes a new set of equilibrium requirements on those that exist in pure water.

Acetic acid reacts with water to produce hydronium ions and acetate ions but in much smaller quantities than the amount of unreacted or undissociated acetic acid molecules, with the equilibrium far to the left.

$$CH_3-C\!\!\underset{OH}{\overset{O}{\diagup}}(aq) + H_2O(l) \rightleftharpoons CH_3-C\!\!\underset{O^-}{\overset{O}{\diagup}}(aq) + H_3O^+(aq)$$

The structural formula for acetic acid is equivalent to the molecular formulas CH_3COOH and $HC_2H_3O_2$. We will use the molecular formula CH_3COOH in reaction equations describing dissociation, because it is a simpler representation.

Ammonia (NH_3) is a weak base because it reacts with water to produce hydroxide ions and because the amount of hydroxide ion produced is much smaller than the total amount of ammonia added to water. The many nitrogen-containing organic compounds that are also weak bases react with water to produce hydroxide ions in the same way as ammonia does. The reaction is

$$NH_3(aq) + H_2O(l) \rightleftharpoons NH_4^+(aq) + OH^-(aq)$$

with the equilibrium far to the left. The basic properties of ammonia and all other bases similar to it depend on the lone pair of electrons on its nitrogen, which bonds with the proton donated by water. This property is best illustrated by using Lewis structures to show the reaction with water:

$$H-\underset{H}{\overset{H}{N}}: + H-\overset{..}{\underset{..}{O}}-H \rightleftharpoons H-\underset{H}{\overset{H}{N^+}}-H + {}^-\overset{..}{\underset{..}{O}}-H$$

The structural feature responsible for basicity—the lone pair of electrons on nitrogen—is illustrated in the margin for ammonia as well as some organic bases.

Many organic bases with similar structural characteristics, such as morphine, codeine, and other physiologically active substances used as medicines, are insoluble in water. Before they can be used medicinally, they are converted into their soluble salts by reaction with HCl.

Quantitative Aspects of Acid–Base Equilibria

To consider weak acids and bases in quantitative terms, we must consider them to be systems in chemical equilibrium, characterized by equilibrium constants. [Remember that pure solids or pure liquids do not appear in the equilibrium-constant expression (Section 8.3).] For example, the equation for the reaction of acetic acid with water is

$$CH_3COOH(aq) + H_2O(l) \rightleftharpoons H_3O^+(aq) + CH_3COO^-(aq)$$

The process is called dissociation, and the equilibrium constant expression for its dissociation is

$$K_a = \frac{[H_3O^+][CH_3COO^-]}{[CH_3COOH]}$$

This equilibrium constant is called an **acid dissociation constant, K_a** ("a" for acid), because the reaction that it describes results in the formation of a hydronium ion.

Similarly, the dissociation of a weak base that results in the production of a hydroxide ion is characterized by a **base dissociation constant, K_b** ("b" for base). Consider the reaction of ammonia with water:

$$NH_3(aq) + H_2O(l) \longrightarrow NH_4^+(aq) + OH^-(aq)$$

The dissociation constant is

$$K_b = \frac{[NH_4^+][OH^-]}{[NH_3]}$$

The values of K_a and K_b are indicative of the relative amounts of products and reactants at equilibrium: the smaller the values of the dissociation constants, the smaller the amount of product relative to reactant, or, in terms of

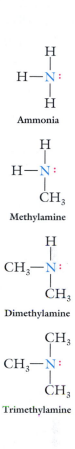

Ammonia

Methylamine

Dimethylamine

Trimethylamine

TABLE 9.4 Values of K_a and pK_a for Some Weak Acids and Values of K_b and pK_b for Some Weak Bases*

Acids				Bases			
Name	Formula	K_a	pK_a	Name	Formula	K_b	pK_b
nitrous acid	HNO_2	4.47×10^{-4}	3.35	dimethylamine	$(CH_3)_2NH$	5.81×10^{-4}	3.24
cyanic acid	HCNO	2.19×10^{-4}	3.66	methylamine	CH_3NH_2	4.59×10^{-4}	3.34
formic acid	HCOOH	1.78×10^{-4}	3.75	ammonia	NH_3	1.75×10^{-5}	4.76
hydrazoic acid	HN_3	1.91×10^{-5}	4.72	cocaine	$C_{17}H_{21}O_4N$	3.90×10^{-9}	8.41
acetic acid	CH_3COOH	1.74×10^{-5}	4.76	aniline	$C_6H_5NH_2$	4.17×10^{-10}	9.38
carbonic acid	H_2CO_3	4.45×10^{-7}	6.35	morphine	$C_{17}H_{19}O_3N$	1.41×10^{-10}	9.85

*All the weak bases listed are derivatives of ammonia.

dissociation, the smaller the degree or extent of dissociation. In other words, the value of this constant provides us with a quantitative measure of just how "weak" a weak acid or weak base is.

Concept check ✓ The smaller the dissociation constant, the weaker the acid or base.

Acid and base dissociation constants are often expressed in the "p" notation mentioned earlier. For weak acids,

$$K_a = 10^{-pK_a} \quad \text{and} \quad -\log K_a = pK_a$$

For weak bases;

$$K_b = 10^{-pK_b} \quad \text{and} \quad -\log K_b = pK_b$$

Table 9.4 lists various weak acids and bases along with their dissociation constants and pK values at 25°C. Note that, reading down each list, as the dissociation constants become smaller, the corresponding pK values become larger.

Box 9.1 shows how acid dissociation constants are used to calculate the pH of weak acids and bases.

9.5 BRØNSTED–LOWRY THEORY OF ACIDS AND BASES

At the end of Section 3.8, we described the reaction between hydrogen chloride and water that produces an H_3O^+ ion as the result of the transfer of a hydrogen ion to water, a reaction that always takes place when compounds called acids react with water. This description is a statement of the **Brønsted–Lowry theory** of acids and bases. In this theory,

- Acids are described as **proton donors.**
- Bases are described as **proton acceptors.**

The reaction of HCl and H_2O is

$$HCl(g) + H_2O(l) \longrightarrow H_3O^+(aq) + Cl^-(aq)$$

Note that both the hydronium ion and the chloride ion are hydrated. In every such reaction, the Brønsted–Lowry theory requires the presence of both a proton donor (the acid) and a proton acceptor (the base; in this case, water). Two facts are noteworthy here:

- Proton transfer requires a pair of substances—a donor and an acceptor.
- Substances other than water can serve as the proton acceptor (base).

9.1 CHEMISTRY IN DEPTH

Acid Dissociation Constants and the Calculation of pH

In the discussion of chemical equilibrium in Chapter 8, we presented a general method (Box 8.2) for determining the concentrations of components of an equilibrium mixture. The same approach is used for quantitative analysis of solutions of weak acids and bases. The objective of the analysis is to determine the hydronium ion or hydroxide ion concentrations in these solutions and then, from these values, the pH. Here we illustrate the general approach by calculating the hydronium ion concentration of a 0.00500 M solution of acetic acid.

The equation for reaction of acetic acid with water is

$$CH_3COOH(aq) + H_2O(l) \rightleftharpoons$$
$$H_3O^+(aq) + CH_3COO^-(aq)$$

Because acetic acid is a weak acid, only a small amount of the acid dissociates to form an unknown number of moles of hydronium ion and an equal amount of acetate ion.

We can use the accounting method developed in Box 8.2 to describe the equilibrium concentrations of the reaction mixture. With the symbol x representing the unknown number of moles of hydronium and acetate ions per liter formed by dissociation, the equilibrium molar concentrations of the reaction mixture are:

	CH_3COOH	H_3O^+	CH_3COO^-
Start	0.00500	0	0
Change	$-[x]$	$+[x]$	$+[x]$
Final	$0.00500 - [x]$	$[x]$	$[x]$

The correct equilibrium constant to use here is the acid dissociation constant (see Table 9.4):

$$K_a = \frac{[H_3O^+][CH_3COO^-]}{[CH_3COOH]} = 1.74 \times 10^{-5}$$

Substituting appropriate values into the dissociation-constant expression from the accounting table gives

$$1.74 \times 10^{-5} = \frac{[x][x]}{0.00500 - [x]} = \frac{[x^2]}{0.00500 - [x]}$$

This quadratic equation in one unknown can be solved by using the standard solution for the general quadratic equation $ax^2 + bx + c = 0$, where

$$x = \frac{-b \pm \sqrt{b^2 - 4ac}}{2a}$$

Note that, although two values of x are obtained by solving this equation, for real substances only the positive value is possible.

We can simplify the calculation further. Because the equilibrium constant is quite small, the value of x is so small that $0.00500 - [x]$ is approximately equal to 0.00500. By using 0.00500 in place of $0.00500 - [x]$, the arithmetic is greatly simplified.

$$1.74 \times 10^{-5} = \frac{[x^2]}{0.00500}$$

$$x^2 = 8.70 \times 10^{-8}$$

$$x = [H_3O^+] = \sqrt{8.70 \times 10^{-8}} = 2.95 \times 10^{-4} \, M$$

$$pH = -\log[H_3O^+] = 3.53$$

The precise value for the hydronium ion concentration obtained by using the quadratic equation is

$$[H_3O^+] = 2.87 \times 10^{-4}$$
$$pH = 3.54$$

The approximated value of $[H_3O^+]$ is about 3% greater than the precise value of $[H_3O^+]$, a discrepancy that is considered acceptable.

This simplified method of obtaining an approximate solution has its limitations. When the equilibrium constant is larger than 1×10^{-4}, the formal quadratic solution must be used because the error in calculating $[H_3O^+]$ becomes unacceptably large—that is, greater than 5% of the precisely calculated value.

The same quantitative analysis can also be used to calculate the pH of solutions of bases, such as ammonia, and solutions of the salts of weak acids or bases. The reaction equation that describes the equilibrium in an aqueous solution of sodium acetate is

$$CH_3COO^-(aq) + H_2O(l) \rightleftharpoons$$
$$CH_3COOH(aq) + OH^-(aq)$$

Because this reaction and those of bases such as ammonia produce hydroxide ions, the quantitative determination of pH requires the use of a base dissociation constant, K_b.

This theory is a powerful tool for understanding the acid-base equilibria for weak acids and bases. Let's revisit these equilibria in the context of the Brønsted–Lowry theory.

Conjugate Acids and Bases

As you learned in Section 9.1, when water reacts with water, one water molecule acts as an acid, donating its proton to the other water molecule, which acts as a base. The result is one molecule of protonated water and one molecule of deprotonated water. The Brønsted–Lowry theory links the two pairs, H_3O^+/H_2O and H_2O/OH^-, and describes them as **conjugate acid–base pairs:**

Conjugate
acid–base pair

$$H_2O(l) + H_2O(l) \rightleftharpoons H_3O^+(aq) + OH^-(aq)$$

Conjugate
acid–base pair

This type of reaction also takes place in other systems such as liquid ammonia and liquid hydrogen sulfide where the ionic forms are solvated by the liquid forms. For example:

$$NH_3(l) + NH_3(l) \rightleftharpoons NH_4^+ + NH_2^-$$

and

$$H_2S(l) + H_2S(l) \rightleftharpoons H_3S^+ + HS^-$$

The reactions between water and a weak acid, such as acetic acid, and between water and a weak base, such as ammonia, are described by the same scheme:

Conjugate
acid–base pair

$$CH_3COOH(aq) + H_2O(l) \rightleftharpoons H_3O^+(aq) + CH_3COO^-(aq)$$

Conjugate
acid–base pair

Conjugate
acid–base pair

$$NH_3(aq) + H_2O(l) \rightleftharpoons OH^-(aq) + NH_4^+(aq)$$

Conjugate
acid–base pair

The central idea is that, as a result of an acid donating its proton, it becomes a base; and, when a base accepts a proton, it becomes an acid. Therefore, the acetate ion is a base, as is the hydroxide ion; and the ammonium ion is an acid, as is the hydronium ion. In general, you will find that a conjugate base (a Brønsted–Lowry base) is the ion produced when its conjugate acid (a Brønsted–Lowry acid) donates a proton to water. For example,

$$HCN(aq) + H_2O(l) \rightleftharpoons CN^-(aq) + H_3O^+(aq)$$
CN^- is the conjugate base of the acid HCN

$$H_2PO_4^-(aq) + H_2O(l) \rightleftharpoons HPO_4^{2-}(aq) + H_3O^+(aq)$$
HPO_4^{2-} is the conjugate base of the acid $H_2PO_4^-$

$$NH_4^+(aq) + H_2O(l) \rightleftharpoons NH_3(aq) + H_3O^+(aq)$$
NH_3 is the conjugate base of the acid NH_4^+

When acetate ion is added to water, it disturbs the existing equilibrium by reacting with water (the acid):

$$CH_3COO^-(aq) + H_2O(l) \rightleftharpoons CH_3COOH(aq) + OH^-(aq)$$

One of the conjugate acid–base pairs is CH_3COOH/CH_3COO^- and the other is H_2O/OH^-. If you prepared an aqueous solution of sodium acetate, the result would be a solution with excess hydroxide ion and therefore a pH above 7.0. Similarly, if you prepared an aqueous solution of ammonium chloride, the result would be a solution with an excess of hydronium ion and therefore a pH below 7. The effects of salts on the pH of aqueous solutions through their hydrolysis reactions (reactions with water) are further considered in Section 9.7.

Relation Between Acid and Base Dissociation Constants

The dissociation constant describing what takes place when acetate ion is added to water,

$$CH_3COO^-(aq) + H_2O(l) \rightleftharpoons OH^-(aq) + CH_3COOH(aq),$$

is a base dissociation constant, K_b, because the product of the equilibrium is an OH^- ion, a base:

$$K_b = \frac{[CH_3COOH][OH^-]}{[CH_3COO^-]}$$

The dissociation constant describing the results of adding acetic acid to water is an acid dissociation constant, K_a, because the product of the reaction is an H_3O^+ ion:

$$K_a = \frac{[H_3O^+][CH_3COO^-]}{[CH_3COOH]}$$

When the two constants are multiplied,

$$K_a \times K_b = \frac{[H_3O^+][CH_3COO^-]}{[CH_3COOH]} \times \frac{[CH_3COOH][OH^-]}{[CH_3COO^-]},$$

and the common terms in the numerator and denominator are canceled, we get

$$K_a \times K_b = [H_3O^+][OH^-]$$

or

$$K_a \times K_b = K_w$$

Although we used specific examples in this derivation, the result can be generalized to the behavior of all conjugate acids and bases in water:

Concept checklist

✓ The acid dissociation constant of a conjugate acid can be calculated from the base dissociation constant of its conjugate base, because their product is always K_w.

✓ The base dissociation constant of a conjugate base can be calculated from the acid dissociation constant of its conjugate acid, because their product is always K_w.

Table 9.5 lists the dissociation constants and pK values for some conjugate acid–base pairs at 25°C.

TABLE 9.5 Values of K_a, pK_a, K_b, and pK_b for Selected Conjugate Acid–Base Pairs

Acids				Bases			
Name	Formula	K_a	pK_a	Name	Formula	K_b	pK_b
nitrous acid	HNO_2	4.47×10^{-4}	3.35	nitrite	NO_2^-	2.24×10^{-11}	10.65
cyanic acid	$HCNO$	2.19×10^{-4}	3.66	cyanate	CNO^-	4.57×10^{-11}	10.34
formic acid	$HCOOH$	1.78×10^{-4}	3.75	formate	$HCOO^-$	5.62×10^{-11}	10.25
acetic acid	CH_3COOH	1.74×10^{-5}	4.76	acetate	CH_3COO^-	5.75×10^{-10}	9.24
carbonic acid	H_2CO_3	4.45×10^{-7}	6.35	hydrogen carbonate	HCO_3^-	2.24×10^{-8}	7.65
dihydrogen phosphate	$H_2PO_4^-$	6.32×10^{-8}	7.20	monohydrogen phosphate	HPO_4^{2-}	1.58×10^{-7}	6.80
ammonium	NH_4^+	5.71×10^{-10}	9.24	ammonia	NH_3	1.75×10^{-5}	4.76
hydrogen carbonate	HCO_3^-	4.72×10^{-11}	10.33	carbonate	CO_3^{2-}	2.19×10^{-4}	3.67
hydrogen phosphate	HPO_4^{2-}	4.84×10^{-13}	12.32	phosphate	PO_4^{3-}	2.07×10^{-2}	1.68

Example 9.11 | **Deriving K_b from K_a for a conjugate acid–base pair**

Using the values in Table 9.5, show how the K_b value for formate is derived from the K_a value of formic acid.

Solution

Table 9.5 shows that the K_a of formic acid is 1.78×10^{-4}. By substituting the known values in the expression

$$K_a \times K_b = [H_3O^+][OH^-] = K_w,$$

we can calculate K_b:

$$1.78 \times 10^{-4} \times K_b = 1.00 \times 10^{-14}$$
$$K_b = 5.62 \times 10^{-11}$$

Problem 9.11 Show how the K_b value for cyanate is derived from the K_a value of cyanic acid.

Relation Between pK_a and pK_b

Once again, the "p" notation can be applied to the dissociation constants of conjugate acid–base pairs, with the result that

$$pK_a + pK_b = pK_w = 14$$

Thus, we can calculate the pK_a of an acid from the pK_b of its conjugate base, and vice versa. For example, given the pK_a for acetic acid of 4.76 (Table 9.4), we can calculate the pK_b—and thus the K_b—for the reaction of acetate ion with water:

$$4.76 + pK_b = 14.00$$
$$pK_b = 14.00 - 4.76 = 9.24$$

Example 9.12 | **Deriving pK_b from pK_a for a conjugate acid–base pair**

Using the values in Table 9.5, show how the pK_b value for formate is derived from the pK_a value of formic acid.

Solution

From Table 9.5, we find that the pK_a of formic acid is 3.75. Rearranging the expression $pK_a + pK_b = pK_w$ and substituting the known values gives

$$pK_b = pK_w - pK_a = 14.00 - 3.75 = 10.25$$

Problem 9.12 Show how the pK_b value for cyanate is derived from the pK_a value of cyanic acid.

We will be applying the Brønsted–Lowry theory of acids and bases in the discussion of polyprotic acids, salt hydrolysis, and buffer systems in the next three sections.

9.6 DISSOCIATION OF POLYPROTIC ACIDS

Among the strong acids listed in Table 9.2 is H_2SO_4 (sulfuric acid) and among the weak acids listed in Table 9.4 is H_2CO_3 (carbonic acid). Both of these acids are **diprotic acids,** meaning that each molecule has two dissociable protons. In Section 4.7, we used two diprotic organic acids, malic and oxaloacetic acids, to illustrate a metabolic redox reaction. They are just two of a large group of similar metabolic products. Phosphoric acid (H_3PO_4), which will be discussed in some detail here, is a **triprotic acid.** Citric acid, an important metabolic intermediate, also is a triprotic acid. All acids with two or more dissociable protons are called **polyprotic acids.** Carbonic acid, phosphoric acid, and their salts are of central importance in maintaining the proper acid–base balance of the blood and tissues of the body.

The Dissociation of Phosphoric Acid

The three sequential dissociations of phosphoric acid and the associated K_a values at 25°C are

$$H_3PO_4(aq) + H_2O(l) \rightleftharpoons H_3O^+(aq) + H_2PO_4^-(aq)$$

$$K_{a1} = \frac{[H_3O^+][H_2PO_4^-]}{[H_3PO_4]} = 5.93 \times 10^{-3}$$

$$H_2PO_4^-(aq) + H_2O(l) \rightleftharpoons H_3O^+(aq) + HPO_4^{2-}(aq)$$

$$K_{a2} = \frac{[H_3O^+][HPO_4^{2-}]}{[H_2PO_4^-]} = 6.32 \times 10^{-8}$$

$$HPO_4^{2-}(aq) + H_2O(l) \rightleftharpoons H_3O^+(aq) + PO_4^{3-}(aq)$$

$$K_{a3} = \frac{[H_3O^+][PO_4^{3-}]}{[HPO_4^{2-}]} = 4.84 \times 10^{-13}$$

The product of the first dissociation step is the dihydrogen phosphate ion, $H_2PO_4^-$; the product of the second is the monohydrogen phosphate ion, HPO_4^{2-}; and the product of the third is the phosphate ion, PO_4^{3-}.

Note that the dissociation constants become smaller for each successive dissociation. This trend is due to the removal, at each successive step, of a positive ion (a proton) from an increasingly negatively charged anion. That is, after each step, the resulting anion exerts a greater force of attraction on the departing proton. Typically, the dissociation constant for each successive step in the dissociation of a polyprotic acid is smaller than the dissociation constant for the preceding step by a factor of about 1×10^{-5}. As a consequence, the hydronium ion in a solution of a polyprotic acid is almost wholly derived from the first dissociation. A method for determining acid dissociation constants is presented in Box 9.2 on the following page.

For example, when H_3PO_4 is added to water, the product of the first dissociation, $H_2PO_4^-$, contributes almost nothing to the total H_3O^+ concentration. All H_3O^+ is derived from the dissociation of the H_3PO_4. When $H_2PO_4^-$ is added to water (in the form of a salt such as KH_2PO_4), the H_3O^+ is derived from the dissociation of the $H_2PO_4^-$, and almost none comes from the dissociation of the product, HPO_4^{2-}.

Each dissociation step has a conjugate acid–base pair: H_3PO_4 can only donate a proton and is thus an acid; PO_4^{3-} can only accept a proton and is thus a base. But both $H_2PO_4^-$ and HPO_4^{2-} can behave either as an acid or as a base: each can both accept and donate a proton. All such substances are called **amphoteric.** You can see that water is also an amphoteric substance.

The $H_2PO_4^-/HPO_4^{2-}$ pair is an important system in cells because it helps to maintain the constant pH necessary for cellular functions (Section 9.8).

The Dissociation of Carbonic Acid

The dissociation steps for carbonic acid at 25°C are

$$H_2CO_3(aq) + H_2O(l) \rightleftharpoons H_3O^+(aq) + HCO_3^-(aq)$$

$$K_{a1} = \frac{[H_3O^+][HCO_3^-]}{[H_2CO_3]} = 4.45 \times 10^{-7}$$

$$HCO_3^-(aq) + H_2O(l) \rightleftharpoons H_3O^+(aq) + CO_3^{2-}(aq)$$

$$K_{a2} = \frac{[H_3O^+][CO_3^{2-}]}{[HCO_3^-]} = 4.72 \times 10^{-11}$$

9.2 CHEMISTRY IN DEPTH

Experimental Determination of Dissociation Constants

You've probably noticed that, in the discussion of the derivation of equilibrium constants and dissociation constants, we never touched on the subject of how one measures or determines the degree or extent of the dissociation or reaction. Let's examine a method that will allow us to make that measurement. We considered the determination of osmolarity in Chapter 7, and now we will show that the measurement of osmolarity can be used to determine the degree of dissociation of weak acids or bases. In Section 7.14, we defined osmolarity as the molarity times i, a factor that specifies the total number of moles as the result of the dissociation of one mole of substance. The substances considered in Section 7.14 were ionic and completely dissociated. There is no reason that we can't deal with substances that only partly dissociate.

Let's consider a molecule of acetic acid in water. When it reacts with water, it forms two new particles—a proton and an acetate ion. We've also learned that, because acetic acid is a weak acid, only a portion of it dissociates into ions. So an aqueous solution of acetic acid consists of a small proportion of ions and a much larger proportion of undissociated acetic acid molecules. The degree of dissociation is the amount of original substance that has dissociated divided by the concentration of the original amount of substance. It can be expressed as a fraction or a percentage. For acetic acid (HAc),

$$\text{Degree of dissociation} = [H^+] / [HAc]$$

If a 1.00 M solution of acetic acid in water is 15% dissociated, 85% of the original concentration remains undissociated. In addition, 15% has dissociated into two particles whose combined concentration is now 0.300 M. Using a modified accounting table developed in Box 9.1, you can see that the i factor is 1.15.

	Acetic acid	H^+	Ac^-	Sum (i)
Start	1.00	0.00	0.00	1.00
Equilibrium	0.85	0.15	0.15	1.15

Now we'll relate the osmolarity factor to the degree of dissociation.

Let's invent a quantity that stands for the degree of dissociation, called α (alpha), which could be 0.10, 0.25, and so forth. Then we can define i as a function of α, for 1 mol of any substance that yields n mol of ions per molecule as

$$i = (1 - \alpha) + (n \times \alpha)$$

Solving for α, we get

$$\alpha = \frac{i - 1}{n - 1}$$

Once we know i, we can calculate α. The equation for osmotic pressure (Section 7.14) contains the i factor:

$$\Pi = i \times M \times R \times T$$

Now let's see how osmotic-pressure measurements can be used to determine dissociation constants.

Example: Calculate i for a 0.0100 M solution of HF for which the osmotic pressure was found to be 0.287 atm at 298 K:

$$0.287 \text{ atm} = i \times M \times R \times T$$
$$= i \times 0.0100 \text{ (mol/L)} \times$$
$$0.0821 \text{ (L} \cdot \text{atm} \cdot \text{K}^{-1} \cdot \text{mol}^{-1}) \times 298 \text{ K}$$
$$i = 1.171, \text{ or, rounded, } i = 1.17$$

Therefore,

$$\alpha = (i - 1) / (n - 1) = 0.17$$

This value of α means that $(0.010 - 0.010 \times 0.17)$ mol of HF dissociates in a 0.010 M aqueous solution and that (0.010×0.17) mol of H^+ and (0.010×0.17) mol of F^- are formed. Then we can calculate the equilibrium constant for the dissociation of HF:

$$K_a = (0.010 \times 0.17) \times$$
$$(0.010 \times 0.17)/(0.010 - 0.010 \times 0.17)$$
$$= 3.5 \times 10^{-4}$$

Note that HCO_3^- (hydrogen carbonate, also known as bicarbonate) is amphoteric and can behave as either acid or base.

The H_2CO_3/HCO_3^- pair is chiefly responsible for keeping the pH of the blood constant at pH = 7.4. The equilibria in this case are more complicated than indicated by the preceding equations because carbonic acid is also in equilibrium with H_2O and CO_2 in solution and with gaseous CO_2 in the lungs. This system is more fully discussed in Section 9.9.

Concept check ✓ The protons of a polyprotic acid dissociate sequentially. Each step of the dissociation creates a new equilibrium system that is characterized by its own dissociation constant.

9.7   SALTS AND HYDROLYSIS

A salt is the product (other than water) of the reaction between an acid and a base. The conjugate bases of weak acids such as acetate, carbonate, and phosphate are the anions of salts such as sodium acetate and potassium dihydrogen phosphate. In aqueous solution, they react with water—a reaction called **hydrolysis**—to produce hydroxide ion (OH^-) and basic solutions. These anions are referred to as basic anions. Similarly, the cation NH_4^+ is the conjugate acid of a weak base; it reacts with water, or **hydrolyzes**, to produce an acidic solution.

There are no basic cations. All Group I and II cations, such as Na^+ and Ca^{2+}, are neutral in aqueous solution. However, salts of Al^{3+}, Pb^{2+}, Sn^{2+}, and the transition metals such as Fe^{2+} and Fe^{3+} form acidic solutions. Water is covalently bound to the metal cation of the salt, and the dissociation of protons from the bound water makes the aqueous solutions of these salts acidic (Figure 9.6). Box 9.3 describes an environmental consequence of the acidity of an iron salt solution.

Figure 9.6 An acidic cation, $Fe(H_2O)_6^{2+}$. The structure of only one of the six bound water molecules is shown.

9.3   CHEMISTRY AROUND US

Acid Mine Drainage

A biologist exploring a wooded area near an abandoned coal mine observes a box turtle moving toward a small pool of water. The turtle slips into the water and, in about 10 to 20 s, it is dead. The biologist notes that the water in the pool has a decidedly red color. She measures the pH with her portable pH meter, which records a pH of 3.3. She determines the density of the water with a hydrometer and finds it to be equivalent to about 4 g of solid per 100 mL of solution, an extraordinarily high solution density. The biologist follows the runoff from the pool down to a small creek, where she finds rocks bordering the stream that are covered with a bright yellow solid (a precipitate) and a small steel bridge that is severely corroded.

What is the explanation of these phenomena? Bituminous coal, the type of coal formerly obtained from this mine, contains significant quantities of an insoluble mineral called marcasite, which is iron(II) sulfide. On exposure to air, the sulfide is oxidized to sulfate, converting the insoluble iron(II) sulfide into water-soluble iron(II) sulfate. The hydrated Fe^{2+} ion is a weak acid, as described in Section 9.7, and produces a significant hydronium ion concentration in its aqueous solution.

$$2\,FeS_2(s) + 7\,O_2(aq) + 2\,H_2O(l) \longrightarrow$$
$$2\,Fe^{2+}(aq) + 4\,SO_4^{2-}(aq) + 4\,H^+(aq)$$

The pH of the solution can be as low as 3.0, depending on the concentration of the Fe^{2+} salt. The reddish pool of water that killed the box turtle consisted of groundwater that had penetrated the old bituminous coal mine and then drained out to the surface heavily laden with the acidic iron(II) sulfate solution.

In the next reaction, ferrous iron is converted into ferric iron. The reaction rate is pH dependant, with the reaction proceeding slowly under acidic conditions (pH 2–3) and several orders of magnitude faster at pH values near 5.0.

Yellow-boy coats the rocks along a creek fed by acid mine drainage from a local abandoned coal mine. (John D. Gordon/Visuals Unlimited.)

$$4\,Fe^{2+}(aq) + O_2(aq) + 4\,H^+(aq) \longrightarrow$$
$$4\,Fe^{3+}(aq) + 2\,H_2O(l)$$

This reaction is followed by the hydrolysis of the ferric ion. The formation of ferric hydroxide precipitate (solid) is pH dependant. Solids form if the pH is above about 3.5, but below pH 3.5 little or no solids will precipitate.

$$4\,Fe^{3+}(aq) + 12\,H_2O(l) \longrightarrow 4\,Fe(OH)_3(s) + 12\,H^+(aq)$$

The ferric hydroxide precipitate is called yellow-boy. After entering the watershed, the acid mine drainage eventually found its way into a larger stream where dilution allowed the pH to rise, resulting in precipitation of the yellow-boy, which coated the rocks bordering the creek. The corrosion of the steel bridge was caused by the acid content of the incoming stream.

Rules for determining the acidity or basicity of a salt solution are:

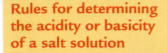

Rules for determining the acidity or basicity of a salt solution

- Salts composed of cations and anions of strong acids and bases form neutral aqueous solutions.
- Salts having anions of weak acids form basic solutions.
- Salts having cations of weak bases form acidic solutions.

When a salt is formed from a strong acid and a weak base or a strong base and a weak acid, the weak base or weak acid determines the pH of the salt solution.

Example 9.13   Predicting the acidity of salt solutions

Predict whether the following salts will produce an acidic, a basic, or a neutral aqueous solution: sodium chloride ($NaCl$), potassium acetate (CH_3COOK), ammonium chloride (NH_4Cl).

Solution

In each case, we must examine both the cation and the anion and determine whether it was formed by neutralization of a strong acid with a weak base or a strong base with a weak acid.

$NaCl$: derived from a strong acid, HCl, and a strong base, $NaOH$.
CH_3COOK: derived from a strong base, KOH, and a weak acid, CH_3COOH.
NH_4Cl: derived from a weak base, NH_3, and a strong acid, HCl.

For each salt, the acidity or basicity of the aqueous solution can be predicted from the rules given in the text.

Figure 9.7 The pH of the solutions of salts considered in Example 9.13 are measured with indicators. Ammonium chloride, on the left, is acidic. Sodium chloride, in the center, is neutral. Sodium acetate, on the right, is basic. (Chip Clark.)

Salt	Cation	Anion	Solution
$NaCl$	neutral	neutral	neutral
CH_3COOK	neutral	basic	basic
NH_4Cl	acidic	neutral	acidic

Problem 9.13   Predict whether the following salts will produce an acidic, basic, or neutral aqueous solution: $MgCl_2$, KNO_3, $SnCl_2$, NH_4NO_3.

Litmus paper tells us whether a solution is acidic or basic. However, many dyes change color over very narrow ranges of pH. In Figure 9.7, the pH of the solutions considered in Example 9.13 were determined by using such dye indicators.

>> More details regarding indicators can be found in Section 9.10.

Example 9.13 illustrates how to predict the acidity of a salt solution by inspection and the application of simple rules. But how would we determine the acidity of a solution of a salt derived from the reaction of a weak base with a weak acid? The pH of such a salt solution depends on which dissociation constant, K_a or K_b, is larger. Recall that K_a describes a reaction in which hydronium ion is produced, and K_b describes a reaction in which hydroxide ion is produced. For example, for ammonium acetate, the K_a for NH_4^+ is 5.71×10^{-10} and the K_b for CH_3COO^- is 5.75×10^{-10} (Table 9.5). In this case, because K_a is about the same value as K_b, the solution will be approximately neutral.

9.8   BUFFERS AND BUFFERED SOLUTIONS

The normal metabolic activity of cells results in the production of acids. Muscle tissue, for example, produces lactic acid. Such acids are a potential danger to proteins, whose structures and physiological functions depend on the maintenance of a proper pH. Even modest changes in pH can result in significant changes in the catalytic activity of enzymes, thereby causing cells to malfunction. The pH of blood must be maintained within very narrow limits, which, if exceeded, can result in death. Fortunately, the body's buffer systems can effectively counter changes in pH.

A **buffered solution** is a solution that resists changes in pH when hydronium or hydroxide ion is added. Such a solution contains a **buffer system** consisting of a conjugate acid–base pair. The $H_2PO_4^-/HPO_4^{2-}$ and H_2CO_3/HCO_3^- systems described in Section 9.6 are just two examples of conjugate acid–base pairs that can be used as buffer systems. Many of the commercially available antacids create a buffer system in the stomach by providing a Brønsted–Lowry base such as citrate. When these systems are added to the low-pH environment of the stomach, they react with hydronium ion to raise the pH.

For optimal buffering action to take place, the concentrations of conjugate acid and conjugate base in the buffer system must be approximately equal. We will see why this is so shortly.

If a small amount of hydroxide ion is introduced into a buffer solution, the conjugate acid will react with the hydroxide ion. If a small amount of hydronium ion is added, the conjugate base will react with the hydronium ion. The conjugate acid behaves as a hydroxide ion sponge, and the conjugate base as a hydronium ion sponge, both constantly ready to absorb the appropriate ion and thus maintain a constant pH.

The greater the concentrations of conjugate acid and base, the greater the capacity of the buffer to absorb changes in hydronium or hydroxide ion concentrations. We will return to this idea of buffering capacity after taking a quantitative look at the conjugate acid–base pair that makes up a buffer system.

Quantitative Aspects of Buffer Systems

Each conjugate acid–base pair has a specific pH that it is able to maintain through its buffering action. This pH can be calculated by using the **Henderson–Hasselbalch equation:**

$$pH = pK_a + \log\left(\frac{[\text{proton acceptor}]}{[\text{proton donor}]}\right)$$

The equation states that the pH of a buffered solution depends on the pK_a of the proton donor of the acid–base pair (not the pK_b of the proton acceptor) and on the logarithm of the proton-acceptor-to-proton-donor ratio. The derivation of the equation is presented in Box 9.4 on the following page.

Table 9.6 on page 247 illustrates how the acceptor-to-donor ratio affects resistance to change in pH in a buffer solution, as calculated from the Henderson–Hasselbalch equation. The effectiveness of a buffer can be seen by noting the change in pH for a wide range of acceptor-to-donor ratios. When the ratio changes from 1.6/1 to 0.6/1, the logarithm of the ratio and therefore the pH varies by only ± 0.2 pH units. Even changes in acceptor-to-donor ratios ranging from 10/1 to 1/10—representing changes of about ± 90% of the original concentrations—result in changes of only ± 1.0 pH unit. However, as you can see, when the ratio is much larger or much smaller than 1.0, the pH changes rapidly.

✓ For maximum buffering effect, the value of the acceptor-to-donor ratio in a buffer solution must be kept as close to 1.0 as possible.

Concept check

Knowing the pK_a of the proton donor and the ratio of acceptor to donor, we can calculate the pH of any buffer solution.

9.4 CHEMISTRY IN DEPTH

The Henderson–Hasselbalch Equation

The Henderson–Hasselbalch equation is derived by considering the dissociation equilibrium of a generalized weak acid. The formula of the weak acid is represented as HA, the anion as A^-. The dissociation process is therefore

$$HA + H_2O \rightleftharpoons H_3O^+ + A^-$$

where A^- is the conjugate base of the acid HA, and the acid dissociation constant expression is

$$K_a = \frac{[H_3O^+][A^-]}{[HA]}$$

Now rewrite the dissociation constant to emphasize the ratio of conjugate base to conjugate acid:

$$K_a = [H_3O^+] \times \frac{[A^-]}{[HA]}$$

To introduce the "p" notation, we first take the negative logarithm of both sides, term by term:

$$-\log K_a = -\log [H_3O^+] - \log \frac{[A^-]}{[HA]}$$

Thus:

$$pK_a = pH - \log \frac{[A^-]}{[HA]}$$

The minus sign is eliminated by rearrangement:

$$pH = pK_a + \log \frac{[A^-]}{[HA]}$$

The relation can be expressed in the more generalized form

$$pH = pK_a + \log \frac{[\text{proton acceptor}]}{[\text{proton donor}]}$$

This form of the equation makes it clear that

1. A buffer system can consist of any conjugate acid–base pair such as $H_2PO_4^-/HPO_4^{2-}$, HCO_3^-/CO_3^{2-}, or NH_4^+/NH_3.
2. The pH of a buffer system depends on two variables:
- The acid dissociation constant (not the base dissociation constant) of the conjugate acid–base pair.
- The ratio of molar concentrations of proton acceptor to proton donor of the conjugate acid–base pair.

Example 9.14 Calculating the pH of a buffer solution

Calculate the pH of an aqueous solution consisting of 0.0080 M acetic acid and 0.0060 M sodium acetate.

Solution

For this type of calculation, we use the Henderson–Hasselbalch equation, substituting the pK_a of acetic acid, 4.76 (Table 9.4), and the concentrations of acceptor and donor. (Note that the pK_a of acetic acid, not the pK_b of acetate, is used.) The acceptor (numerator) is acetate, and the donor (denominator) is acetic acid.

$$pH = 4.76 + \log \frac{[0.0060]}{[0.0080]}$$
$$= 4.76 - 0.13$$
$$pH = 4.63$$

Problem 9.14 Calculate the pH of an aqueous solution consisting of 0.0060 M acetic acid and 0.0080 M sodium acetate.

The information provided by the Henderson–Hasselbalch equation has other uses besides the calculation of pH. It tells us how best to select a buffer system that will buffer at a specific pH—a common requirement in laboratory work. For example, a particular experiment on a biological reaction may require a buffer that maintains a solution at pH 7.0 ± 0.3.

TABLE 9.6	Effect on pH of the Ratio of Proton Acceptor to Proton Donor Forms of an $HPO_4^{2-}/H_2PO_4^-$ Buffer ($pK_a = 7.2$)	
Ratio	Logarithm	pH
1000/1	3.00	10.20
100/1	2.00	9.20
10/1	1.00	8.20
5/1	0.70	7.90
1.6/1	0.20	7.40
1/1	0.00	7.20
0.6/1	− 0.20	7.00
1/5	− 0.70	6.50
1/10	− 1.00	6.20
1/100	− 2.00	5.20
1/1000	− 3.00	4.20

The key to selecting a buffer system is to consider what the Henderson–Hasselbalch equation reveals when the concentrations of conjugate base and acid are equal; that is, when [acceptor] = [donor]. When that is the case,

$$pH = pK_a + \log 1.0$$

But log 1.0 = 0; so, when [acceptor] = [donor],

$$pH = pK_a$$

✓ To prepare a buffer of specific pH, select a conjugate acid–base pair with a pK_a as close as possible to the desired pH.

Concept check

For a solution that buffers at pH 7.0 ± 0.3, this acid–base pair is $H_2PO_4^-/HPO_4^{2-}$, with a pK_a of 7.2 (see Table 9.5). We can prepare the buffer by dissolving equal molar quantities of KH_2PO_4 and K_2HPO_4 in water.

Buffering Capacity

The ratio of the molar concentrations of a conjugate acid–base pair is independent of the concentrations themselves, but the buffering capacity of the system is not. The ratio of concentrations can have the value of 1.0 whether the concentrations of donor and acceptor are equal at 0.001, 0.01, or 0.1 *M*. If more acid is added to a buffer solution than the conjugate base can absorb, then the pH will drastically change. Because the conjugate acid and base behave like sponges, the amount of hydronium ion or hydroxide ion that can be buffered depends on the amount (concentration) of the conjugate acid–base pair present.

✓ The buffering capacity of a buffer increases with increasing concentration of the buffer's conjugate acid–base pair.

Concept check

9.9 BUFFER SYSTEM OF THE BLOOD

Carbon dioxide produced by cells as an end product of metabolism diffuses into the erythrocytes of the blood. There, it is converted into bicarbonate ion in a reaction catalyzed by the enzyme carbonic anhydrase. This process can be represented by a two-step process:

$$CO_2(aq) + H_2O(l) \rightleftharpoons H_2CO_3(aq)$$
$$H_2CO_3(aq) + H_2O(l) \rightleftharpoons H_3O^+(aq) + HCO_3^-(aq)$$

You can see that each molecule of CO_2 added to the blood results in the formation of an H_3O^+ ion. The buffer system appears to consist of H_2CO_3/HCO_3^-, whose pK_a is 6.35, a value far lower than the blood pH of 7.4. That value is a full pH unit away from blood pH; so how can the H_2CO_3/HCO_3^- system act as the blood's buffer? This apparent discrepancy can be cleared up by noting that the concentration of H_2CO_3 depends on the concentration of dissolved CO_2.

In the Henderson–Hasselbalch equation, the proton-donor (acid) component is the sum of the concentrations of dissolved carbon dioxide and carbonic acid ($CO_2 + H_2CO_3$), and the pH is given by

$$pH = pK_a + \log\left(\frac{[HCO_3^-(aq)]}{[CO_2(aq) + H_2CO_3(aq)]}\right)$$

Substituting the values for blood pH (7.4) and pK_a (6.35) gives

$$7.4 - 6.35 \approx 1 = \log\left(\frac{[HCO_3^-(aq)]}{[CO_2(aq) + H_2CO_3(aq)]}\right)$$

Because the log of the ratio is about 1, the ratio of proton acceptor to proton donor is approximately 10/1. The acid component (the proton donor) would appear to be at such a low concentration that the introduction of a small quantity of base should drastically change the pH. What saves the day is the reservoir of dissolved CO_2.

Carbon dioxide is constantly renewed by metabolic processes, and so the carbonic acid reservoir, $CO_2 + H_2CO_3$, resists depletion and the H_2CO_3/HCO_3^- system effectively buffers the blood.

Carbon Dioxide Maintains the Acid–Base Balance of the Blood

As CO_2 is transported from metabolizing cells to the lungs for excretion to the air, it is accompanied by the protons generated by metabolic activity. Bicarbonate and protons are formed from CO_2 in erythrocytes at the tissue level, and those same protons are eliminated at the lungs by recombination with bicarbonate:

$$\text{Tissue level:} \quad CO_2 + H_2O \rightleftharpoons H^+ + HCO_3^-$$
$$\text{At the lungs:} \quad H^+ + HCO_3^- \rightleftharpoons CO_2 + H_2O$$

This abbreviated scheme emphasizes the role of CO_2 in maintaining the acid–base balance of the blood. It is continually removed by diffusion because its concentration at the tissue level is greater than that at the lungs. Remember that metabolizing tissues and the lungs rapidly communicate through the transport of blood by the vascular system. The buffering system is an open system, not at equilibrium, and there are serious consequences if the CO_2 is removed too rapidly or too slowly at the lungs. The rate of CO_2 removal depends on the rate of breathing, which we call the **ventilation rate.**

Acidosis and Alkalosis

In certain circumstances—for example, in the lung disease emphysema or when a person is under anesthesia—the ventilation rate may be too low, and CO_2 is not removed from the lungs rapidly enough. Consequently, the bicarbonate buffer system will "back up," hydrogen ion will not be removed by reaction with HCO_3^-, and the blood pH will fall. This condition is called **respiratory acidosis.** Immediate treatment consists of intravenous bicarbonate infusion. Conversely, if CO_2 is removed from the lungs faster than it arrives, the blood pH will rise. This condition, called **respiratory alkalosis,** develops,

A PICTURE OF HEALTH
Normal pH of Some Body Fluids

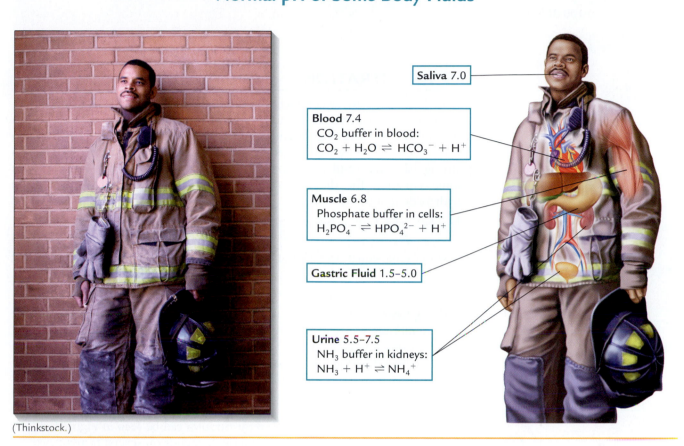

Saliva 7.0

Blood 7.4
CO_2 buffer in blood:
$$CO_2 + H_2O \rightleftharpoons HCO_3^- + H^+$$

Muscle 6.8
Phosphate buffer in cells:
$$H_2PO_4^- \rightleftharpoons HPO_4^{2-} + H^+$$

Gastric Fluid 1.5–5.0

Urine 5.5–7.5
NH_3 buffer in kidneys:
$$NH_3 + H^+ \rightleftharpoons NH_4^+$$

(Thinkstock.)

for example, under circumstances of great excitement (rapid breathing). First aid for this condition is to have the patient breathe into a paper bag, an action that increases the CO_2 content of the lungs.

Metabolic acidosis is a lowering of the blood pH as a result of a metabolic disorder, rather than as a result of a failure of the blood buffer system. For example, a large and serious decrease in pH can occur as a result of uncontrolled diabetes. The blood pH may fall from the normal 7.4 to as low as 6.8. The increased H^+ concentration is due to the large amounts of acidic compounds produced in the liver. The bicarbonate buffer system attempts to compensate for the excess H^+, producing excess CO_2, which must be eliminated at the lungs. However, so much CO_2 is lost by ventilation that the absolute concentration of the buffer system decreases. The capacity of the buffer system is severely compromised and cannot reduce the metabolically produced excess H^+. In such cases, buffer capacity can be temporarily restored by intravenous administration of sodium bicarbonate.

An increase in blood pH as a result of a metabolic disorder is called **metabolic alkalosis.** This condition can arise when excessive amounts of H^+ ion are lost—for example, during excessive vomiting. H^+ ion is then borrowed from the blood, and the blood pH will rise. Intake of excessive amounts of antacids also will cause the blood pH to rise. Immediate treatment consists of intravenous administration of ammonium chloride.

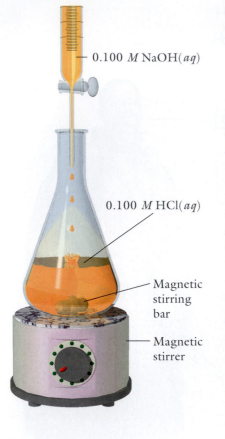

— 0.100 M NaOH(aq)

0.100 M HCl(aq)

— Magnetic stirring bar

— Magnetic stirrer

Figure 9.8 A titration setup. NaOH (aq) is added drop by drop from a buret into a constantly stirred solution of acid until the end point is neared. The end point is identified by a change in color of the indicator.

9.10 TITRATION

Acid–base chemistry as conducted in the laboratory, including the clinical laboratory, often requires the measurement of the concentration of an acid or a base in solution. This measurement is done with a titration using a neutralization reaction. Neutralization is the reaction between a strong acid and a strong base that results in the formation of a salt and water. The complete ionic equation for the reaction between hydrochloric acid and sodium hydroxide in water is

$$H^+(aq) + Cl^-(aq) + Na^+(aq) + OH^-(aq) \rightleftharpoons$$
$$H_2O(l) + Na^+(aq) + Cl^-(aq)$$

The net ionic equation for the neutralization of any strong acid by any strong base is

$$H^+(aq) + OH^-(aq) \rightleftharpoons H_2O(l)$$

and emphasizes the formation of water. The ionic reaction takes place because the equilibrium constant for the reaction is 1.00×10^{14}. Nitric acid, HNO_3, or any strong acid added to a solution of calcium hydroxide, $Ca(OH)_2$, or any strong base is described by the same net ionic equation.

The first step in the **titration** procedure is to accurately measure a volume of the solution containing the acid. A basic solution of known concentration is then added to the acid solution in small measured amounts until the acid is neutralized. The experimental setup for titration can be seen in Figure 9.8.

The success of this analytical technique depends on several factors:

- The reaction must go to completion (have a very large equilibrium constant).

- The reaction must be fast.

- There must be a way of detecting when the reaction is complete—that is, when neutralization has occurred.

A reaction that is stoichiometrically complete is said to be at its **equivalence point**. In a titration, the experimentally determined point of completion is called the **end point**. End points are detected by using dyes called **indicators** (Table 9.7), which change color near the pH of the equivalence point (Figure 9.9 and Box 9.5 on page 252) or by using a pH meter. The goal of titration is to get the stoichiometrical equivalence point and the experimentally determined

www.whfreeman.com/bleiodian2e

TABLE 9.7 | Indicator Dye pK_as and Titration End Points of Some Weak Acids and Bases

Indicator	pK_a	Weak acid or base	End-point pH
methyl red	4.6	NH_3	4.5
β-naphtholphthalein	7.6	HCOOH	7.9
phenolphthalein	8.7	CH_3COOH	8.4
thymol blue	9.6	$H_2PO_4^-$	9.6

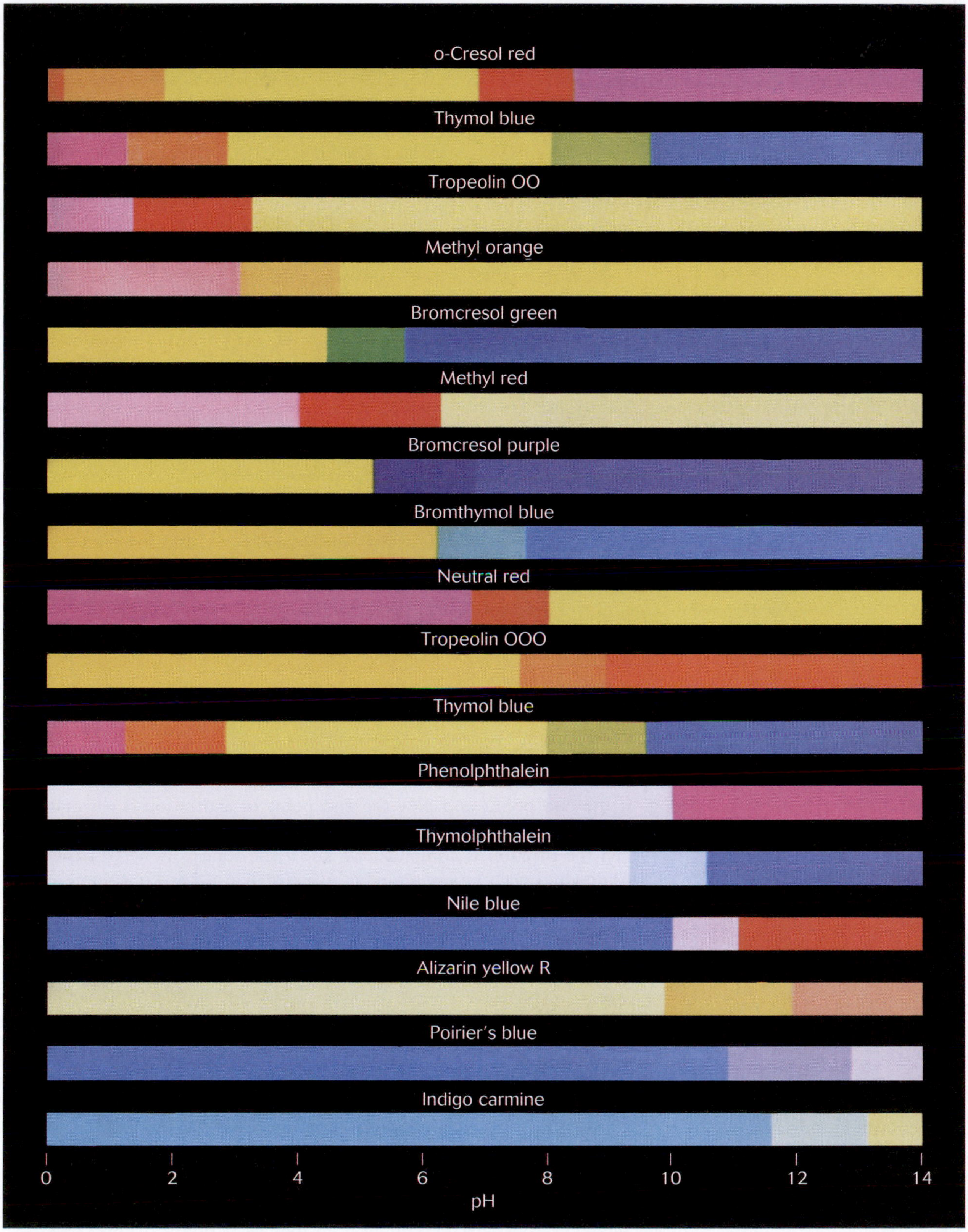

Figure 9.9 The colors of indicators used to determine end points of titrations over a wide range of pH values.

9.5 CHEMISTRY IN DEPTH

Use of Indicators in Determining pH

Not all titrations have end points at neutral pH—that is, pH 7—because, as we have seen, not all salt solutions are neutral. The salt of a weak acid forms a basic solution, and the salt of a weak base forms an acidic solution. For example, a solution of 0.1 M sodium acetate can have a pH of about 8.4. Therefore, the end point of a titration of acetic acid with sodium hydroxide consists of a solution of sodium acetate at a pH of 8.4, not pH 7. To detect the end point of a titration, we must be able to determine when the pH of the solution coincides with the predicted pH. This measurement can be done with a pH meter, but most work of this type is accomplished with indicator dyes (Sections 9.7 and 9.10).

A typical indicator dye is itself a weak acid and, like any weak acid, it reacts with water. What makes the indicator so useful is that a solution of the undissociated acid and a solution of the dissociated acid, or anion, have different colors. The difference is usually so marked that it is a simple matter to determine the transition from one color to another. How can this color change be used to measure pH?

First, let's write the acid dissociation reaction, calling the acid form of the indicator HIn and its anion In⁻:

$$HIn + H_2O \rightleftharpoons H_3O^+ + In^-$$

We can now express the equilibrium in terms of the Henderson–Hasselbalch equation (see Box 9.4):

$$pH = pK_a + \log \frac{[In^-]}{[HIn]}$$

From many years of experience, chemists have found that the eye can detect a color change when 10% of an indicator has undergone dissociation. In other words, they can "see" the end point of a titration to well within ± 1 pH unit of the pK_a of the indicator. However, for the accurate and rapid detection of the end point of a titration, they must choose an indicator with a pK_a very close to the calculated end point. Table 9.7 lists a number of indicator dyes, the pK_a of each dye, and the weak acid whose titration end point coincides with each pK_a. Figure 9.9 shows the colors of many other indicators used to determine end points over a wide range of pH values.

Most indicator dyes used in the laboratory are generally added directly to solutions, but they are also available as dye-impregnated paper strips that will determine pH to within about 0.5 pH unit. Litmus paper, by comparison, will reveal only whether the pH is above or below 7.

Figure 9.10 Aqueous solutions at pH 2 (red), pH 5 (orange), and pH 10 (yellow). The acid–base indicator is the dye methyl red. (Chip Clark.)

end point to coincide. The behavior of an indicator in solutions at the beginning, at the end point, and past the end point of a titration is illustrated in Figure 9.10.

When the titration is complete, we know the total volume of basic solution of known concentration required to neutralize a known volume of the acid. We can then calculate the concentration of the acid by a method very similar to the one used to quantitate dilution in Section 7.7. In Example 9.15, we will use this method to calculate the concentration of a solution of hydrochloric acid.

Example 9.15 Calculating acid concentration from a neutralization reaction

The neutralization of 25.0 mL of a solution of HCl of unknown concentration required 16.7 mL of a 0.0101 M KOH standard solution. (A **standard solution** is one of precisely known concentration.) Calculate the concentration of the acid.

Solution

As in dilution calculations (Section 7.7), the number of moles of added base in a neutralization reaction must be equal to the number of moles of acid:

$$\frac{Moles_{base}}{Liter} \times volume_{base} = \frac{moles_{acid}}{liter} \times volume_{acid}$$

$$M_{base} \times volume_{base} = M_{acid} \times volume_{acid}$$

The only restriction on units is that they must be the same on both sides of the equation:

$$0.0101 \times 16.7 \text{ mL} = M_{acid} \times 25.0 \text{ mL}$$
$$M_{acid} = 0.00675 \; M$$

Problem 9.15 What volume of a 0.0200 M KOH standard solution is required to neutralize 35.0 mL of a 0.0150 M solution of HNO_3?

In Section 7.10, we pointed out that reactions between ions can take place if one of the products is insoluble and forms a precipitate. We know that a neutralization reaction results in the formation of water. Now we can look at neutralization reactions in which not only water but also a gas is formed.

The reaction between sodium bicarbonate, taken orally to relieve the distress of excess stomach acid, and HCl results in the formation of a gas. The neutralization products are water and carbonic acid (H_2CO_3). Carbonic acid is unstable and rapidly decomposes to form CO_2 and H_2O. The reaction between HCl and $CaCO_3$ is shown in Figure 9.11. The equation for the formation of carbonic acid is

$$NaHCO_3(aq) + HCl(aq) \rightleftharpoons H_2CO_3(aq) + NaCl(aq)$$

The net ionic equation is

$$H^+(aq) + HCO_3^-(aq) \rightleftharpoons H_2CO_3(aq)$$

The gas-formation step consists of the decomposition of the unstable carbonic acid:

$$H_2CO_3(aq) \rightleftharpoons H_2O(l) + CO_2(g)$$

The net ionic equation showing the overall result is

$$H^+(aq) + HCO_3^-(aq) \rightleftharpoons CO_2(g) + H_2O(l)$$

Figure 9.11 Carbon dioxide is one of the products of the reaction of metal carbonates with acids. Here, an eggshell, largely composed of calcium carbonate, reacts with hydrochloric acid. The bubbles on the eggshell surface are carbon dioxide. (Chip Clark.)

9.11 NORMALITY

A unit of concentration that we have not yet introduced is **normality, N,** defined as **equivalents per liter.** Although less frequently used than molarity, this unit is often encountered in clinical chemistry. Its meaning rests on a quantity called equivalent mass.

Equivalent mass and **equivalents** are related to formula mass and moles. These units are used in acid–base chemistry to express the mass in grams of a base such as KOH that is equivalent to the mass in grams of an acid required for neutralization.

For example, consider the stoichiometry of the reaction between HCl and KOH: 1 mol of KOH is equivalent to 1 mol of HCl. Therefore, 56.1 g of KOH is equivalent to 36.5 g of HCl. We define 36.5 g of HCl as 1.00 equivalent of HCl, and 56.1 g of KOH as 1.00 equivalent of KOH. For either HCl or KOH, 1.00 equivalent contained in 1.00 L of solution is 1.00 N in concentration.

In the calculation of the normality for the solution of a polyprotic acid, the stoichiometry of its neutralization reaction is of key importance. Consider the reaction between H_2SO_4 and KOH: 1 mol of H_2SO_4 contains 2 mol of H^+; so $\frac{1}{2}$ mol of H_2SO_4 is equivalent to 1 mol of KOH. To obtain a mass of H_2SO_4 equivalent to 1 mol of KOH in this neutralization reaction, we must divide the molar mass of H_2SO_4 by 2: $98/2 = 49$ g/mol. The equivalent mass of H_2SO_4 is 49 g/mol; that is, 49 g of H_2SO_4 is equivalent to 56 g of KOH. A 1.0 N solution of H_2SO_4 contains 49 g/L (see Example 4.11).

These concepts are best explained in the context of a typical laboratory problem.

Example 9.16 Preparing a solution of known normality

How would you prepare a 1.00 N solution of phosphoric acid (H_3PO_4)?

Solution

We first write a balanced equation for the neutralization of H_3PO_4 by a base such as KOH:

$$H_3PO_4(aq) + 3\ KOH(aq) \rightleftharpoons K_3PO_4(aq) + 3\ H_2O(l)$$

Then we rewrite the equation so that the acid reacts with only 1 mol of base.

$$\tfrac{1}{3} H_3PO_4(aq) + KOH(aq) \rightleftharpoons \tfrac{1}{3} K_3PO_4(aq) + H_2O(l)$$

The modified stoichiometry shows that $\tfrac{1}{3}$ mol of H_3PO_4 is equivalent to 1 mol of KOH in this neutralization reaction.

To prepare a 1.00 N solution of the acid, $\tfrac{1}{3}$ mol of H_3PO_4, $98.0/3 = 32.7$ g (or 1.00 equivalent), should be dissolved in sufficient water to make 1.00 L of aqueous solution.

Problem 9.16 How would you prepare a 1.00 N solution of the diprotic acid $(CH_2)_2(COOH)_2$ (succinic acid)?

Concept check

✓ The equivalent mass of an acid is the formula mass of the acid divided by the number of reacting H^+ ions per mole of acid.

The normality of an acid solution used in neutralization is calculated by the following relation:

$$N = \frac{\text{equivalents}_{\text{acid}}}{\text{L}} = \left(\frac{\text{mol}_{\text{acid}}}{\text{L}}\right)\left(\frac{\text{moles dissociable } H^+}{\text{mol}_{\text{acid}}}\right)$$

The normality of a basic solution used for neutralization is calculated in the same way; simply replace H^+ with OH^-.

The meaning of equivalent mass or equivalents can be expanded to include the equivalent mass of charged species such as Na^+ or Mg^{2+}. In this case, an equivalent mass of a charged species is the mass that contains an Avogadro number of charges. For example, the equivalent mass of Na^+ ion is its atomic mass, that of Ca^{2+} is one-half its atomic mass, and that of Al^{3+} is one-third its atomic mass. The ion concentrations in blood plasma and other physiological fluids are frequently recorded in units of normality, specifically in **milliequivalents,** or **meq,** per liter, where 1000 meq = 1 equivalent. We will put these ideas to work in the next example.

Example 9.17 Calculating molar concentrations of ions from milliequivalents per liter (normality)

The concentrations of sodium and calcium ions in a sample of blood plasma are recorded as $Na^+ = 142$ meq/L and $Ca^{2+} = 5.0$ meq/L. Calculate the molar concentrations of these ions.

Solution

To convert equivalents into moles, we use the following conversion factor:

$$\frac{\text{mol}}{\text{L}} = \left(\frac{\text{equivalents}}{\text{L}}\right)\left(\frac{\text{mol}}{\text{equivalent of charge}}\right)$$

$$Na^+:\ \left(\frac{142\ \text{meq}}{\text{L}}\right)\left(\frac{1.00\ \text{eq}}{1000\ \text{meq}}\right)\left(\frac{1.00\ \text{mol}}{1.00\ \text{eq of charge}}\right) = 0.142\ M$$

$$Ca^{2+}:\ \left(\frac{5.0\ \text{meq}}{\text{L}}\right)\left(\frac{1.0\ \text{eq}}{1000\ \text{meq}}\right)\left(\frac{1.0\ \text{mol}}{2.0\ \text{eq of charge}}\right) = 0.0025\ M$$

Problem 9.17 The concentrations of potassium and magnesium ions in a sample of blood plasma are given as $K^+ = 5.0$ meq/L and $Mg^{2+} = 3.0$ meq/L. Calculate the molar concentrations of these ions.

Summary

Water Reacts with Water

- Water reacts with water to form two ions: H_3O^+, the hydronium ion, and OH^-, the hydroxide ion.

- The equilibrium constant for this reaction is called the ion-product of water, K_w.

- In pure water and in neutral solutions, $[H_3O]^+ = [OH]$.

- In acidic solutions, $[H_3O]^+ > [OH]$.

- In basic solutions, $[H_3O]^+ < [OH]$.

Strong and Weak Acids and Bases

- Strong acids and bases become completely dissociated on reaction with water.

- In solutions of strong acids or bases, the concentration of hydronium or hydroxide ion is equal to the acid or base solution concentration.

- Only a small percentage of the molecules of a weak acid or base react with water to form hydronium or hydroxide ions.

A Measure of Acidity: pH

- Acidity, the molar concentration of hydronium ion, is expressed by an acidity scale called pH.

- The three possible conditions in aqueous solutions with respect to acidity are

$$pH = 7 \text{ in a neutral solution}$$
$$pH < 7 \text{ in an acidic solution}$$
$$pH > 7 \text{ in a basic solution}$$

Quantitative Aspects of Acid–Base Equilibria

- Equilibrium constants describing the reaction of weak acids or bases with water are called dissociation constants.

- The smaller the dissociation constant, the weaker the acid or base.

Brønsted–Lowry Theory of Acids and Bases

- The Brønsted–Lowry theory describes acids as proton donors and bases as proton acceptors.

- When an acid donates a proton, it becomes a base, and, when a base accepts a proton, it becomes an acid.

- A weak acid together with its deprotonated form, now a base, is called a conjugate acid–base pair.

The Dissociation of Polyprotic Acids

- Acids with two or more dissociable protons are called polyprotic acids.

- The protons of a polyprotic acid dissociate sequentially, and each step of the dissociation process is characterized by its own dissociation constant.

- The intermediary products of the dissociation can behave as both proton donors and proton acceptors, and such substances are called amphoteric.

Salts and Hydrolysis

- In aqueous solution, the anions of weak acids—that is, conjugate bases—react with water to produce basic solutions.

- Cations that are weak acids (for example, ammonium ion) react with water to produce acidic solutions.

Buffers and Buffered Solutions

- A buffered solution contains a conjugate acid–base pair that causes the solution to resist changes in pH.

- The ratio of concentrations of proton acceptor (base) to proton donor (acid) controls the pH.

Titration

- Titration is a neutralization reaction used to determine the concentrations of acids or bases in solutions.

- Experimentally determined neutralization end points are detected by using indicator dyes or an electronic pH meter.

Normality

- In acid–base reactions, the equivalent mass of an acid is its formula mass divided by the number of its dissociable protons.

- The equivalent mass of a charged species such as Na^+ or Mg^{2+} is its atomic mass divided by its charge.

- Normality is concentration defined as equivalents per liter.

Key Words

Exercises

Water Reacts with Water

9.1 The concentration of hydronium ion and of hydroxide ion in pure water at 25°C is 1.00×10^{-7} M. What is the value of the ion-product for the ionization of water at that temperature?

9.2 The ion-product of pure water, K_w, is 2.51×10^{-14} at body temperature (37°C). What are the molar H_3O^+ and OH^- concentrations in pure water at body temperature?

9.3 What is meant by an amphoteric substance?

9.4 Can water be considered to be an amphoteric substance?

Strong Acids and Bases

9.5 What is the definition of a strong base? Give two examples.

9.6 What is the definition of a strong acid? Give two examples.

9.7 Calculate the molar H_3O^+ and Cl^- concentrations in a 0.30 M aqueous solution of HCl.

9.8 Calculate the molar H_3O^+ and ClO_3^- concentrations of a 0.28 M solution of perchloric acid ($HClO_3$).

9.9 Calculate the $[OH^-]$ and $[Tl^+]$ concentrations in a 0.25 M aqueous solution of TlOH.

9.10 Calculate the $[OH^-]$ and $[Ca^{2+}]$ of a 0.025 M aqueous $Ca(OH)_2$ solution.

9.11 Calculate the $[OH^-]$ in a 0.350 M aqueous solution of HCl.

9.12 Calculate the $[H_3O^+]$ and $[K^+]$ in a 0.16 M aqueous solution of KOH.

9.13 Calculate the hydronium ion concentration of a 0.020 M aqueous solution of $Ca(OH)_2$.

9.14 Calculate the hydroxide ion concentration in a 0.285 M aqueous solution of HCl.

A Measure of Acidity: pH

9.15 What is the pH of pure water at 25°C?

9.16 What is the pH of pure water at 37°C? The ion-product of pure water, K_w, is 2.51×10^{-14} at 37°C.

9.17 What is the pH of an aqueous solution with $[H_3O^+] = 0.010$ M?

9.18 What is the pH of an aqueous solution with $[H_3O^+] = 0.0026$ M?

9.19 What is the pH and the pOH of an aqueous solution with a hydroxide ion concentration of 0.015 M?

9.20 What is the pH and the pOH of an aqueous solution with a hydrogen ion concentration of 0.027 M?

9.21 Calculate the pH of the following aqueous solutions at 25°C: (a) 0.0033 M HCl; (b) 0.025 M HNO_3; (c) 0.00073 M HBr; (d) 0.074 M HI.

9.22 Calculate the pH of the following aqueous solutions at 25°C: (a) 0.0041 M KOH; (b) 0.00094 M $Ca(OH)_2$; (c) 0.035 M $Ba(OH)_2$; (d) 0.0084 M TlOH.

Weak Acids and Bases

9.23 What is the definition of a weak acid? Give two examples.

9.24 What is the definition of a weak base? Give two examples.

9.25 Is an aqueous solution containing equal concentrations of acetic acid (CH_3COOH) and ammonia (NH_3) acidic, basic, or neutral?

9.26 Is an aqueous solution containing equal concentrations of formic acid ($HCOOH$) and aniline ($C_6H_5NH_2$) acidic, basic, or neutral?

The Brønsted–Lowry Theory of Acids and Bases

9.27 Identify the conjugate acid–base pairs in the following equations:

(a) $HNO_2(aq) + H_2O(l) \rightleftharpoons$
$$NO_2^-(aq) + H_3O^+(aq)$$

(b) $H_2PO_4^-(aq) + H_2O(l) \rightleftharpoons$
$$HPO_4^{2-}(aq) + H_3O^+(aq)$$

9.28 Identify the conjugate acid–base pairs in the following equations:

(a) $HCOOH(aq) + NH_3(l) \rightleftharpoons$
$$HCOO^-(aq) + NH_4^+(aq)$$

(b) $H_2O(l) + H_2O(l) \rightleftharpoons OH^-(aq) + H_3O^+(aq)$

9.29 Using Table 9.5, identify the equilibrium constant for the reaction of acetate ion with water (hydrolysis) at 25°C.

9.30 Using Table 9.5, identify the equilibrium constant for the reaction of nitrous acid with water at 25°C.

9.31 Compute the pK_a for the dissociation of acetic acid and the pK_b for the hydrolysis of acetate ion in water at 25°C.

9.32 Compute the pK_a for the dissociation of nitrous acid and the pK_b for the hydrolysis of nitrite ion in water at 25°C.

9.33 What is the sum of the pK_a for the dissociation of acetic acid and the pK_b for the hydrolysis of acetate ion at 25°C in water?

9.34 What is the sum of the pK_a for the dissociation of nitrous acid and the pK_b for the hydrolysis of nitrite ion at 25°C in water?

Polyprotic Acids

9.35 Write equations for the complete dissociation of carbonic acid (H_2CO_3) in water.

9.36 Write equations for the complete dissociation of phosphoric acid (H_3PO_4) in water.

9.37 Write equilibrium-constant expressions for the complete dissociation of carbonic acid in water.

9.38 Write equilibrium-constant expressions for the complete dissociation of phosphoric acid in water.

9.39 In an aqueous solution of H_2CO_3, is the proportion of the H_3O^+ contributed by HCO_3^- of major or minor significance?

9.40 In an aqueous solution of H_3PO_4, is the proportion of the H_3O^+ contributed by $H_2PO_4^-$ of major or minor significance?

Salts and Hydrolysis

9.41 Predict whether each of the following salts will produce an acidic, basic, or neutral aqueous solution: (a) $CaCl_2$ (calcium chloride); (b) Na_2CO_3 (sodium carbonate); (c) $FeCl_3$ [iron(III)chloride]; (d) NH_4Cl (ammonium chloride); (e) $MgSO_4$ (magnesium sulfate).

9.42 Predict whether aqueous solutions of each of the following salts are acidic, neutral, or basic: (a) NH_4NO_3 (ammonium nitrate); (b) $Al_2(SO_4)_3$ (aluminum sulfate); (c) KI (potassium iodide); (d) $NaHCO_3$ (sodium hydrogen carbonate); (e) K_2HPO_4 (potassium hydrogen phosphate).

9.43 Predict whether aqueous solutions of the following salts are acidic, neutral, or basic: (a) $Ca(NO_3)_2$ (calcium nitrate); (b) $NaNO_2$ (sodium nitrite); (c) KCN (potassium cyanide); (d) CH_3COONa (sodium acetate); (e) $(HCOO)_2Mg$ (magnesium formate).

9.44 Predict whether aqueous solutions of the following salts are acidic, neutral, or basic: (a) $Fe(NO_3)_3$ [iron(III) nitrate]; (b) $Mg(NO_2)_2$ (magnesium nitrite); (c) KBr (potassium bromide); (d) $(CH_3COO)_2Ca$ (calcium acetate); (e) $MgCO_3$ (magnesium carbonate).

Buffers and Buffer Solutions

9.45 Why does a solution of a conjugate acid–base pair behave as a buffer solution?

9.46 Why must the concentrations of the conjugate acid and conjugate base in a buffer system be comparable?

9.47 What is the pH of an aqueous solution consisting of 0.0060 M acetic acid and 0.0080 M sodium acetate?

9.48 What is the pH of an aqueous solution consisting of 0.070 M formic acid ($HCOOH$) and 0.070 M sodium formate ($HCOONa$)?

9.49 Calculate the pH of an aqueous solution consisting of 0.075 M K_2HPO_4 and 0.050 M KH_2PO_4.

9.50 Calculate the pH of an aqueous solution consisting of 0.050 M K_2HPO_4 and 0.075 M KH_2PO_4.

Titration

9.51 The neutralization of a 20.0-mL aqueous solution of HCl of unknown concentration required 16.2 mL of a 0.0210 M KOH standard solution. Calculate the concentration of the acid.

9.52 What was the molar concentration of 25.0 mL of an aqueous solution of HNO_3 that required 32.0 mL of a 0.0180 M standard solution of KOH for complete neutralization?

9.53 What was the molar concentration of 40.0 mL of an aqueous H_2SO_4 solution that required 33.2 mL of a 0.0410 M standard solution of KOH for complete neutralization?

9.54 What was the molar concentration of 20.0 mL of an aqueous H_3PO_4 solution that required 42.3 mL of a 0.0850 M standard solution of KOH for complete neutralization?

Normality

9.55 How would you prepare a 0.500 N aqueous solution of phosphoric acid (H_3PO_4)?

9.56 How would you prepare an aqueous 0.200 N solution of sulfuric acid (H_2SO_4)?

9.57 What is the normality of a 0.024 M aqueous solution of H_3PO_4?

9.58 What is the normality of a 0.016 M aqueous solution of H_2SO_4?

9.59 Calculate the molar concentrations of the following ions, whose concentrations are recorded in units of normality: (a) Na^+ = 0.125 eq/L; (b) K^+ = 0.035 eq/L; (c) Ca^{2+} = 0.072 eq/L; (d) Mg^{2+} = 0.028 eq/L.

9.60 Compute the normality of the following solution with respect to each ionic species. The solution contains (a) 27.4 mmol of K^+, (b) 7.81 mmol of Ca^{2+}, and (c) 6.08 mmol of Al^{3+} ions, all in 54.7 mL.

Unclassified Exercises

9.61 Calculate the pH of the following aqueous solutions: (a) 0.0031 M HNO_3; (b) 1.0 M HCl; (c) 0.0069 M HI; (d) 0.019 M HBr; (e) 0.023 M $HClO_3$.

9.62 Calculate the pOH of the solutions in Exercise 9.61.

9.63 Calculate the pOH of the following aqueous solutions: (a) 0.0062 M KOH; (b) 0.0041 M $Ca(OH)_2$; (c) 0.028 M $Ba(OH)_2$; (d) 1.0 M KOH: (e) 0.010 M NaOH.

9.64 Calculate the pH of the solutions in Exercise 9.63.

9.65 Are aqueous solutions of the following salts acidic, neutral, or basic? (a) $Fe_2(SO_4)_3$; (b) NaBr; (c) $NaNO_2$; (d) NH_4NO_3; (e) $Mg(CN)_2$.

9.66 Calculate the pH of an aqueous solution that is 0.050 M in H_2CO_3 and 0.075 M in $KHCO_3$.

9.67 What is the molar ratio of $HPO_4^{2-}/H_2PO_4^-$ in a buffered aqueous solution of pH 8.20?

9.68 Compute the molarity of 15.0 mL of an aqueous nitric acid solution that required 35.9 mL of 0.0480 M KOH for complete neutralization.

9.69 Calculate the normality of 25.0 mL of an aqueous $Ca(OH)_2$ solution that required 62.0 mL of a 0.0250 M HCl solution for complete neutralization.

9.70 The pH of an aqueous HCl solution is 3.20. Calculate the concentration of the HCl solution.

9.71 Calculate the pOH and the hydroxide ion concentrations of the following aqueous solutions: (a) 0.0034 M HCl; (b) 0.025 M HNO_3; (c) 0.00073 M HBr; (d) 0.074 M HI.

9.72 What is the definition of a polyprotic acid?

9.73 Predict whether CH_3COONH_4 (ammonium acetate) will form an acidic or a basic aqueous solution.

9.74 Give some examples of buffer systems.

9.75 What are the components of a buffered solution?

Expand Your Knowledge

Note: The icons ⚗ ⚗ ⚗ denote exercises based on material in boxes.

9.76 The ion-product of water increases as temperature increases. Is the dissociation of water an exothermic or an endothermic reaction?

9.77 What is a conjugate acid–base pair?

9.78 What is the meaning of pH?

9.79 Suppose the capacity of your stomach to be 2.00 L and its pH to be 1.50. How many grams of the antacid sodium bicarbonate, $NaHCO_3$, would be necessary to bring the pH of your stomach to 7.00?

9.80 How would you prepare a buffer to maintain the pH of an aqueous solution as close to pH 5.00 as possible? (Consult Table 9.5.)

9.81 Calculate the pH of a 0.010 M aqueous solution of sodium acetate at 25°C (see Box 9.1).

9.82 Calculate the pH of a 0.010 M aqueous solution of ammonium chloride at 25°C (see Box 9.1).

9.83 Use the ionization constant in Table 9.4 to calculate the osmotic pressure of a 0.0050 M aqueous solution of formic acid at 25°C.

9.84 The osmotic pressure of a 0.00500 M aqueous solution of hydrazoic acid at 298 K is 98.2 torr. Calculate its ionization constant. Compare its value with that given in Table 9.4 (see Box 9.2).

9.85 The Henderson–Hasselbalch equation is often used to calculate the pH of a buffer solution, but it also describes the progress of the titration of a weak acid such as acetic acid with a strong base such as sodium hydroxide. Describe the situation when 50% of the acid has been neutralized:

what species are present in the solution, and what is the relation of the pH to the pK_a of acetic acid?

9.86 Would the pH at the equivalence point of the titration in Exercise 9.85 be 7.00? Explain your answer.

9.87 Arrange the pH values of the following 0.010 M aqueous solutions in the order highest to lowest: Na_3PO_4, Na_2HPO_4, NaH_2PO_4.

9.88 Your laboratory instructor provides you with four solutions. She tells you that they are equimolar solutions of monoprotic weak acids. You are to measure the pH of each solution and correlate them with their dissociation constants, smallest to largest. The list with their pH values is (1) 6.65; (2) 3.41; (3) 4.82; (4) 2.85.

9.89 What can you say about the relative basic strengths of the anions produced by the step-by-step dissociation of diprotic or triprotic acids such as carbonic acid or phosphoric acid?

9.90 Write equations that describe the acid–base reactions of each of the following pairs of Brønsted–Lowry acids and bases. (a) H_2O and NH_2^-, (b) H_2O and HS^-, (c) HCNO and NH_3, (d) HNO_2 and CH_3COO^-.

9.91 Your laboratory instructor gives you two solutions identified as A and B. She tells you that one of them has a pH of 2 and the other has a pH of 12. Both solutions are colorless. She adds a few drops of Bromcresol green to each solution. Solution A turns yellow, and solution B turns blue. Which solution has a pH of 2, and which has a pH of 12 (see Box 9.5)?

9.92 The indicator Nile blue is added to an aqueous solution of pH 12. What will be the color of the resulting solution (see Box 9.5)?

CHAPTER 10

CHEMICAL AND BIOLOGICAL EFFECTS OF RADIATION

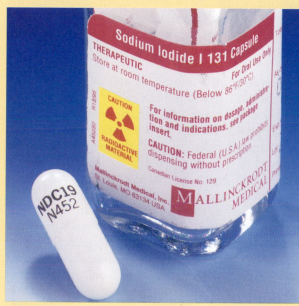

(Richard Megna/Fundamental Photographs.)

Chemistry in Your Future

Less than a month ago, when Ms. Jensen came to the radiation laboratory where you work as a technician, she was experiencing extreme nervousness, insomnia, and weight loss. Her physician had discovered that Ms. Jensen's thyroid gland was enlarged and, after getting the results of some blood tests, determined that the problem was Graves' disease, a condition characterized by an overactive thyroid gland. Now, after several visits in which you administered the prescribed series of radioactive iodine capsules, Ms. Jensen's symptoms are gone and she feels like herself again. This chapter identifies the radioactive isotope that was used and explains why it was chosen for her therapy.

For more information on this topic and others in this chapter, go to www.whfreeman.com/bleiodian2e

Learning Objectives

- Define radioactivity and describe the principal kinds of nuclear emissions.

- Use nuclear equations to explain transmutation.

- Describe radioactive decay and the half-life concept.

- Describe the chemical effects of high-energy radiation.

- Describe the characteristics of radioisotopes used in diagnosis and in therapy.

- Explain how CAT scans, PET scans, and MRI are used in diagnosis.

- Describe the origin of nuclear energy and its effect on the biosphere.

We are immersed in a sea of electromagnetic radiation—cosmic rays, heat from the sun, radio frequencies. The energies of certain frequencies of radiation are great enough to break chemical bonds and to ionize atoms, with serious consequences for biological systems. It is therefore important for us to examine the origin and properties of radiation and its interaction with matter. We will find that radiation also plays a significant role in medical diagnosis and therapy and can be harnessed as an important source of energy.

10.1   ELECTROMAGNETIC RADIATION REVISITED

Before we launch into the principal topic of this chapter, we have to return to the topic of electromagnetic radiation introduced in Chapter 2. There we stressed the relation between energy and frequency. The reason was that our goal was the understanding of atomic structure, which required an understanding of energy states and the transitions between them. However, the relation between energy and frequency is not the whole story about electromagnetic radiation. Further consideration of the properties of electromagnetic radiation will allow a more complete presentation of the effects of radiation on biological tissue.

Electromagnetic radiation travels from a source through space to a detector. As an example, heat from an incandescent bulb can be felt by your hand at some distance from the bulb. What form does the radiation take as it travels through space? Recall that, in Chapter 2, we described how the vibration of an electrical charge generated a disturbance that traveled through space with a frequency dependent on the speed of the vibration. If you drop a stone into still water, the vertical disturbance will generate a circular series of waves with the disturbance as the point of origin. A vibrating-point charge produces a similar disturbance, but a medium such as water is not needed for the disturbance to travel through space. It is characterized as a disturbance traveling through space in the form of a wave that has a characteristic frequency.

Figure 10.1 illustrates the form of waves of the same intensity generated by the vibration of charged particles at various frequencies. Notice that the distances between wave crests are different for different frequencies. The higher the frequencies, the closer the wave crests. The distance between the wave crests is called the wavelength. A frequency is the number of times that an event takes place over some fixed period of time—for example, 6 times per second or 42 times per minute. The units of frequencies are 1/time (for example,

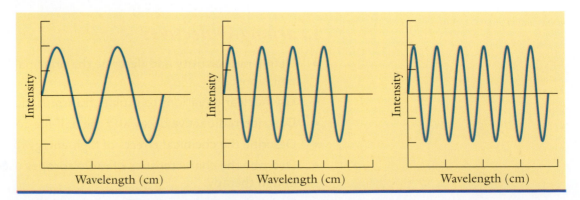

Figure 10.1 The forms of electromagnetic waves of the same intensity, generated by the vibration of charged particles at three different frequencies, which increase from left to right.

seconds^{-1}, minutes^{-1}) and the units of wavelengths are length (for example, centimeters, meters). Distance and time are related by a quantity called speed:

$$\text{Speed} = \text{distance}/\text{time}$$

Because frequency is 1/time,

Speed of radiation = wavelength of radiation $\times$ frequency of radiation

The traditional symbols used for these quantities are c for speed, the Greek letter λ (lambda) for wavelength, and the Greek letter v (nu) for frequency.

Speed of radiation = wavelength of radiation $\times$ frequency of radiation

becomes

$$c = \lambda \times v$$

It is a fact of life that the speed of electromagnetic radiation in our universe is a constant, 3×10^8 m/s. Knowing the frequency of radiation allows us to calculate the wavelength of any radiation. For my favorite FM radio station at 91.3 MHz,

$$3 \times 10^8 \text{ m/s} = \lambda \times 91.3 \times 10^6 \text{ s}^{-1}$$
$$\lambda = 3.3 \text{ m}$$

From the list of radiation frequencies in Table 2.4, let's calculate the wavelength of X-radiation.

$$3 \times 10^8 \text{ m/s} = \lambda \times 1 \times 10^{18} \text{ s}^{-1}$$
$$\lambda = 3 \times 10^{-10} \text{ m}$$

These examples demonstrate the inverse relation between frequency and wavelength. But they say nothing about the energy of electromagnetic radiation.

One of the most revealing experiments demonstrating the energy of electromagnetic radiation was called the **photoelectric effect.** Irradiation of a metal by light of a particular frequency was found to cause an electric current to flow in the metal. Albert Einstein proposed that light could be considered to consist of particles called **photons,** whose energy depended on the frequency of the light. He was eventually awarded a Nobel Prize for this discovery.

The discovery began a great controversy in physics regarding the question of how something can be both a particle and a wave at the same time. The answer to that question still troubles many people. The explanation accepted today is that it all depends on the experiment that is done. Some experiments demonstrate the particle nature of light, and others demonstrate the wave nature of light. Whether one accepts one model or the other, the fact of the matter is that these models have led to the creation of the modern electronic world.

Since Einstein's analysis, the particle–wave duality has been shown to be characteristic of all subatomic particles such as electrons. The relative energies of electromagnetic energy listed in Table 2.4 were calculated with the mathematical relation between energy and frequency proposed by a physicist named Max Planck, also a Nobel Prize winner for this work. The relation is

$$E = h \times v$$

where E is the photon's energy, v, is the photon's frequency, and h is called the Planck constant in honor of its discoverer.

Remember, you should not be surprised that, depending on the nature of the experiment, the same radioactive emission is called either a beta (β)-particle or a β-ray or that X-rays and gamma (γ)-rays are considered to be photons of very high energy.

10.2 RADIOACTIVITY

In 1895, Wilhelm Roentgen discovered that, by exposing metals to high electrical voltages, he could generate radiation that penetrated solid objects. He called this mysterious radiation **X-rays.** Within months of their discovery, they were being used in forensic applications and to explore pathologies of bone. Shortly after that, Henri Becquerel discovered penetrating radiation that emanated spontaneously from uranium salts. After Becquerel's discovery, Marie and Pierre Curie found that two new elements that they discovered—polonium and radium—also emitted penetrating radiation. Many other natural emitters of radiation were found in the ensuing years and came to be called radioactive substances.

By 1912, Ernest Rutherford had shown that radioactivity was the result of processes taking place in the atomic nucleus. He proposed that the emission of high-energy particles from the nucleus allowed unstable radioactive elements to achieve stability. Today, the origin of the instability is believed to lie in the fact that an atom's nucleus contains protons in extraordinarily close contact, with the resulting strong repulsive forces reduced by the presence of neutrons. As atomic number increases, the number of neutrons increases to compensate for the increasing repulsive forces within the nucleus. Nevertheless, above atomic number 83, the atomic nuclei become so unstable that all elements above that atomic number are radioactive. Some of the isotopes of the lighter elements are radioactive as well. One of the radioactive isotopes of potassium is of particular importance because it is naturally present within our bodies and is a constant internal source of radiation.

Radioactive emissions often result in changes in mass number or atomic number or both, and these changes can be accounted for only by noting the precise number of **nucleons** (the name for either the protons or the neutrons in the nucleus) present before and after radioactive emissions. For this reason, in considering nuclear processes, we must always be careful to refer to a specific isotope. We will use the notation developed in Section 2.5 for characterizing isotopes, in which one isotope is distinguished from another in either of two ways. One method is to write the mass number as a superscript and the atomic or proton number as a subscript in front of the symbol for the element. An example is $^{17}_{8}O$. The alternate method is to write the name of the element followed by a hyphen and the atomic mass number. For example, oxygen-17 or potassium-38, with the atomic number implied by the element's name.

The word isotope is used to denote atoms of the same element possessing different mass numbers. The word **nuclide** is a more general term used for referring to isotopes of either the same or different elements. A radioactive isotope is called a **radionuclide.** For example, potassium-40 and oxygen-17 are both radionuclides.

10.3   RADIOACTIVE EMISSIONS

The major types of radioactive emissions are alpha(α)-particles, beta(β)-particles, gamma(γ)-rays, and positrons. **Alpha-particles** have been shown to consist of helium nuclei—that is, helium atoms with no electrons. **Beta-particles** are energetic electrons originating in the nucleus and emitted from it. **Positrons** have all the properties of electrons but with a positive rather than a negative charge. **Gamma-rays** are not particles; they are very high energy photons. Table 10.1 lists the emissions, their symbols, and the nuclear consequence of their loss.

Because the α-particle is a helium nucleus, its symbolic notation as a nuclear particle is $^{4}_{2}He$. Considered a helium nucleus—that is, a helium atom that has lost its electrons—it is also an ion and appears in a chemical equation as He^{2+}. When an element loses an α-particle, $^{4}_{2}He$, its mass undergoes a change of 4 amu, and its atomic number is reduced by two units. It is therefore transmuted into a new element of lower atomic number with a reduced mass. **Transmutation** is a nuclear process in which a nuclide is transformed into a

| TABLE 10.1 | Summary of the Properties and Results of Nuclear Emissions | | |

| Emission | Symbol | Change in nucleus | |
		Mass number	Atomic number
α	^4_2He	decreases by 4	decreases by 2
β	$^0_{-1}\text{e}$	no change	increases by 1
positron	^0_1e	no change	decreases by 1
γ	$^0_0\gamma$	no change	no change

new element either spontaneously or artificially, the latter by interaction with a high-energy nuclear particle.

Transmutation is described by a **nuclear equation.** For example, to show that radium-226 emits an α-particle and forms radon-222, we write

$$^{226}_{88}\text{Ra} \longrightarrow {}^{222}_{86}\text{Rn} + {}^4_2\text{He}$$

Notice that the sum of the subscripts (numbers of protons) on the right-hand side is equal to the subscript on the left-hand side and that the sum of the superscripts (protons + neutrons) on the right is equal to the superscript on the left.

As shown in Table 10.1, the symbol for a β-particle, $^0_{-1}\text{e}$, has a superscript of zero, denoting the infinitesimal mass of an electron relative to a nucleon. The subscript of -1 refers to the electron's charge. The loss of a β-particle from a nucleus causes a transmutation by increasing the positive charge of the nucleus (the atomic number) by one unit.

The symbol for a positron, ^0_1e, indicates that it has no mass but does possess a positive charge. Its loss decreases the positive charge of the nucleus (the atomic number) and leads to a transmutation of the element into the next element that is one atomic number lower. The symbol for the γ-ray, $^0_0\gamma$, indicates no mass and no charge; thus γ-ray emission does not lead to transmutation.

Example 10.1   Writing a nuclear equation: I. Emission of an alpha–particle

What is the effect of the radioactive emission of an α-particle by an atom of uranium-238?

Solution

When a $^{238}_{92}\text{U}$ nucleus emits a helium nucleus, ^4_2He, the mass number decreases by four, and the atomic number decreases by two. The change in atomic number reveals that a new element has been produced. The nuclear equation that describes the ejection of a helium nucleus from uranium-238 is

$$^{238}_{92}\text{U} \longrightarrow {}^{234}_{90}\square + {}^4_2\text{He}$$

Check to be sure that the sum of the mass numbers of the product superscripts (protons plus neutrons) equals the mass number of the uranium-238 and that the sum of the product subscripts (protons) is equal to the atomic number of uranium-238. The new element, $^{234}_{90}\square$, is identified by referring to the periodic table to find the element with atomic number 90. It is thorium. The new nuclide is therefore thorium-234.

Problem 10.1   Radium-223 is a radioactive α-emitter. Write the nuclear equation for this emission event, and identify the product.

The following example shows how to predict the results of β-emission.

| Example 10.2 | Writing a nuclear equation: II. Emission of a beta-particle |

What is the effect of the emission of a β-particle on an atom of thorium-234.

Solution

As in Example 10.1, we first write the nuclear equation for this emission event. Radioactive thorium-234 emits a β-particle as

$$^{234}_{90}\text{Th} \longrightarrow {}^{234}_{91}\text{Pa} + {}^{0}_{-1}\text{e}$$

The mass number of the product is the same as that of thorium, because the β-particle has no mass. But notice that one negative nuclear charge has been lost. So the positive nuclear charge and consequently the atomic number have each been increased by one. Examination of the periodic table indicates that the element possessing atomic number 91 is protactinium.

Problem 10.2 Radium-230 is a radioactive β-emitter. Write the nuclear equation for the event, and identify the product.

We have said that a β-particle is equivalent to an electron. How, then, can it be emitted from a nucleus, which contains only protons and neutrons? The answer is that the β-particle is the result of the conversion within the nucleus of a neutron ($^{1}_{0}\text{n}$) into a proton:

$$^{1}_{0}\text{n} \longrightarrow {}^{1}_{1}\text{H} + {}^{0}_{-1}\text{e}$$

The energetic β-particle is ejected, whereas the proton, $^{1}_{1}\text{H}$, remains in the nucleus to increase the atomic number by one.

In an analogous process, the emission of a positron is the result of the conversion of a proton into a neutron:

$$^{1}_{1}\text{H} \longrightarrow {}^{1}_{0}\text{n} + {}^{0}_{1}\text{e}$$

If a positron comes into contact with an electron, the positron is rapidly annihilated to produce γ-rays:

$$^{0}_{-1}\text{e} + {}^{0}_{1}\text{e} \longrightarrow 2\,{}^{0}_{0}\gamma$$

Note that the reaction of a positron with an electron produces two γ-rays. This process is the basis for positron emission tomography (PET), a medical imaging technique that will be described later in this chapter. Example 10.3 illustrates a nuclear reaction equation for positron emission.

| Example 10.3 | Writing a nuclear equation: III. Emission of a positron |

A radioactive isotope of potassium, $^{38}_{19}\text{K}$, is a positron emitter. Describe the transmutation that takes place as a result, and identify the product(s).

Solution

The nuclear equation for the process is

$$^{38}_{19}\text{K} \longrightarrow {}^{38}_{18}\text{Ar} + {}^{0}_{1}\text{e}$$

The product has the same mass number as that of potassium but, because the positive nuclear charge has been reduced by one, its atomic number has decreased by one. The periodic table reveals the new element to be argon.

Problem 10.3 Sodium-21 is a radioactive positron emitter. Write the nuclear equation for the event, and identify the product(s).

The emission of an α- or a β-particle often leaves the nucleus of the product in an excited state. Recall from Section 2.7 that, when an atom is in an excited state, it can return to its ground state by losing the energy of excitation as emitted light—that is, as a photon. The photons emitted by excited atomic nuclei, however, are far more energetic than the photons emitted by the return of excited electrons to the ground state. Moreover, they are invisible. These photons are the γ-rays, or γ-radiation, emitted by radioactive nuclei. Because γ-ray emission does not alter the mass of a radioactive nuclide, it is usually not included in nuclear equations.

Radioactivity can be induced in nonradioactive elements by bombarding them with high-energy nuclear particles. The first artificial conversion of one nucleus into another was performed by Ernest Rutherford in 1919. He bombarded nitrogen-14 (the target) with α-particles (the projectiles) emitted by radium and produced oxygen-17. The reaction is

$$^{14}_{7}N + ^{4}_{2}He \longrightarrow ^{17}_{8}O + ^{1}_{1}H$$

Since that time, hundreds of radioisotopes have been produced in the laboratory.

In order for charged particles such as α-particles to penetrate a positively charged nucleus or even the electron cloud surrounding the nucleus, they must be moving at extremely high velocities. Such velocities are achieved in machines called particle accelerators such as the cyclotron, the Van de Graaf generator, and the betatron. All the elements above atomic number 92 have been created in this manner. None of them occur naturally.

Most radioisotopes used in medicine have been produced with the use of neutrons as subatomic projectiles. For example, cobalt-60, a γ-emitter used in radiation therapy for cancer, is prepared by bombarding iron-58 with neutrons (obtained from a nuclear reactor; Section 10.9). The successive reactions are

$$^{58}_{26}Fe + ^{1}_{0}n \longrightarrow ^{59}_{26}Fe$$

$$^{59}_{26}Fe \longrightarrow ^{59}_{27}Co + ^{0}_{-1}e$$

$$^{59}_{27}Co + ^{1}_{0}n \longrightarrow ^{60}_{27}Co$$

10.4 RADIOACTIVE DECAY

Radioactive decay is the loss of radioactivity when a radioactive element emits nuclear radiation. The decay of radioisotopes found in nature results in the formation of products called **daughter nuclei,** which may or may not be radioactive. If the resulting products are nonradioactive (stable), the decay stops. However, if the products are radioactive, decay will continue to produce new elements until a stable state is achieved.

A series of nuclear reactions that begins with an unstable nucleus and ends with the formation of a stable one is called a **nuclear decay series** or **nuclear disintegration series.** Three such series are found in nature. One series begins with uranium-238, proceeds through 14 decay steps, and terminates with lead-206. One of the daughters of this decay series is radium-226, which decays to a number of radon radionuclides. Of these radionuclides, radon-222 is a major health hazard and comprises almost 50% of the background radiation on Earth. Background radiation will be more fully discussed in Box 10.4 on page 277. Another series begins with uranium-235 and terminates with lead-207. The third begins with thorium-232 and terminates with lead-208.

The rate at which a particular radionuclide decays is as characteristic of that nuclide as the melting point is of a pure compound. Some radionuclides decay in millionths of a second, others in millions of years. Decay is independent of the variables, such as temperature, pressure, and concentration, that affect chemical reactions. It is an intrinsic property of the unstable nucleus and is outside of human control. For this reason, the disposal of nuclear wastes

Figure 10.2 Relation between
radioactive decay and half-life.
Fifty percent of the radioactivity
of a sample is lost at the end of
every half-life.

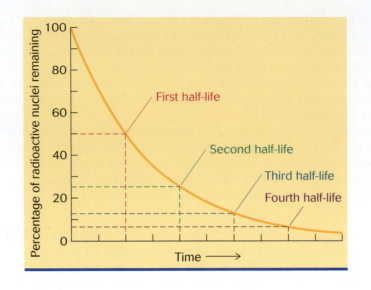

is very difficult and becomes one of the considerations in the selection of
radioisotopes used for diagnosis and therapy.

The rate of radioactive decay can be described as the loss of a constant
percentage of the original element per unit time. Chemists measure it in
terms of the time required for 50% of the original material to decay. This
period of time is called the **half-life**, $t_{1/2}$. Because a constant percentage of
the element decays in a given period of time, 50% of the original amount is
lost in the first half-life, 50% of the remainder will decay in the second half-
life, 50% of what remains after that will decay in the third half-life, and so
forth. The relation between half-life and radioactive decay is illustrated in
Figure 10.2.

Example 10.4   Determining the relation between half-life and percentage of decay

Estimate how much of a radioisotope will be left after four half-lives.

Solution
The simplest way to arrive at this estimate is to put the problem into tabular form.

End of half-life	Amount left (%)
0	100.00
1	50.00 (half the original amount)
2	25.00 (half the preceding remainder)
3	12.50 (half the preceding remainder)
4	6.25 (half the preceding remainder)

At the end of four half-lives, 6.25% of the original amount of the radioisotope
will be left.

Problem 10.4   Estimate how much of a radioisotope will be left after
six half-lives.

We can show that the percentages arrived at in Example 10.4 and Fig-
ure 10.2 are the result of repeated multiplications of the factor $\frac{1}{2}$, once for
the first half-life, twice to get the second half-life result, three times to get the
third half-life result, and so forth.

$$\frac{1}{2} = 0.5$$

$$\frac{1}{2} \times \frac{1}{2} = \left(\frac{1}{2}\right)^2 = \frac{1}{4} = 0.25$$

$$\frac{1}{2} \times \frac{1}{2} \times \frac{1}{2} = \left(\frac{1}{2}\right)^3 = \frac{1}{8} = 0.125$$

$$\frac{1}{2} \times \frac{1}{2} \times \frac{1}{2} \times \frac{1}{2} = \left(\frac{1}{2}\right)^4 = \frac{1}{16} = 0.0625$$

These repeated multiplications can be summarized in a convenient equation. If the amount of radioisotope at time zero (before decay begins) is defined as N_0, and the amount remaining after n half-lives is N, then the fraction of isotope remaining after n half-lives is

$$\frac{N}{N_0} = \left(\frac{1}{2}\right)^n$$

and the percentage of isotope remaining after n half-lives is

$$\frac{N}{N_0} \times 100\% = \left(\frac{1}{2}\right)^n \times 100\%$$

Conversion of the fractions remaining after one, two, three, and four half-lives into percentages gives

$$0.5 \times 100\% = 50\%$$
$$0.25 \times 100\% = 25\%$$
$$0.125 \times 100\% = 12.5\%$$
$$0.0625 \times 100\% = 6.25\%$$

Example 10.5 Determining the amount of radioisotope remaining after an integral number of half-lives

Calculate the percentage of radioisotope remaining after 7.0 half-lives.

Solution
The percentage of radioisotope left after any number of half-lives can be calculated by using the following relation:

$$\frac{N}{N_0} \times 100\% = \left(\frac{1}{2}\right)^n \times 100\%$$

where N = amount left after a number, n, of half-lives, and N_0 is the amount of isotope at the start of the process. You can calculate $(0.5)^n$ on your calculator in several ways. There may be a special function key labeled Y^x. To use it, enter 0.5 (Y equals 0.5), press the Y^x key, enter n, press the enter key, and read the result. In the absence of a special key, enter 0.5, and multiply it n times by 0.5.

The percentage of a radioisotope left after seven half-lives is:

$$\frac{N}{N_0} \times 100\% = \left(\frac{1}{2}\right)^{7.0} \times 100\% = \frac{1}{128} \times 100\% = 0.78\%$$

Problem 10.5   Calculate the percentage of radioisotope remaining after 5.0 half-lives.

Example 10.6 Determining the number of half-lives from the amount of remaining radioisotope

A nurse locates the iodine-131 prescribed for measurement of thyroid activity and finds that it was produced 21 days before. What percentage of the isotope is left in the sample?

Solution

This problem is solved by calculating how many half-lives have taken place since the isotope's manufacture. We find the half-life of the isotope in Table 10.2 to be 8.05 days. Therefore the number of half-lives is

$$21 \text{ days} \left(\frac{\text{half life}}{8.05 \text{ days}} \right) = 2.6 \text{ half-lives}$$

As in Example 10.5, the percentage of radioisotope left after any number of half-lives can be calculated by using the relation

$$\frac{N}{N_0} \times 100\% = \left(\frac{1}{2} \right)^n \times 100\%$$

Substituting the value for half-lives into this equation, we get

$$\frac{N}{N_0} \times 100\% = \left(\frac{1}{2} \right)^{2.6} \times 100\%$$

We can solve this equation as we did in Example 10.5:

$$\left(\frac{1}{2} \right)^{2.6} \times 100\% = 17\%$$

Problem 10.6   Calculate the percentage of radioisotope remaining after 3.8 half-lives.

Table 10.2 lists the half-lives of some biomedically useful radioisotopes. We will learn more about their practical applications in Section 10.6. In addition, radioisotopes such as carbon-14, hydrogen-3 (tritium), phosphorus-32, and sulfur-35 are important in biochemical research. They are incorporated into biochemical intermediates by synthesis. The resultant radioactivity, or "label," of the intermediates makes it possible to map their progress through the metabolic pathways in which they participate. An understanding of radioactive isotopes and their half-lives has also enabled scientists to assess the ages of ancient rocks, fossils, and human artifacts. Radiocarbon dating is discussed in Box 10.1.

TABLE 10.2   Half-Lives and Biomedical Applications of Some Radioisotopes

Nuclide	Half-life	Emission	Application
barium-131	11.6 days	gamma	detect bone tumors
carbon-11	20.3 min	positron	PET brain scan
chromium-51	27.8 days	gamma	determine blood volume and red-blood-cell lifetime
gold-198	64.8 h	beta	assess kidney activity
iodine-131	8.05 days	beta	measure thyroid uptake of iodine
iron-59	45 days	beta	assess blood-iron metabolism
krypton-79	34.5 h	gamma	assess cardiovascular function
phosphorus-32	14.3 days	beta	detect breast carcinoma
selenium-75	120 days	beta	measure size and shape of pancreas
technetium-99	6.0 h	gamma	detect blood clots
indium-111	2.8 days	gamma	label blood platelets
gallium-67	78 h	gamma	diagnose lymphoma and Hodgkin's disease
chromium-51	27.8 days	gamma	diagnose gastrointestinal disorders

*The m stands for metastable.

The reason for advising people to stay as far away as practicable from a radiation source is that radiation of all kinds follows an inverse-square law with respect to distance:

$$I \propto \frac{1}{d^2}$$

where I, the intensity of the radiation, is proportional to the reciprocal of the square of the distance d from the origin of the radiation. The radiation intensity falls off very rapidly as the distance from the source increases. The effect of this law is calculated in Example 10.7.

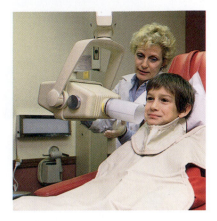

A technician positions a device to take an X-ray of a patient's teeth. (D. Wray/The Image Works.)

Example 10.7   Calculating the effect of increasing distance on radiation intensity

If the distance between a source and target is 2 m, how far should the source be moved from the target to decrease the radiation intensity to one-fourth (1/4) of its current value?

Solution

A ratio of two intensities can be constructed by using the equation introduced earlier:

$$\frac{I_{new}}{I_{old}} = \frac{\left(\dfrac{1}{d^2_{new}}\right)}{\left(\dfrac{1}{d^2_{old}}\right)}$$

This equation simplifies to

$$\frac{I_{new}}{I_{old}} = \frac{d^2_{old}}{d^2_{new}}$$

Substituting appropriate values into these ratios gives

$$\frac{1}{4} = \frac{4\ \text{m}^2}{d^2_{new}}$$

$$d^2_{new} = 16\ \text{m}^2$$

$$d_{new} = 4\ \text{m}$$

Irrespective of the specific values of distance and intensity, a doubling of the distance causes the radiation intensity to decrease by a factor of four.

Problem 10.7   If the distance between a source and target is 6 m, how far should the source be moved from the target to decrease the radiation intensity to one-fourth of its current value?

10.6   DETECTION OF RADIOACTIVITY

Nuclear emissions, as already stated, are quite energetic and leave a trail of ions behind them as they penetrate through matter. These ions serve as evidence that radiation is present. Many techniques have been devised for detecting them.

The most common and inexpensive method of detection is by the use of photographically sensitive film that becomes progressively darker as its exposure to radiation increases. Film badges are routinely worn by scientists and technicians who work with radioactive materials or around radiation sources. The badges can differentiate between different types of radiation. Gamma-rays penetrate much more matter than do β-particles, which in turn penetrate more matter than do α-particles before they are stopped; so covering photographic film with different thicknesses of material makes it possible to identify whatever type of radiation is present.

An ionizing smoke detector uses the basic principle of the Geiger counter. When smoke particles enter the chamber of an ionizing smoke detector, however, they attract the ionized air molecules already present and reduce the electric current. (Michael Dalton/ Fundamental Photographs.)

Amplifier
and counter

Window

Argon gas

Positive
electrode

Negative
electrode
(cylinder case)

Figure 10.4 Diagram and
photograph of a Geiger counter.
Radiation enters through a thin
window in the cylinder, ionizing
the argon gas inside. The ions then
carry electrical current between the
cathode (negative electrode) and the
anode (positive electrode), which is
detected by the amplifer and counter.
(Yoav Levy/Phototake.)

Figure 10.5 A scintillator, such as
those in the photograph, emits a flash
of light when a charged particle passes
through it. The color of the emitted
light depends on the type of material.
(Courtesy of BICRON, a division of Saint-
Gobain Industrial Ceramics, Inc.)

Other devices, such as the **Geiger counter,** use electronics to detect the
ions created when radiation passes through matter. The counter, shown in
Figure 10.4, consists of a tube fitted with a thin window and filled with argon
gas. A positively charged electrode runs down the center of the tube, and the
tube itself serves as the negative electrode. When a photon or particle passes
through the window into the tube's interior, it creates positive argon ions in
the gas, which are collected by the negative electrode. The resulting current
can be measured on a meter or heard audibly as a click. Each burst of current
signals the presence of an ionization event. The same ionization effect is used
in one type of smoke detector.

In devices called **scintillation counters,** radiation generates flashes of
light: the more flashes of light per unit time, the greater the dosage of radiation in the environment. A scintillation counter uses a device called a scintillator, which contains substances that emit a flash of light when struck by an
energetic photon or particle (Figure 10.5). These substances—sodium and

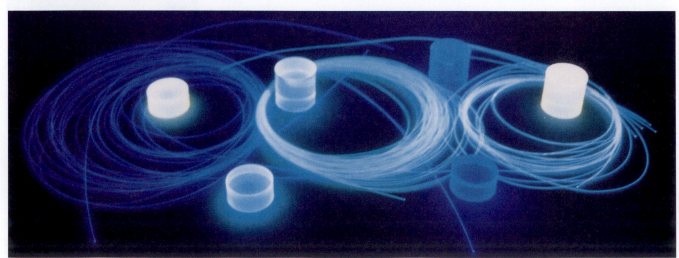

thallium iodides, zinc sulfide, and others—are similar to the substances that cover the inner surface of television tubes. The light is then converted into an electrical signal by a phototube.

10.7 MEASURING RADIOACTIVITY

The general approach to measuring the radioactivity of a sample is to measure the disintegrations per second (dps)—that is, the number of atoms that decay per second. The most commonly used unit, the **curie (Ci),** represents the number of radioactive disintegrations per second in a 1.0-g sample of pure radium. It is named after the discoverer of radium, Marie Sklodowska Curie.

$$1 \text{ Ci} = 3.7 \times 10^{10} \text{ dps}$$

This level of radioactivity is very intense; so it is usually more convenient to report a measurement in millicuries (mCi), microcuries (μCi), or picocuries (pCi). The SI unit of radioactivity is the becquerel (Bq), after the scientist credited with discovering radioactivity.

$$1 \text{ Bq} = 1 \text{dps}$$

Radiation can also be quantified by the number of ions produced by the radiation passing through air. The unit describing this mode of detection is the roentgen (R), named after the discoverer of X-rays. It is used to measure the exposure to X-rays or γ-rays and provides an estimate of exposure to high-energy radiation.

$$1 \text{ R} = 2.1 \times 10^9 \text{ units of charge/cm}^3 \text{ of dry air}$$

Table 10.4 lists these types of radiation units and others, as well as the situations in which they are appropriately applied.

TABLE 10.4	Units of Radiation, Their Symbols, and Definitions	
Unit	**Symbol**	**Definition**
curie	Ci	An older non-SI unit that describes the activity of a radioactive sample. It is the number of disintegrations per second (dps) in 1.0 g of pure radium: $1 \text{ Ci} = 3.7 \times 10^{10} \text{ dps}$ This activity is extremely high. More commonly, activity is reported in milli- or microcuries: $1 \text{ μCi} = 3.7 \times 10^4 \text{ dps}$
becquerel	Bq	An SI unit that describes the activity of a radioactive sample: $1 \text{ Bq} = 1 \text{ dps}$ $3.7 \times 10^{10} \text{ Bq} = 1 \text{ Ci}$
roentgen	R	A non-SI unit of exposure to γ- or X-radiation based on the amount of ionization produced in air. 1 R will produce 2.1×10^9 ions in 1.0 cm³ of dry air at 0°C and 1.0 atm pressure.
rad	rad	A measure of the energy absorbed in matter as a result of exposure to any form of radiation. It does not reveal the biological effects of the radiation. $1 \text{ rad} = 0.01 \text{ J absorbed per kilogram of matter}$
gray	Gy	An SI unit that describes the energy absorbed by tissue. Like the rad, it does not reveal the biological effects of radiation. $1 \text{ Gy} = 1 \text{ J absorbed per kilogram of tissue}$ $1 \text{ Gy} = 100 \text{ rad}$
rem	rem	A measure of the energy absorbed in matter as a result of exposure to specific forms of radiation. This unit takes into account the fact that different forms of radiation produce different biological effects. 1 rem is equal to 1 rad times a factor dependent on the type of radiation.
sievert	Sv	An SI unit that, like the rem, takes into account the different biological effects caused by different forms of radiation. $1 \text{ Sv} = 100 \text{ rem}$

10.3 CHEMISTRY WITHIN US

Radiation Dosimetry and Wristwatches

In the late 1930s, many cases of severe radiation sickness were discovered among workers in a factory that produced watch faces on which the numerals glowed in the dark. Workers painted numerals by hand with paint containing very low concentrations of a radium salt and a phosphorescent material. The radiation from radium caused the phosphor to glow in the dark. The work required precise location of the radioactive paint, and so the workers used very thin brushes and kept the brush ends pointed by licking them with their tongues. In this way, α-emitting solids came to be embedded in the tissue of their mouths and tongues, causing radiation damage that eventually led to cancer.

In 1941, Robley Evans, a physician, completed a study of these workers and others whose jobs entailed exposure to radiation. It led to the development of standards for maximum allowable lifetime exposure that are still in use today. Such standards are crucial if scientists are to safely conduct experiments with materials such as the radioactive isotopes routinely used in medicine for diagnosis and therapy.

What if we wish to predict the amount of biological damage likely to be caused by a given amount of radiation? The number of disintegrations per unit time of a radioactive sample or the number of ionizations in a volume of air will not tell us much in this regard. We saw earlier that biological damage depends to a great extent on the type of radiation to which the organism is exposed. Two units, the rad and the rem, have been devised to take this correlation into account.

As stated earlier, X-rays and γ-rays are particularly effective in penetrating human tissue, α-particles are stopped by skin, and β-particles may penetrate to a depth of 5 to 10 mm. Alpha-radiation and β-radiation from external sources are therefore not ordinarily considered dangerous. Nevertheless, when solid particles emitting these radiations enter the body and are not excreted quickly, they can do a great deal of damage in the tissue surrounding the emitting particle (Box 10.3). The experience of workers in the plutonium-purifying industry has shown that, if particles become lodged in the lungs, cancer is a very likely result.

The **rad** is a measure of the energy absorbed in matter as a result of exposure to any form of radiation. To convert this information into a measure of the biological damage likely to be induced by radiation, the rad is multiplied by a factor known as the **RBE (relative biological effectiveness)**. The exact value of the RBE depends on the type of tissue irradiated, the dose rate, and the total dose, but it is approximately 1 for X-rays, γ-rays, and β-rays and 10 for α-particles, protons, and neutrons. The product is the **rem (roentgen equivalent man)**.

$$\text{rads} \times \text{RBE} = \text{rems}$$

Note that a rad of α-rays will produce more damage than a rad of β-particles. An SI unit called the **sievert (Sv)** measures a similar property—the effect of absorbed radiation on different kinds of tissues and under different situations. One sievert = 100 rem. The clinical effects of short-term exposure to radiation are listed in Table 10.5.

When we weigh the relative hazards posed by different types of radiation, it is important to remember that all life forms on Earth are continuously exposed to low levels of **background radiation** from natural sources such as cosmic rays and radioactive elements in soil and rocks. A typical exposure to this radiation is about 350 mrem per year (1 millirem = 1×10^{-3} rem). About 50% of this background radiation is from radon, a naturally occurring radioactive noble gas formed from the decay of radioactive

TABLE 10.5	Effects of Short-Term Exposure to Radiation
Dose (rem)	Clinical effect
0–25	undetectable
25–50	temporary decrease in white-blood-cell count
100–200	large decrease in white-blood-cell count; nausea; fatigue
200–300	immediate nausea, vomiting; delayed appearance of appetite loss, diarrhea; probable recovery in 3 months
500	within 30 days of exposure, death for 50% of the exposed population

radium-226 (Box 10.4). Additional background radiation from manufactured sources—X-rays, radiotherapy, radioisotopes—is about 65 mrem per year. Table 10.5 indicates that background radiation is too weak to be considered dangerous.

10.4 CHEMISTRY WITHIN US

Radon: A Major Health Hazard

The U.S. Environmental Protection Agency (EPA) estimates that from 5000 to 20,000 deaths from lung cancer yearly are caused by radon-222 gas. Radon-222 is a naturally occurring radionuclide derived from the radioactive decay of uranium ores. It is one of the daughters arising in the 14-member decay series from uranium to lead. Because uranium ores are present in many types of rocks and soil, radon-222 and its decay products also are commonly found components of our environment.

The normal average background radon concentration is about 0.2 pCi (1 picocurie = 1×10^{-12} curie, see Table 10.4) per liter of air, constituting about 50% of natural background radiation. The EPA's maximum recommended level is 4 pCi/L of air. However, when radon-222 first came to the attention of authorities, the reason was that it was found at concentrations of more than 2500 pCi/L of air in a house in Pennsylvania. It was detected there because it was the home of an engineer who worked at a nearby nuclear energy plant. He kept setting off the radiation alarms when he reported for work because his clothes were contaminated with radon-222 and its decay products, which had been picked up from the air in his house.

Because radon is a noble gas, most radon-222 is exhaled, unchanged, after spending only a short time in the lungs. However, two of its decay daughters—polonium-218 and polonium-214—are radioactive α-emitters like their parent, with the insidious difference that they are solids and can remain in contact with lung tissue. Thus polonium-218 and polonium-214 are able to remain in the lungs to irradiate tissue, damage cells, and possibly lead to cancer.

(Blair Seitz/Photo Researchers.)

The radon-222 hazard is higher inside houses than outside because the ventilation indoors is restricted. The gas can be detected with inexpensive commercial test kits and then, as long as the house is not constructed of uranium-ore-containing granite, its concentration can be significantly reduced by adequate ventilation in the basement and careful sealing of any cracks and openings in the foundation.

10.8   APPLICATIONS

Current knowledge of the properties of radiation and its interaction with matter has led to the development of a wide variety of applications, particularly in medical and biological science. Much of the recent research in molecular biology, metabolism, and cell biology would have been impossible without radioisotopes.

X-Rays and Medical Imaging

The main use of radiation is in medical imaging. Because at least 80% of medical imaging is done with X-rays, we will consider their fundamental properties—that is, what they are, where they come from, and how they are used for medical imaging. Let's begin with how they are produced.

X-rays are generated in a vacuum tube containing a coiled wire filament, which serves as an anode, and a metal target, which serves as the cathode. When the filament is heated, electrons are literally boiled off and, with a potential difference between the metal target and the filament, the electrons will be attracted and accelerated to the target. The collision between the incoming electrons and the atoms of the target results in the accelerated electron's loss of energy, and that lost energy is converted into X-rays. In Section 10.1, we said that the energy of an electron is characterized by a frequency and, in this case, by very high frequencies (Table 2.4).

We may therefore say that X-rays are high-energy photons generated by the collisions of high-energy electrons with positively charged target atoms. It is important to note that less than 1% of the electron-beam energy is transferred to X-ray energy. Therefore, most of the energy lost from the accelerated electrons appears as heat, which must be dissipated at the cathode. Consequently, tungsten, with its high melting point, is the usual choice as the target material.

Two types of X-rays are generated when high-energy electrons collide with a metal target. First, incoming electrons can strike an electron in the inner $1s$ shell of the target atom. If the incoming electron's energy is great enough, the impact can cause the $1s$ electron's ejection from the atom (Chapter 2). Consequently, an atomic electron from a higher orbit will fill the vacancy, and the difference in energy between the upper-level atomic electron and the inner atomic shell will be radiated as an X-ray of very high energy. These X-rays are characteristic of the material used for the target and are called **K-shell X-rays.** They are too energetic for medical radiology and are typically used for identification of the metals used as targets.

Second, when the incoming electrons strike the target, they may closely approach the nuclei of the target atoms but not strike any atomic electrons. In that case, the incoming electrons are only attracted to the positively charged nuclei. This attraction changes the electrons' trajectories and slows them down. When this occurs, they lose energy, and this energy is emitted as X-rays. These X-rays are called **Bremsstrahlung,** a German word that means "braking radiation," radiation that is the result of slowing down or "braking" the incoming electrons. Bremsstrahlung is the only radiation used for X-ray imaging because its energy can be controlled. Figure 10.6 illustrates the results of altering the path of incoming electrons by atomic nuclei.

Bremsstrahlung emerges from an X-ray tube scattered in all directions and possesses a wide range of energies. The X-rays cannot be allowed to flood the room, and so the tube is heavily shielded to prevent radiation damage. In addition, to be useful for imaging, a small window is left open in the direction of a detector. However, much of the bremsstrahlung passing through that window possesses energies too high for exposure to tissue. Therefore, an additional filter, usually aluminum, is placed in the window to allow the passage of only those X-rays of energies compatible with tissue.

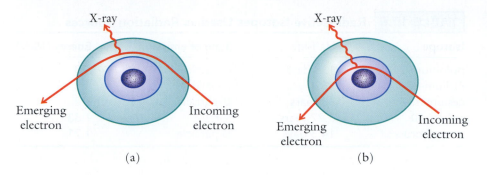

X-ray

Emerging
electron

Incoming
electron

(a)

X-ray

Emerging
electron

Incoming
electron

(b)

Figure 10.6 Incoming electrons are deflected from their original path by attractions of highly positively charged atomic nuclei of the metal target. The electron in part *b* that comes closer to the nucleus is slowed more than the electron in part *a*. The electron in part *b* therefore loses more energy and, as a result, produces more-energetic X-rays.

The energies of X-rays produced in the X-ray tube depend on the electron velocities, which in turn depend on the potential difference between the heated filament and the target cathode. The greater the potential difference, the greater the electron velocities. By varying that potential difference, the amount and energy of the resultant bremsstrahlung can be regulated, as illustrated in Figure 10.7

The ultimate goal of X-ray imaging is to produce sharply defined images of the hidden structures of a patient's anatomy. It begins with the passage of X-rays through tissue where they meet structures of different densities. Because of this variation in density, the X-rays are absorbed differentially. This differential absorption is the origin of the images produced when the X-rays emerging from tissue reach a detector.

Photographic plates have been the main detector used for recording X-ray images ever since Wilhelm Roentgen produced an X-ray image of his wife's hand complete with wedding band in 1895. Since that time, we have learned that exposure to X-rays can have serious biological consequences; so a short radiation exposure time is critical. In addition, a short radiation exposure time allows us to freeze the motion of internal organs. One of the most important advances in reducing exposure time is the use of solid-state detectors of great sensitivity for recording images. Solid-state detectors work in much the same way as a television tube does. They produce digital records that have the advantages of being immediately available for reading, electronically transmissible, enlargeable, and easily stored. These new types of detectors have been coupled with computer methods to expand medical radiology in many new directions to be discussed in the subsection titled Radiation and Computer Imaging Methods.

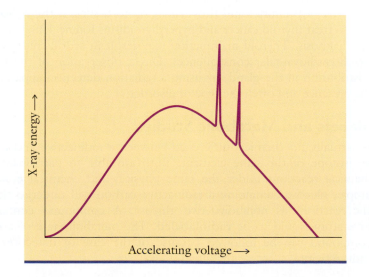

X-ray energy →

Accelerating voltage →

Figure 10.7 The intensity of filtered X-radiation as a function of increasing accelerating voltage. K-shell radiation appears as spikes above the bremsstrahlung.

TABLE 10.6	Radioactive Isotopes Used as Radiation Sources		
Isotope	Half-life	Type of emission	Energy (MeV)
polonium-210	138 days	α-particle	5.3
radium-226	1620 years	α-particle	4.8
cesium-137	30 years	β-particle	1.2
cobalt-60	5.27 years	γ-ray	1.33
phosphorus-32	14.3 days	β-particle	1.71

Radioisotopes

Radioisotopes introduced into the body are used for diagnosing pathological conditions as well as for therapy. The characteristic emission of a radioactive isotope is the decisive factor in its choice either as a diagnostic or as a therapeutic tool.

- Diagnostic use requires that the radiation have significant penetrating power to be accurately detected; that is, it should be primarily a γ-emitter.

- Therapeutic use requires intentional damage to abnormal (cancerous) tissue; therefore, the isotope should be an α- or β-emitter.

Another key factor in a physician's decision to make use of a radionuclide for either of these purposes is that the benefits of the radiation must outweigh the risks. Important considerations include the intensity of radiation—the shorter the half-life, the more intense the radiation—and the rate of excretion. If the rate of excretion is too low, unwanted radiation damage will occur. If it is too high, the isotope will not remain at the irradiation site long enough to have its desired effect. Moreover, any decay products should not be radioactive.

Some isotopes have a natural tendency to concentrate in particular cells of the body. In such cases, the isotope can be administered as a solution, whether orally or by injection. For example, the thyroid gland secretes thyroxin, an iodine derivative of the amino acid tyrosine. As a result, radioactive iodine, which is often used to treat thyroid disorders, will have a tendency to concentrate there. When an isotope is not one that selectively concentrates in the target organ, it can instead be inserted into the subject tissue as a package in a metal or plastic tube. Table 10.2 lists radionuclides used in diagnosis, and Table 10.6 lists some radionuclides used solely in therapy, along with their half-lives and the types of radiation emitted.

Iodine-131, a β-emitter and thus damaging to tissues, is the radionuclide often used to treat thyroid cancer and hyperthyroidism. Iodine-123, a γ-emitter, is used in diagnosis. Thyroid malfunction may result in serious metabolic disorders, with cardiovascular consequences. Direct observation of the radionuclide's distribution in the gland, by using a radiation-detection apparatus, will reveal the presence and often the type of disorder.

Radiolabels and Metabolic Studies

There is virtually no chemical difference between a radioactive and a nonradioactive isotope of the same element. Both are able to participate in the same chemical reactions under the same physical and chemical conditions. For example, glucose containing a radioactive carbon will undergo the same metabolic reactions as nonradioactive glucose. A compound containing a radioactive atom is said to be **radiolabeled.** The metabolic fate of a radiolabeled compound can be followed because of the ease of detection of the label's radioactivity.

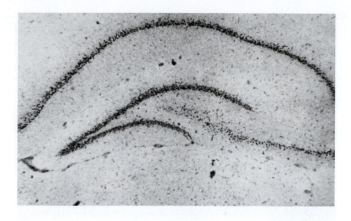

Figure 10.8 An autoradiograph showing the localization of steroid receptors in the brain of a rat injected with radioactive estradiol. The hormone was localized in the hippocampal region of the brain. (Reprinted with the permission of Bruce McEwen of the Rockefeller Institute.)

One technique used to follow the fate of a radiolabeled compound is called **autoradiography.** This technique recognizes radiolabeled substances by their effects on photographic film. Melvin Calvin made use of it in the research that led to his discovery of the chemical steps of glucose photosynthesis. Calvin fed radiolabeled bicarbonate ($H^{14}CO_3^-$) to actively growing algae. After short periods of growth, the cells were killed and broken, and chromatography was used (see Box 1.2) to separate all the different molecular components on a single sheet of paper. Each component ended up in a unique position on the paper. When the paper was covered with photographic film, the radiolabeled compounds revealed their positions by darkening the film wherever they were located. The identity of each labeled compound was learned by comparing its position on the chromatogram with reference chromatograms, which had been made by using known compounds. Calvin was awarded a Nobel Prize in 1961 for this work. The same techniques are used today to study the physiological distribution and metabolic fate of drugs.

Autoradiography also uses photographic film to locate radiolabeled compounds in undisrupted cells and tissues. In a typical experiment, living tissue is exposed to a radiolabeled compound and then allowed to continue its normal metabolic activity for a while. When enough time has passed, a sample of tissue is removed, fixed, and mounted on slides. (To fix a tissue sample means to treat it with a preparation that stops all further chemical action and preserves the structures of the cell.) Next, the slides are covered with photographic film. The location of the label within the tissue is revealed by the darkening of the corresponding location on the film. An image obtained from such an experiment can be seen in Figure 10.8.

Radiation and Computer Imaging Methods

Advances in radiochemistry and computer technology have combined to produce powerful new scanning methodologies for the diagnosis of injury and disease. The general approach common to many of these techniques is to direct radiation through the body and then detect the degree of attenuation of the radiation when it emerges. The information captured in this way is translated by a computer into an image of the body's inner structures.

In one of these methods, called a **CAT scan (computer-assisted tomography),** short bursts of X-rays are directed toward some part of the body. The X-rays are absorbed differentially, and the emerging radiation is detected by a circular array of detectors surrounding the body. The detected X-rays are then transformed by a computer program into an image of the internal structures.

In **magnetic resonance imaging,** or **MRI,** the body is placed in the field of a strong electromagnet and simultaneously exposed to a radiofrequency field. Protons, as well as many other atomic nuclei, behave as tiny magnets and

Figure 10.9 PET scan of a normal brain and that of a patient with Alzheimer's disease. The differences in brain scans are easily detected. (N. I. H/ Science Source/Photo Researchers.)

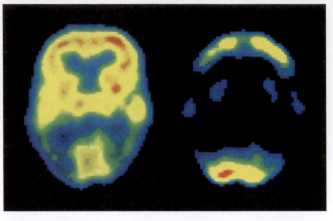

Normal   Alzheimer's disease

will tend to become aligned in an intense magnetic field. These tiny magnets also interact with energy from the radio signals and absorb it, to a degree that depends on the protons' local molecular environment. The signals detected during MRI (by detectors arrayed about the body, similar to the arrangement used in CAT scans) correspond to the differences in energy absorption, and a computer is able to transform these differences into an image of soft-tissue structures within the body. The energies needed for MRI imaging are much smaller than the energies need for a CAT scan and therefore allow much longer scan times.

In the technique known as a **PET scan (positron emission tomography),** a solution of positron-emitting isotope is injected into the body and produces γ-radiation that is detected outside the body by arrays of detectors similar to those used in X-ray tomography (the CAT scan). When an emitted positron encounters a nearby electron, both are annihilated, producing two γ-rays traveling in opposite directions. The most commonly used isotopes are O-13, N-13, C-11, and F-18. The half-lives of these positron emitters range from 2 to 110 min. Thus, the key obstacle in the use of this technique has been the need to develop rapid chemical syntheses to incorporate the isotopes into metabolically useful compounds.

Positron emission tomography has been extensively used to study disorders in the brain. Glucose, labeled with the positron-emitting radionuclide carbon-11, is often used in such studies. After the glucose is fed intravenously to the subject (and enters the brain), the positrons that it emits quickly combine with electrons from any nearby molecules (for example, water) to create γ-rays. Because the brain uses glucose exclusively for energy, the internally generated γ-rays produce brain scans (like the one in Figure 10.9) that can reveal disorders in metabolism indicative of a variety of pathologies. Recent work has centered on the metabolism of neurotransmitters—compounds such as acetylcholine and norepinephrine that effect the transfer of nerve impulses between neurons. PET scans provided the first evidence that acetylcholine concentrations are significantly lower than normal in the brains of patients with Alzheimer's disease.

10.9 NUCLEAR REACTIONS

In 1938, scientists found that, when uranium is bombarded with neutrons, its atoms fragment into smaller nuclei—such as barium, krypton, cerium, and lanthanum—and enormous amounts of energy are released. The general explanation is that the uranium nucleus absorbs the high-energy neutrons, becomes destabilized as a result, and responds by spontaneously decomposing into more-stable products with the evolution of much heat. The new process was given the name **nuclear fission.**

There are many possible products of uranium fission. More than 200 isotopes of 35 different elements have been found as a result of the fission of uranium-235. Two possible outcomes are

$$^{235}_{92}U + {}^{1}_{0}n \longrightarrow {}^{139}_{56}Ba + {}^{94}_{36}Kr + 3\,{}^{1}_{0}n$$

$$^{235}_{92}U + {}^{1}_{0}n \longrightarrow {}^{144}_{54}Xe + {}^{90}_{38}Sr + 2\,{}^{1}_{0}n$$

These equations provide a key to the power and danger of nuclear fission. Note that, when one neutron is absorbed by one uranium nucleus and destabilizes it, at least two neutrons are released as a result. These fission-produced neutrons, in their turn, can each destabilize another uranium nucleus and result in the release of still more fission-produced neutrons. In other words, the multiple yield of neutrons results in an exothermic chain reaction that grows exponentially—explosively—and produces extremely large amounts of energy. This is the operating principle for the atomic bomb.

Nuclear reactors, which use fission to produce energy, keep the reaction to manageable levels with the use of control rods made of cadmium, an efficient absorber of neutrons. The control rods are arranged and deployed as shown in Figure 10.10. A nuclear power plant that produces electricity is diagrammed in Figure 10.11. The heat released in the fission reaction is used to drive steam turbines that in turn generate electricity. Although we have stopped building nuclear plants in the United States, other countries—for example, France and Japan—use nuclear power extensively to generate electricity.

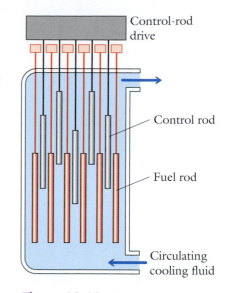

Figure 10.10 Diagram of the control mechanism in a nuclear reactor. The fuel rods contain fissionable material. The control rods are made from materials that are good absorbers of neutrons. The density of neutrons in the reactor core can be controlled by raising or lowering the control rods. Controlling the neutron density controls the rate of energy production.

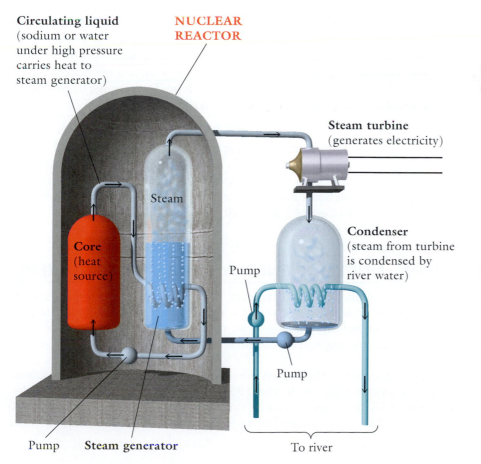

Figure 10.11 Heat transfer in a nuclear reactor. The heat produced in the core is transferred by a closed loop of liquid sodium metal or liquid water to a steam generator. The steam produced in the generator runs a steam turbine that produces electricity. The steam from the turbine must be cooled and pumped back into the generator. Cooling can be accomplished by using cooling towers (evaporative cooling) or by using nearby sources such as a river.

Nuclear fusion, a process that seems very much like the opposite of nuclear fission, also is able to produce large amounts of energy. In **nuclear fusion,** the nuclei of two light elements collide at velocities high enough to overcome the mutual repulsion of the nuclear protons and fuse to form a nucleus of higher atomic number. Nuclear fusion is the source of the energy continuously generated in stars such as our sun, where protons, deuterons ($^2_1$H), and helium nuclei collide successively as follows:

$$^1_1\text{H} + ^1_1\text{H} \longrightarrow ^2_1\text{H} + ^0_1\text{e}$$

$$^1_1\text{H} + ^2_1\text{H} \longrightarrow ^3_2\text{He}$$

$$^3_2\text{He} + ^3_2\text{He} \longrightarrow ^4_2\text{He} + 2\,^1_1\text{H}$$

$$^3_2\text{He} + ^1_1\text{H} \longrightarrow ^4_2\text{He} + ^0_1\text{e}$$

Nuclear fusion reactions require a temperature of about 4×10^7 K to overcome the mutual repulsions of the various nuclei. When fusion occurs, it releases vast quantities of energy, more than enough to sustain stellar fusion processes (see Section 10.10).

Research on the use of nuclear fusion is in progress in the hope that it will be a major source of energy in the foreseeable future. The chief problem is to design a system that will produce and maintain the extraordinarily high temperatures required in a controlled manner.

The energy changes in nuclear reactions are millions of times as great as those in chemical reactions. The reason is that there are changes in mass in nuclear processes. Mass and energy are related by the Einstein equation:

$$E = mc^2$$

where m is the mass and c is the velocity of light, 3×10^8 m/s. As you can see, the value of the speed of light is so large that even a small change in mass produces an enormous change in energy.

The masses of nuclei are always less than the sum of the masses of the constituent nucleons. For example, the mass of a helium-4 nucleus is 4.00150 amu. The sum of two individual protons and two individual neutrons is 4.03190 amu. The difference in mass of 0.03040 amu represents the energy released when protons and neutrons fuse into a helium-4 nucleus. That change in energy can be calculated by using the Einstein relation in the form

$$E = \Delta m \times c^2$$

where Δm is the change, or difference, in mass. The disintegration of 1 mol of helium-4 requires an energy input of 2.73×10^{12} J/mol. The fusion of protons and neutrons to form a helium nucleus will produce the same amount of energy. It is 1 million times the energy released in chemical reactions, which at the most may reach values of 1 to 2×10^6 J/mol.

10.10   NUCLEAR ENERGY AND THE BIOSPHERE

The dangers of uncontrolled exposure to high-energy radiation have been understood for many years. Because we use nuclear reactors to generate electricity and to manufacture weapons, we must confront the possibility of widespread damage through the accidental or deliberate release of radioactive substances.

The first concern about such possibilities was prompted by atmospheric testing of atomic weapons in the 1950s. Two of the more dangerous byproducts of nuclear fission released through bomb testing are strontium-90 and cesium-137. The properties of strontium are very similar to those of calcium. Therefore, if radioactive strontium enters the body, it will be incorporated where calcium is found, in bone, particularly growing bone, such as the bones of children. When incorporated into bone, the radioisotope becomes a lifetime burden and can produce cancer of the bone or leukemia or both. Cesium, on the other hand, is an alkali metal; so its properties are similar to those of potassium, whose cation concentrates inside cells. The presence of radioactive cesium in cells can inflict damage in a wide variety of tissues. However, cesium cations are quite soluble and are rapidly eliminated from the body. There is no pharmacological treatment for exposure to these radioactive substances. This awareness led countries in possession of atomic weapons to enter into a treaty that banned atmospheric testing.

Today, a more acute concern is the possibility of the accidental release of radioactivity from the nuclear reactors that are being used to produce electricity. In 1986, when a serious accident occurred in Ukraine at the plant in Chernobyl, the fallout of radioactivity was intense. One of the more dangerous of the radiation products was iodine-131, for which, fortunately, a pharmacological antidote is available.

Because iodine becomes concentrated in the thyroid gland, high concentrations of its I-131 isotope can produce serious radiation damage. For this reason, in the wake of the Chernobyl accident, anyone living in a region where iodine-131 was known to have been deposited was encouraged to use table salt enriched with nonradioactive iodine-127. The thyroid gland cannot discriminate between iodine-127 and iodine-131, so flooding the body with the nonradioactive iodine ensured that the body's concentration of iodine-131 would be significantly diluted and its rate of uptake sharply reduced compared with the uptake of iodine-127.

We can never fully escape exposure to high-energy radiation. We are constantly exposed to cosmic rays from space and to radiation from naturally occurring radioisotopes within our bodies (such as potassium-40) and in the surrounding rocks and soil (such as radon). Add to that the occasional medical or dental X-ray session. For residents of the United States, these sources of exposure add up to about 350 mrem per year, a small amount compared with the standard for safe maximum exposure, which is about 5000 mrem per year. Although this exposure varies from place to place, background radiation does not seem to pose a serious problem.

Summary

Radioactivity

- Radioactivity is spontaneous nuclear decomposition, in which high-energy subatomic particles are ejected from the nucleus.

- The major types of radioactive emissions are α-particles, β-particles, γ-rays, and positrons.

- Alpha-, β-, and positron radiation result in changes in atomic number or mass or both, to create a new element, but γ-ray emission does not lead to the formation of a new element.

Radioactive Decay

- Radioactive decay occurs when a radioactive element is transformed into another element by emitting nuclear radiation.

- The rate of radioactive decay is the time required for 50% of the original material to decay, and that period of time is called the half-life ($t_{1/2}$).

Effects of Radiation

- High-energy radiation produces ions and free radicals along its penetration tracks.

- The consequences of irradiation depend on the kind of tissue that is damaged.
- Protection from high-energy radiation is effected by using an absorbing barrier, increasing the distance from the radiation source, and limiting exposure time.

Detection of Radiation

- High-energy radiation is detected by the use of photographically sensitive film, Geiger counters, or scintillation counters.
- Measurements of radiation intensity, the dose rate, and the total dose can be used to predict the likelihood of biological damage.
- With the same amount of energy absorbed, the likelihood of damage depends on the type of tissue irradiated and is different for different types of radiation.

Applications

- X-radiation has been used since 1895 to generate images of the body's inner structures.

- Radioisotopes are used for the diagnosis of pathological conditions and for therapy.
- In diagnostic applications, radioisotopes are primarily γ-emitters and, in therapeutic use, they are either α- or β-emitters.
- In newer methods, radiation originating inside or passing through the body is detected externally and transformed by computer to generate an image of the body's inner structures.
- Recent applications include the CAT scan, produced with X-rays, the PET scan, produced by positron decay of a metabolite within the brain, and MRI, produced by the absorption of radio signals.

Nuclear Reactions

- Nuclear fission and nuclear fusion yield a million times the energy released in chemical reactions.
- The energy changes in these nuclear reactions are the result of mass changes in the nucleus.

Key Words

α-particle, p. 262
background radiation, p. 276
β-particle, p. 262
bremsstrahlung, p. 278
computer assisted tomography (CAT scan), p. 281
free radical, p. 270
γ-ray, p. 262
Geiger counter, p. 274

half-life, p. 266
magnetic resonance imaging (MRI), p. 281
nuclear equation, p. 263
nuclear fission, p. 283
nuclear fusion, p. 284
nucleon, p. 262
photoelectric effect, p. 261
photon, p. 261

positron, p. 262
positron emission tomography (PET scan), p. 282
radioactivity, p. 261
radiation chemistry, p. 270
relative biological effectiveness (RBE), p. 276
transmutation, p. 262
X-ray, p. 261

Exercises

Radioactivity

10.1 What are the characteristics of α-radiation?

10.2 What are the properties of β-radiation?

10.3 Describe the important characteristics of γ-radiation.

10.4 How do positrons compare with other subatomic particles?

10.5 Specify the missing component of the following nuclear equation:
$$^{240}_{95}\text{Am} + \square \longrightarrow ^{243}_{97}\text{Bk} + ^{1}_{0}\text{n}$$

10.6 Specify the missing component in the following nuclear equation:
$$^{238}_{92}\text{U} + ^{12}_{6}\text{C} \longrightarrow ^{244}_{98}\text{Cf} + \square$$

10.7 Complete the following nuclear equation:
$$^{40}_{19}\text{K} \longrightarrow ^{40}_{20}\text{Ca} + \square$$

10.8 Complete the following nuclear reaction:
$$^{0}_{-1}\text{e} + \square \longrightarrow 2\,\gamma$$

10.9 What is induced radioactivity?

10.10 Give some examples of induced radioactivity.

10.11 What is a radioactive decay series?

10.12 What is a daughter nucleus?

10.13 The amount of C-14 remaining in a charcoal sample from an archeological site in New Mexico was determined to be 23.3% of the original quantity. The half-life of C-14 is 5730 years. How old is the site?

10.14 Paleontologists found that 2.75 half-lives of C-14 had passed since a bone knife had been carved at a site in Alaska. What was the percentage of C-14 left in that sample?

Effects of Radiation

10.15 Which is more effective in penetrating tissue, α-radiation or β-radiation?

10.16 Describe the pattern of radiation produced in tissue by β-radiation.

10.17 What happens to X-radiation as it passes through tissue?

10.18 What happens to γ-radiation as it passes through tissue?

10.19 What are the immediate results of the interaction of γ-radiation with water?

10.20 Do any chemical events follow the immediate results of γ-irradiation of water?

10.21 Which has the most long-term consequences, a primary radiation event or its secondary chemical processes?

10.22 Describe how biochemical radiation damage may have a deleterious biological consequence.

10.23 What is radiation sickness?

10.24 What is background radiation?

10.25 A radiation technician was standing 4 feet from a radiation source. She determined that she must move away to reduce the intensity of the radiation to 1/4 of its value. How much farther did she move?

10.26 If the technician in Exercise 10.25 could not work farther away from the radiation source, is there anything else that she could do to protect herself?

Detection

10.27 What is the basis of detection of radiation by the Geiger counter?

10.28 How does a scintillation counter detect radiation?

Units

10.29 What is a rad?

10.30 What is the difference between the rad and the rem?

10.31 Does 1 rad of α-radiation have the same effect on tissue as 1 rad of γ-radiation?

10.32 Does 1 rad of β-radiation have the same effect on tissue as 1 rad of γ-radiation?

Applications

10.33 Can α-emitting isotopes be used for diagnosis? Explain your answer.

10.34 Can γ-emitting isotopes be used for diagnosis? Explain your answer.

10.35 Can cobalt-60 be used for cancer therapy? Explain.

10.36 Can radium-226 be used for cancer therapy? Explain.

10.37 Can positrons be detected directly? Explain.

10.38 Can β-particles be detected directly? Explain.

10.39 Would iodine-131, a β-emitter, be used for scanning the thyroid for disorders?

10.40 Would technetium-99m, a γ-emitter, be used for scanning for disorders of the vascular system?

10.41 What is a CAT scan used for?

10.42 Describe how a PET scan works.

Unclassified Exercises

10.43 Technetium-99m is a radioisotope used to assess heart damage. Its half-life is 6.0 h. How much of a 1.0-g sample of technetium-99m will be left after 30 h?

10.44 Iodine-126, a γ-emitter, has a half-life of 13.3 h. If a diagnostic dose is 10 ng, how much is left after 2.5 days?

10.45 A sample of $Na_3{}^{32}PO_4$ had an activity of 6.3 mCi. What does that mean in dps?

10.46 The radiation in rads absorbed by tissue from an α-emitter was the same as that absorbed from a β-emitter. The radiation absorbed from the β-emitter was determined to be 15.2 mrem. What was the absorbed dose in rems from the α-emitter?

10.47 The radioisotope fluorine-18 is used for medical imaging. It has a half-life of 110 min. What percentage of the original sample activity is left after 4 h?

10.48 The intensity of X-rays at 2 m from a source was 0.8 roentgen (R). How far from the source should the target be to reduce the intensity to 0.2 R?

10.49 What is radioactivity?

10.50 What are the primary chemical events when high-energy radiation penetrates tissue?

Expand Your Knowledge

Note: The icons ▲▲▲ denote exercises based on material in boxes.

10.51 Does radioactive decay always result in transmutation?

10.52 Would you expect infrared light to initiate chemical events? Explain.

10.53 What kinds of tissue are most sensitive to radiation damage?

10.54 Is a radioisotope useful if its decay products are equally radioactive? Explain your answer.

10.55 What is the difference between nuclear fission and nuclear fusion?

10.56 What are X-rays?

10.57 How are X-rays produced?

10.58 What are two problems associated with the production of X-rays, and how have they been solved?

10.59 Do all X-rays have a common origin?

10.60 Is all X-radiation suitable for medical radiology?

10.61 How does X-radiation produce images of internal organs?

10.62 How are X-rays detected?

10.63 What is the origin of stratospheric ozone (see Box 10.2)?

10.64 You want to purchase a house located in a city where radon is known to be a natural product emanating from the local geological strata. Are there steps that you can take to minimize the risks associated with radon gas (see Box 10.4)?

Answers to Problems Following In-Chapter Worked Examples

Chapter 1

1.1 (a) 0.002 s; (b) 0.05 m; (c) 0.100 liters

1.2 7.068×10^{-4}

1.3 7.311×10^{-4}

1.4 27 or 2.7×10^1

1.5 4.8×10^5

1.6 6.1×10^2 cm^3

1.7 40.35

1.8 87.3 m

1.9 0.0741 L

1.10 1.659 lb

1.11 6.048×10^5

1.12 2.87×10^3 g

1.13 3.97×10^3 mL

1.14 118.2 g

1.15 36°C

1.16 212°F

1.17 37.0°C and 310 K

1.18 0.20 J/g × °C

1.19 49 J

1.20 1412 kcal/day

Chapter 2

2.1 C = 58.54%, H = 4.071%, N = 11.38%, O = 26.02%

2.2 30 oranges, 10 cantaloupes

2.3 0

2.4 3+

2.5 Titanium, 48 amu

2.6 $^{14}_{7}$N and $^{15}_{7}$N

2.7 (a) protons = 7, neutrons = 7
(b) protons = 7, neutrons = 8

2.8 24.3 amu

2.9 Four subshells, s, p, d, and f

2.10 $1s^2 2s^2 2p^6 3s^2 3p_x^2 3p_y^1 3p_z^1$

2.11

Element	Atomic number	1s	2s	2p
Boron	5	↑↓	↑↓	↑

2.12

Element	Atomic number	1s	2s	2p$_x$	2p$_y$	2p$_z$
Nitrogen	7	↑↓	↑↓	↑	↑	↑

2.13 $[\text{Ne}]3s^2 3p^4$

2.14 Na· $\longrightarrow$ Na$^+$ + e

:Cl̈· + e $\longrightarrow$:C̈l:$^-$

Chapter 3

3.1 (a) K$^+$; (b) Ca^{2+}; (c) In^{3+}; (d) P^{3-}; (e) S^{2-}; (f) Br$^-$

3.2 (a) KBr; (b) GaF$_3$; (c) Ca$_3$P$_2$

3.3 (a) KCl; (b) Mg$_3$P$_2$; (c) Ga$_2$O$_3$

3.4 (a) Titanous chloride; (b) mercuric chloride; (c) ferrous chloride; (d) plumbic oxide

3.5 (a) Ammonium hydrogen sulfite; (b) calcium carbonate; (c) magnesium cyanide; (d) potassium hydrogen carbonate; (e) ammonium sulfate

3.6 (a) Nitrogen triiodide; (b) diphosphorus pentoxide; (c) disulfur dichloride; (d) sulfur trioxide

3.7 (a) CH$_4$; (b) PCl$_3$; (c) CO$_2$

3.8

:C̈l—P̈—C̈l:
        |
      :C̈l:

3.9 :C̈l—S̈—C̈l:

3.10

```
     H   H   H
     |   |   |
 H—C—C—C—H
     |   |   |
     H   H   H
```

3.11 :Ö=C=Ö:

3.12

```
 H—N̈—N̈—H
     |   |
     H   H
```

3.13 CCl$_4$ will have a tetrahedral shape.

3.14 NI$_3$ will have the shape of a trigonal pyramid.

3.15 OF$_2$ will have a bent shape.

Chapter 4

4.1 (a) 257.4 amu; (b) 231.6 amu; (c) 18.02 amu; (d) 98.09 amu; (e) 88.09 amu

4.2 (a) 142.1 g; (b) 119.0 g; (c) 40.31 g; (d) 86.17 g

4.3 1.251 mol

4.4 1.998×10^{-3} mol

4.5 (a) 1 mol each of Zn, H, and P atoms, 4 mol O atoms; (b) 2 mol H atoms, 1 mol S atoms, 4 mol O atoms; (c) 2 mol Al atoms, 3 mol O atoms; (d) 3 mol Ca atoms, 2 mol P atoms

4.6 Li, 1.153×10^{-23} g; N, 2.326×10^{-23} g; F, 3.155×10^{-23} g; Ca, 6.656×10^{-23} g; C, 1.994×10^{-23} g

4.7

Number of calcium atoms	Mass in grams
10	6.656×10^{-22}
100	6.656×10^{-21}
100000	6.656×10^{-18}
1.0×10^{15}	6.656×10^{-8}
6.022×10^{20}	0.04008
6.022×10^{23}	40.08

4.8 MgO

4.9 $MgSO_4$

4.10 P_2O_5

4.11 $3\, Ca(OH)_2 + 2\, H_3PO_4 \longrightarrow Ca_3(PO_4)_2 + 6\, H_2O$

4.12 $N_2H_4 + 4\, H^+ + 2\, I^- \rightarrow 2\, NH_4^+ + I_2$

4.13 (a) +1; (b) +3; (c) +4; (d) +6

4.14 $C_4H_{10} + \frac{13}{2} O_2 \longrightarrow 4\, CO_2 + 5\, H_2O$
$2\, C_4H_{10} + 13\, O_2 \longrightarrow 8\, CO_2 + 10\, H_2O$

4.15 $C_3H_6O_3 + 3\, O_2 \longrightarrow 3\, CO_2 + 3\, H_2O$

4.16 $\dfrac{3 \text{ mol } Mg(OH)_2}{2 \text{ mol } FeCl_3}$ $\dfrac{3 \text{ mol } Mg(OH)_2}{2 \text{ mol } Fe(OH)_3}$ $\dfrac{3 \text{ mol } Mg(OH)_2}{3 \text{ mol } MgCl_2}$

$\dfrac{2 \text{ mol } FeCl_3}{2 \text{ mol } Fe(OH)_3}$ $\dfrac{2 \text{ mol } FeCl_3}{3 \text{ mol } MgCl_2}$ $\dfrac{2 \text{ mol } Fe(OH)_3}{3 \text{ mol } MgCl_2}$

plus the reciprocal of each.

4.17 6.400 mol

4.18 1.000 mol

4.19 62.4 g $BaCl_2$, 60.1 g $Ba_3(PO_4)_2$

Chapter 5

5.1 (a) 0.741 atm; (b) 707 mmHg, or 707 torr

5.2 $V = 407$ mL

5.3 $P = 0.950$ atm

5.4 $V = 1.14 \times 10^3$ mL

5.5 $T = 619$ K, or 346°C

5.6 $P = 1.73$ atm

5.7 $T = 913$ K, or 640°C

5.8 $V = 1.11$ L

5.9 $CO_2 = 0.191$ g

5.10 $M = 32.2$ g/mol

5.11 $M = 40.1$ g/mol

5.12 $P_{\text{hydrogen}} = 494$ torr; $P_{\text{oxygen}} = 247$ torr

5.13 $V = 8.14$ L

5.14 0.0152 mL N_2/mL H_2O

5.15 0.0122 mL N_2/mL H_2O

Chapter 6

6.1 (c) $H-C\equiv N$

6.2 c and e.

6.3 $d > c > b > a$

6.4 No. Liquids having strong attractive forces will not mix with liquids possessing weak attractive forces.

6.5 (a) The secondary forces between CCl_4 molecules are weak London forces, and those between H_2O molecules are strong H-bonds. Solutions between substances possessing such very different secondary forces are not possible.
(b) The secondary forces between C_5H_{12} molecules and between CCl_4 molecules are London forces. Because they are the same, solutions of these two substances are possible.

6.6 Ethanol, 200 torr; hexane, 390 torr; chloroform, 500 torr.

6.7 Methyl alcohol, CH_3OH, in the liquid state is hydrogen bonded, and methyl iodide, CH_3I, is not.

6.8 Ammonia can form hydrogen bonds; phosphine cannot. Therefore ammonia has the higher boiling point.

6.9 $b > a > d > c$

Chapter 7

7.1 Dissolve 11.0 g of gelatin in 189 g of water.

7.2 0.0020 g/g of solution

7.3 438 g of solution

7.4 Add 4.8 mL of acetone to sufficient water so that the final volume of solution equals 100 mL.

7.5 6.6 g of sodium chloride

7.6 Dissolve 18.7 g of KH_2PO_4 in 500 mL of solution.

7.7 0.918 mol

7.8 1.67 L

7.9 0.300 M

7.10 0.0200 mmol/mL

7.11 1.80×10^3 mL, or 1.80 L

7.12 503 mL, or 0.503 L

7.13 9.0×10^{-5} g/mL

7.14 (a)
$3\, Ca^{2+}(aq) + 6\, Cl^-(aq) + 6\, Na^+(aq) + 2\, PO_4^{3-}(aq) \longrightarrow$
$\qquad Ca_3(PO_4)_2(s) + 6\, Cl^-(aq) + 6\, Na^+(aq)$
(b) $3\, Ca^{2+}(aq) + 2\, PO_4^{3-}(aq) \longrightarrow Ca_3(PO_4)_2(s)$

Chapter 8

8.1 $\Delta H_{\text{rxn}} = +6$ kJ. The reaction is endothermic.

8.2 $K_{\text{eq}} = \dfrac{[HI]^2}{[I_2][H_2]}$

8.3 $K_{\text{eq}} + \dfrac{[CO]^2}{[O_2]}$

8.4 $[A] = [B]$

8.5 $K_{\text{eq}} = 1.1 \times 10^{-1}$

8.6 $K_{\text{eq}} = 1 \times 10^2$

8.7 Shift of mass to the left (reactant side).

8.8 (a) Shift of mass to the right; (b) shift of mass to the left.

Chapter 9

9.1 0.050 M

9.2 0.050 M

9.3 0.060 M

9.4 2.00×10^{-12} M

9.5 2.00×10^{-12} M

9.6 $[H_3O^+] = 1.0 \times 10^{-12}$ M; pH = 12.00

9.7 pH = 2.43

9.8 pH = 10.00

9.9 pOH = 11.30

9.10 $[HI] = 3.5 \times 10^{-3}$

9.11 $K_{\text{b(cyanate)}} = K_{\text{w}}/K_{\text{a(cyanic acid)}}$

9.12 $pK_{\text{b(cyanate)}} = pK_{\text{w}} - pK_{\text{a(cyanic acid)}}$

9.13 $MgCl_2$ and KNO_3, neutral; $SnCl_2$ and NH_4NO_3, acidic.

9.14 pH = 4.89

9.15 26.3 mL

9.16 Dissolve 59.1 g of succinic acid in 1.00 L of water.

9.17 $[K^+] = 0.0050$ M; $[Mg^{2+}] = 0.0015$ M

Chapter 10

10.1 $^{223}_{88}Ra \rightarrow ^{219}_{86}Rn + ^{4}_{2}He$

10.2 $^{230}_{88}Ra \rightarrow ^{230}_{89}Ac + ^{0}_{-1}e$

10.3 $^{21}_{11}Na \rightarrow ^{21}_{10}Ne + ^{0}_{1}e$

10.4 1.56%

10.5 3.1%

10.6 7.2%

10.7 12 m

Answers to Odd-Numbered Exercises

Chapter 1

1.1 (a) Chemical; (b) physical; (c) physical; (d) chemical

1.3 (a) Mixture; (b) pure substance; (c) pure substance; (d) pure substance

1.5 Length, meter; mass, kilogram; temperature, kelvin; time, second; amount of substance, mole

1.7 (a) 0.630 km; (b) 1.440 s; (c) 65 mg; (d) 1.3 mg

1.9 (a) 3; (b) 4; (c) 3; (d) 3

1.11 (a) 6; (b) 3; (c) 1; (d) 4; (e) 3

1.13 (a) 8.39×10^{-3}; (b) 8.3264×10^4; (c) 3.72×10^2; (d) 2.08×10^{-5}

1.15 (a) 9.368×10^5; (b) 1.638×10^3; (c) 5.68×10^{-5}; (d) 9.17×10^{-3}

1.17 (a) 48.4 cm^2; (b) 8.4

1.19 (a) 14.02; (b) 28.615

1.21 (a) 1.29 km; (b) 1.29×10^3 m; (c) 1.29×10^5 cm; (d) 1.29×10^6 mm

1.23 1.42×10^4 mL

1.25 6.214 miles

1.27 28.35 g/oz

1.29 (a) 0.27 kg; (b) 2.7×10^2 g; (c) 2.7×10^5 mg

1.31 10 cm

1.33 79 g

1.35 20°C

1.37 300 K

1.39 0.069

1.41 1.130

1.43 1.10 lb

1.45 1.04 g/mL

1.47 125.1 cm^3

1.49 (a) Chemical; (b) physical; (c) physical; (d) chemical

1.51 (a) 9.62×10^6 kg; (b) 5.487×10^4 days; (c) 2.53×10^{-1} s; (d) 2.74×10^{-4} km

1.53 1.62×10^{-3} oz

1.55 (a) 3.83×10^4; (b) 5.48×10^{-2}; (c) 7.33×10^{-3}

1.57 1.0×10^3 cm^3

1.59 6.3×10^5 kJ/24 h

1.61 The unspecified solution contains two components, NaCl and water. You could obtain pure NaCl by evaporating all the water from the mixture. Five grams of the pure NaCl could then be weighed on a laboratory balance, and the specified mixture of salt and water could then be prepared.

1.63 You must modify your original hypothesis and retest it.

1.65 The "uk" location stands for United Kingdom (England), a country that uses the SI metric system. In the United States, the Fahrenheit system is still in popular use; so 175°C = 347°F.

1.67 (a) 3.94 yd; (b) 66.0 in.; (c) 5.90×10^9 miles/yr; (d) 2.26 oz (Hint: Assume the density of water to be 1.00 g/mL.)

1.69 The price per milliliter of the active ingredient in the recommended product is $7.03; that for the generic is $6.06. The generic product is more cost effective.

1.71 8.10%

1.73 Density of box = 0.0375 g/cm^3
Amount of water to be added = 9.63×10^5 mL

1.75 You may have rounded off your figures before you finished the calculation.

Chapter 2

2.1 No mass is lost as a result of a chemical reaction.

2.3 C = 40.00%, H = 6.671%, O = 53.33%

2.5 Atomic mass O = 16.0000 amu

2.7 Atomic mass N = 14.0 amu

2.9 Proton: mass, 1.00728 amu; charge, + 1
Neutron: mass, 1.00867 amu; charge, 0
Electron: mass, 0.0005486 amu; charge, − 1

2.11 Atomic mass includes the mass of neutrons.

2.13 It would be a cation; charge, 3+

2.15 Atomic mass, Sr = 87.62 amu

2.17 $^{16}_{8}O$, $^{17}_{8}O$; $^{24}_{12}Mg$, $^{25}_{12}Mg$; $^{28}_{14}Si$, $^{29}_{14}Si$

2.19 (a) Lithium, (b) calcium, (c) sodium, (d) phosphorus, (e) chlorine

2.21

Element	Group	Period
Li	I	2
Na	I	3
K	I	4
Rb	I	5
Cs	I	6

2.23 Li, Na, K, Rb, and Cs are metals.

2.25 Main-group elements

2.27 Group VII

2.29 No. The ground state is stable; no energy can be lost.

2.31 Shells identified by a principal quantum number. Subshells within shells identified by the letters *s, p, d*, etc. Orbitals within subshells.

2.33 An orbital is used to define the probability of finding an electron in a region of space around an atomic nucleus.

2.35 One orbital in an *s* subshell

2.37 No, an *s* subshell is spherically symmetrical.

2.39 (a) Magnesium; (b) sulfur; (c) potassium

2.41 Sodium ion

2.43 Group I. All possess one outer electron.

2.45 Group II. All possess two outer electrons.

2.47 True

2.49 Its energy increases.

2.51 They have virtually identical chemical properties.

2.53 Metal

2.55 $^{16}_{8}O$, $^{18}_{8}O$

2.57 0.022%

2.59 Blue light

2.61 The octet rule

2.63 You need to know the natural abundances of each of the isotopes.

2.65 Dalton's atom was indestructible, but the modern atom can be decomposed into subatomic particles. In addition, the discovery of isotopes showed that the masses of the atoms of an element are not identical.

2.67 Each element has a characteristic spectrum. An element's presence in a sample will be detected by its characteristic spectral lines.

2.69 (a) A discrete bundle of light energy; (b) a number describing an energy state of an atom; (c) the lowest energy state of an electron in an atom; (d) an allowed energy state that is higher in energy than the ground state

2.71 The repulsion between two similarly charged particles

2.73 (a) Ca: $[Ar]4s^2$; (b) Br: $[Ar]4s^23d^{10}4p^5$; (c) Ag: $[Kr]5s^14d^{10}$; (d) Zn: $[Ar]4s^23d^{10}$

2.75 (a) $1s^22s^22p_x^12p_y^12p_z^1$; (b) $1s^22s^22p_x^12p_y^12p_z^1$; (c) $1s^22s^22p_x^22p_y^12p_z^1$; (d) $1s^22s^22p_x^22p_y^22p_z^2$

Chapter 3

3.1 Chemical-bond formation is the result of a process that allows the reacting elements to achieve stability.

3.3 By losing, gaining, or sharing electrons

3.5 (a) Potassium sulfate; (b) manganese(II) hydroxide; (c) iron(II) nitrate; (d) potassium dihydrogen phosphate; (e) calcium acetate; (f) sodium carbonate

3.7 (a) Ammonium; (b) nitrate; (c) sulfate; (d) phosphate (e) acetate

3.9 (a) $AlCl_3$; (b) Al_2S_3; (c) AlN

3.11 (a) Li_2O; (b) $CaBr_2$; (c) Al_2O_3; (d) Na_2S

3.13 1 (Group I) : 1 (Group VII)

3.15 1 (Group II) : 2 (Group VII)

3.17 (a) CaO; (b) CaS; (c) CaSe

3.19 (a) Al_2O_3; (b) Al_2S_3; (c) Al_2Se_3

3.21 (a) AlF_3; (b) $AlCl_3$; (c) $AlBr_3$; (d) AlI_3

3.23

$$
\begin{array}{c}
\ddot{\ddot{Cl}} \\
| \\
:\ddot{Cl}-C-\ddot{Cl}: \\
| \\
:\ddot{Cl}:
\end{array}
$$

3.25 $:\ddot{F}-\ddot{O}-\ddot{F}:$

3.27 $\ddot{O}=C=\ddot{O}$

3.29 The four fluorine atoms are situated at the corners of a tetrahedron with the boron atom at its center.

3.31 (a) A tetrahedron, no dipole moment; (b) a tetrahedron, no dipole moment; (c) linear, no dipole moment

3.33 A chemical bond that consists of two electrons with paired spins.

3.35 (a) Silver nitrate; (b) mercury(II) chloride; (c) sodium carbonate; (d) calcium phosphate

3.37 (a) Magnesium chloride; (b) magnesium sulfide; (c) magnesium nitride

3.39 (a) Barium oxide; (b) barium sulfide; (c) barium selenide

3.41 (a) Potassium nitride; (b) potassium phosphide; (c) potassium arsenide

3.43 The atoms in $TeBr_2$ are oriented toward the corners of a tetrahedron. The molecule is bent.

3.45 (a) Covalent; (b) covalent; (c) covalent; (d) ionic; (e) covalent

3.47 LiF; LiCl; LiBr; LiI

3.49 Yes. Hydrogen requires only two electrons, and beryllium and boron require four and six electrons, respectively, to complete their valence shells.

3.51 Test its ability to conduct electricity when dissolved in water. An electrolyte is an ionic substance.

3.53 $H-\ddot{O}-\ddot{O}-H$

3.55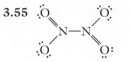

3.57 (a) Si—F > Ge—F > C—F; (b) Si—F > Sn—F > C—F; (c) Al—Br > Ga—Br; (d) Sb—Br > As—Br

3.59 (a) $:\ddot{Cl}:\ddot{P}:\ddot{Cl}:$

$:\ddot{Cl}:$

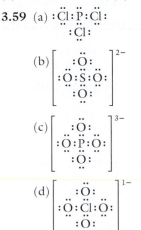

3.61 (a) $[Ar]4s^24p^1 + 3\,[He]2s^22p^5 \longrightarrow [Ar]^{3+} + 3\,[Ne]^{1-}$
(b) $[Kr]4d^{10}5s^1 + [Ne]3s^23p^5 \longrightarrow [Kr]4d^{10}]^{1+} + [Ar]^{1-}$
(c) $3\,[He]2s^1 + [He]2s^22p^3 \longrightarrow [He]^{1-} + [Ar]^{3-}$

3.63 (a) $TlCl_3$; (b) Zn_2N_3; (c) CdS; (d) GaO

3.65 (a) $:\ddot{Cl}:\ddot{S}:\ddot{Cl}:$

(b)
$$:\ddot{Cl}:$$
$$:\ddot{Cl}:\ddot{Ge}:\ddot{Cl}:$$
$$:\ddot{Cl}:$$

(c) $:\ddot{Br}:\ddot{As}:\ddot{Br}:$
$:\ddot{Br}:$

(d) $H:\ddot{P}:H$
$\quad H$

3.67 (a)
$$
\begin{array}{cc}
H & H \\
\ddot{}C::C\ddot{} \\
H & H
\end{array}
$$

(b)
$$
\begin{array}{c}
\ddot{Cl}: \\
:\ddot{O}::C\ddot{} \\
\ddot{Cl}:
\end{array}
$$

3.69 (a) Linear; (b) bent; (c) bent

3.71 (a) Trigonal planar; (b) bent

Chapter 4

4.1 (a) 156.1; (b) 58.33; (c) 189.7; (d) 262.0; (e) 60.10

4.3 (a) 474.0; (b) 204.1; (c) 271.5; (d) 254.5; (e) 391.5

4.5 0.4221 mol

4.7 101 g

4.9 1.21×10^{23}

4.11 280.7 g

4.13 $ZnCrO_4$

4.15 (a) $N_2 + 3 Br_2 \rightarrow 2 NBr_3$
(b) $2 HNO_3 + Ba(OH)_2 \rightarrow Ba(NO_3)_2 + 2 H_2O$
(c) $HgCl_2 + H_2S \rightarrow HgS + 2 HCl$

4.17 (a) $4 P + 5 O_2 \rightarrow 2 P_2O_5$
(b) $FeCl_2 + K_2SO_4 \rightarrow FeSO_4 + 2 KCl$
(c) $HgCl + 2 NaOH + 2 NH_4Cl \rightarrow Hg(NH_3)_2Cl +$
$2 NaCl + 2 H_2O$

4.19 $C_6H_6 + 7.5 O_2 \rightarrow 6 CO_2 + 3 H_2O$
$2 C_6H_6 + 15 O_2 \rightarrow 12 CO_2 + 6 H_2O$

4.21 $C_5H_{12} + 8 O_2 \rightarrow 5 CO_2 + 6 H_2O$
$2 C_8H_{18} + 25 O_2 \rightarrow 16 CO_2 + 18 H_2O$
$C_{10}H_{20} + 15 O_2 \rightarrow 10 CO_2 + 10 H_2O$

4.23 $CH_4 + 2 O_2 \rightarrow CO_2 + 2 H_2O$

4.25 $\dfrac{2 Al(OH)_3}{3 H_2SO_4}$ $\dfrac{1 Al(OH)_3}{3 H_2O}$ $\dfrac{1 H_2SO_4}{2 H_2O}$

$\dfrac{2 Al(OH)_3}{1 Al_2(SO_4)_3}$ $\dfrac{3 H_2SO_4}{1 Al_2(SO_4)_3}$ $\dfrac{1 Al_2(SO_4)_3}{6 H_2O}$

and the inverse of each of these unit-conversion factors.

4.27 0.83 mol

4.29 3.3 mol

4.31 53.48 g

4.33 (a) 0.44 mol; (b) 2.90 mol; (c) 9.4 g; (d) 8.500 g

4.35 (a) $Ca_3(PO_4)_2 + 4 H_3PO_4 \rightarrow 3 Ca(H_2PO_4)_2$
(b) $FeCl_2 + (NH_4)_2S \rightarrow FeS + 2 NH_4Cl$
(c) $2 KClO_3 \rightarrow 2 KCl + 3 O_2$
(d) $3 O_2 \rightarrow 2 O_3$
(e) $2 C_5H_6 + 13 O_2 \rightarrow 10 CO_2 + 6 H_2O$

4.37 (a) 0.5200 mol; (b) 0.7507 mol; (c) 1.500 mol; (d) 0.800 mol; (e) 0.36 mol

4.39 1.80

4.41 1.1×10^{24}

4.43 $AgCl$

4.45 C_2H_6O

4.47 (a) $AgNO_3 + KCl \rightarrow AgCl + KNO_3$
(b) $Ba(NO_3)_2 + Na_2SO_4 \rightarrow BaSO_4 + 2 NaNO_3$
(c) $2 (NH_4)_3PO_4 + 3 Ca(NO_3)_2 \rightarrow$
$Ca_3(PO_4)_2 + 6 NH_4NO_3$
(d) $3 Mg(OH)_2 + 2 H_3PO_4 \rightarrow Mg_3(PO_4)_2 + 6 H_2O$

4.49 684.4 g

4.51 Carbon: 67.33%; hydrogen: 6.931%; nitrogen: 4.620%; oxygen: 21.12%

4.53 (a) 45 g of H_2SO_4; (b) 133 g of $Ca(OH)_2$; (c) 35 g of C_2H_6O; (d) 110 g of C_4H_{10}; (e) 211 g of $FeCl_2$

4.55 P_4O_{10}

4.57 (a) 0.219 mol; (b) 0.0850 mol

4.59 $\$9.4 \times 10^{13}$ per person

4.61 (a) 5.68×10^{-22} g; (b) 2.52×10^{-22} g

4.63 $CoCl_2$

4.65 200.6 amu

4.67 1.60×10^4 amu

4.69 (a) $3 NaHCO_3 + C_6H_8O_7 \longrightarrow$
$3 CO_2 + 3 H_2O + Na_3C_6H_5O_7$
(b) 152 mg
(c) 105 mg

4.71 $5 Fe^{2+}(aq) + MnO_4^-(aq) + 8 H^+(aq) \longrightarrow$
$5 Fe^{3+}(aq) + Mn^{2+}(aq) + 4 H_2O(l)$

4.73

	Oxidized	Reduced
(a)	I_2	I_2
(b)	Al	OH^-

Chapter 5

5.1 (a) 364 torr; (b) 483 torr; (c) 675 torr; (d) 735 torr

5.3 (a) 735 torr; (b) 0.967 atm

5.5 (a) 912 torr; (b) 342 mm Hg; (c) 1.12 atm

5.7 T and n

5.9 V and n

5.11 2.5 L

5.13 1.00 L

5.15 348°C, or 621 K

5.17 3.25 atm

5.19 2.25 L

5.21 1.52 L

5.23 6.24×10^4 (torr × mL)/(K × mol)

5.25 3.59 L

5.27 5.60 L

5.29 1.01 g of H_2, 4.25 g of NH_3

5.31 0.920 atm

5.33 0.243 mol

5.35 All volumes are the same.

5.37 200 torr

5.39 40.4 L

5.41 153°C

5.43 2.0 L

5.45 0.900 L

5.47 0.0906 g/L

5.49 4.49×10^3 mL

5.51 (a) 570 torr; (b) 935 torr; (c) 722 torr; (d) 1.330×10^3 torr

5.53 (a) 27.7 L; (b) 23.1 atm; (c) 329 K; (d) 1.00 mol

5.55 2.04 mol

5.57 0.031 mL CO_2/mL arterial blood

5.59 0.0361 mL

5.61 675.8 g

5.63 1.17×10^{13} molecules in 1.0 mL

5.65 $d = 0.000588$ g/mL

5.67 C_4H_8

5.69 3.39×10^2 g used

5.71 Ratio of $O_2/CH_4 = 1/1$.

5.73 Every dive adds nitrogen to the blood. Making the first of a series of dives the deepest maximizes the calculation of blood-nitrogen accumulation. Making it the last dive fools the diver into a false sense of security and could possibly bring on a serious case of the bends on surfacing.

Chapter 6

6.1 All matter would exist only in the gaseous state. No matter would be present in either liquid or solid form.

6.3 The melting of a solid is the transition from a structure consisting of a highly ordered array of molecules to a disordered structure with the molecules still in contact but now free to move about. This transition is effected by adding energy to the system to overcome the strong attractive forces in the solid.

6.5 True. The vapor pressure of a liquid is a measure of the escaping tendency of its molecules. Increasing the temperature increases the kinetic energy of its molecules, thus increasing their escaping tendency.

6.7 Because the molecules are in contact.

6.9 The molar heat of vaporization is always greater than the molar heat of fusion.

6.11 True. The transformation between the solid state and the liquid state takes place at the same temperature. If the transformation is approached by raising the temperature of the solid until the liquid forms, this temperature is called the melting point. If the transformation is approached by lowering the temperature of the liquid until the solid forms, it is called the freezing point.

6.13 9.7 kcal of heat will be released.

6.15 Through the weak London forces (temporary dipoles)

6.17 Only b

6.19 The greater the number of electrons (atomic number) in a molecule, the larger the temporary dipole (London force). Thus, in general, the greater the molecular mass, the greater the strength of London forces exerted between molecules.

6.21 The four C–Cl polar bonds are arranged symmetrically (tetrahedrally) with carbon at the center of the tetrahedron. The symmetrical arrangement of dipoles results in a net dipole moment of zero.

6.23

	London force	Dipole–dipole	Hydrogen bond
CH_4	yes	no	no
$CHCl_3$	yes	yes	no
NH_3	yes	yes	yes

6.25 Compound a has the greater heat of vaporization, because it can form hydrogen bonds between molecules, whereas compound b cannot.

6.27 The molecules of a liquid have acquired sufficient kinetic energy to partly overcome the intermolecular forces holding them in fixed positions. Although they are free to assume new positions under an applied force (for example, gravity), they still remain in contact.

6.29 In an open system, vapor molecules will continuously escape into the surroundings. Therefore the rate of evaporation will always exceed the rate of condensation, and evaporation will occur continuously.

6.31 When a solid melts, the ordered solid-state structure becomes disordered, but the molecules in the liquid state remain in close contact. Therefore the volume of the liquid is close to that of the solid.

6.33 The word vapor is used to describe the gaseous state of a substance whose liquid and gaseous states are present at the same time. Under normal conditions of room temperature, the liquid states of oxygen or nitrogen are not possible.

6.35 c < a < b

6.37 No. Vapor pressure is a balance between evaporation and condensation. In an open container, the rate of evaporation must be greater than the rate of condensation, and equilibrium cannot be established.

6.39 Equilibrium describes a situation that does not change over time.

6.41 The temperature at which vapor bubbles form throughout a liquid at an external pressure of 1 atmosphere

6.43 Both molecules are polar, but, in addition, water can form hydrogen bonds; so the secondary attractive forces in water are greater than those in chloroform. Therefore water will have the greater surface tension of the two.

6.45 Acetic acid molecules form hydrogen bonds both among themselves and with water. Pentane molecules can interact only by London forces. The attractive forces between acetic acid and water are similar, and therefore acetic acid will dissolve in water. The attractive forces between pentane molecules and water are very different and will therefore not form solutions.

6.47 An amorphous solid has no particular shape, melts over a range of temperatures, and if it can be shattered, forms pieces that have round or smooth surfaces resembling a liquid.

6.49 $CH_4 < C_2H_5OH < NaCl$

6.51 68%

6.53 H_2O, HF, and $CHCl_3$

6.55 H_2O can form two hydrogen bonds per molecule, while HF can only form one hydrogen bond per molecule. Therefore, more thermal energy is required to break two hydrogen bonds than one. This results in a higher boiling point.

6.57 Vapor pressure measures the tendency of molecules to escape from liquid into the gas phase. The greater the secondary forces, the lower the escaping tendency and the smaller the vapor pressure.

6.59 The Lewis dot structure of dimethyl ether shows that the oxygen has a free pair of electrons. This free pair of electrons acts to force the molecule into a bent configuration like water. Thus, dimethyl ether has a dipole.

6.61 Both water and hydrogen sulfide can form dipole–dipole interactions and London forces. However, water can also form hydrogen bonds, but hydrogen sulfide cannot. The strong hydrogen bonds cause water to be a solid at the same temperature at which hydrogen sulfide exists as a gas.

6.63 The temperature on the outside of an uninsulated glass of ice and water is about 0°C, the freezing point of water. Water vapor in the air will therefore condense on the outside of the glass.

6.65 (a) HF forms hydrogen bonds, HCl cannot.

(b) $TiCl_4$ condenses by London forces, the liquid form of LiCl consists of ions.

(c) HCl condenses through both dipole–dipole interactions and London forces, the sum of which is weaker than the ionic forces in liquid LiCl.

(d) The hydrogen bonds formed in ethanol are stronger than the secondary forces in ethyl ether.

6.67 When a substance melts, the intermolecular forces are only loosened so that the crystalline order is disrupted. Vaporization requires all secondary forces to be broken, which requires more energy.

6.69 Both a liquid and a gas are disordered compared with a solid.

6.71 The condensation of steam (gaseous water) releases much more heat than does the cooling of liquid water from 100°C to skin temperature.

6.73 Without lung surfactant, the surface tension at the surface of alveoli would cause the alveoli to collapse and therefore prevent the absorption of oxygen and the diffusion of carbon dioxide.

Chapter 7

7.1 (a) 1.07 g; (b) 8.19 g

7.3 1.50 g

7.5 87.1 mL of solution

7.7 125 g

7.9 448 mL

7.11 1.21 mg %

7.13 0.0042% w/v

7.15 1.0 mol

7.17 (a) 2.0 M; (b) 3.38 M

7.19 (a) 2.78 mol; (b) 0.384 mol

7.21 (a) 1.50 L; (b) 3.83 L

7.23 9.09×10^{-4} M

7.25 3.75×10^3 mg

7.27 5.40×10^2 mL

7.29 0.020 L

7.31 (a) 1.50 L; (b) 325 mL

7.33 9.0% w/v

7.35 0.500 M

7.37 No. The data show that the solubility of strontium acetate decreases with increase in temperature. If the solution had been saturated, strontium acetate would have crystallized out.

7.39 6.71 M

7.41 (a) 0.40 osmol/L; (b) 0.02 osmol/L

7.43 The dextrose solution is hypertonic; erythrocytes will lose water and shrink.

7.45 (a) 8.90% w/w; (b) 7.04% w/w; (c) 4.0% w/w

7.47 (a) 2.20 g; (b) 13.1 g

7.49 (a) 9.70% w/v; (b) 1.04% w/v; (c) 4.00% w/v

7.51 (a) 0.0030 mol; (b) 0.078 mol; (c) 0.124 mol; (d) 1.09 mol

7.53 (a) 0.559 mol; (b) 2.52 mol; (c) 0.960 mol; (d) 0.679 mol

7.55 Add water to 13.3 g of KI to a final volume of 500.0 mL of solution.

7.57 (a) 5.66% w/w; (b) 4.83% w/w; (c) 1.77% w/w

7.59 (a) 69.5 g NaCl; (b) 69.5 g $Mg(NO_3)_2$; (c) 98.7 g K_2CO_3

7.61 15.0 L

7.63 9.0% w/v

7.65 (a) Precipitate forms: $Ag^+(aq) + Cl^-(aq) \longrightarrow AgCl(s)$
(b) Precipitate forms:
$Mg^{2+}(aq) + 2\ OH^-(aq) \longrightarrow Mg(OH)_2(s)$
(c) Precipitate forms: $Pb^{2+}(aq) + 2\ Cl^-(aq) \longrightarrow PbCl_2(s)$
(d) Precipitate forms:
$3\ Ca^{2+}(aq) + PO_4^{3-}(aq) \longrightarrow Ca_3(PO_4)_2(s)$

7.67 Propylene glycol is the solvent, and water is the solute.

7.69 Water moves out of the arterial blood.

7.71 0.30 osmol/L (see Box 7.3)

7.73 10.1% w/w

7.75 25.3 m

7.77 18.4 M

7.79 (a) $Na_2SO_4(aq) + BaCl_2(aq) \longrightarrow 2\ NaCl(aq) + BaSO_4(s)$
$Ba^{2+}(aq) + SO_4^{2-}(aq) \longrightarrow BaSO_4(s)$
(b) $NaCl(aq) + AgNO_3(aq) \longrightarrow NaNO_3(aq) + AgCl(s)$
$Ag^+(aq) + Cl^-(aq) \longrightarrow AgCl(s)$

7.81 At solution concentrations greater than the critical micelle concentration, surface-active substances associate through secondary forces to form macromolecular structures called micelles.

Chapter 8

8.1 2.25 mol/min

8.3 The proportion of molecules undergoing reactive collisions doubled.

8.5 Reaction rate a is greater than reaction rate b.

8.7 $E_{back} = +58$ kJ

8.9 No. The reaction takes place in an open system, the CO_2 escapes into the atmosphere, and the system can never come to equilibrium.

8.11 Forward reaction: $CaCO_3(s) \longrightarrow CaO(s) + CO_2(g)$
Back reaction: $CaO(s) + CO_2(g) \longrightarrow CaCO_3(s)$

8.13 $K_{eq} = [CO_2]$

8.15 $K_{eq} = 0.099$

8.17 Equilibrium shifts to the right (to product).

8.19 (a) Color shift to pink; (b) color shift to blue.

8.21 The square brackets around chemical components in an equilibrium-constant expression denote their molar concentrations.

8.23 (a) $K_{eq} = \dfrac{[CO][H_2]^3}{[CH_4][H_2O]}$; (b) $K_{eq} = \dfrac{[CO_2]^8[H_2O]^8}{[O_2]^{12}}$

8.25 (a) $K_{eq} = [CO_2][NH_3]^2$; (b) $K_{eq} = \dfrac{1}{[O_2]^{0.5}}$

8.27 (a) $COCl_2(g) \rightleftharpoons CO(g) + Cl_2(g)$
(b) $CH_4(g) + H_2O(g) \rightleftharpoons CO(g) + 3\ H_2(g)$

8.29 (a) $K_{eq} = \dfrac{[NO_2][NO_3]}{[N_2O_5]}$; (b) $K_{eq} = \dfrac{[N_2]^2[H_2O]^6}{[NH_3]^4[O_2]^3}$

8.31 (a) 1000/1; (b) 1/100; (c) 7.1/1; (d) 12.0/1

8.33 Some of the solid would dissolve.

8.35 $E_{forward} = +72$ kJ

8.37 E_a of the forward reaction was larger.

8.39 If a system in an equilibrium state is disturbed, the system will adjust to neutralize the disturbance and restore the system to equilibrium.

8.41 As the reaction proceeds, the total number of moles decreases. Because all components are gaseous, follow the course of the reaction by measuring the decrease in pressure as a function of time.

8.43 Pulse rate at 72°F is 40 ± 5 beats per minute.

8.45 The easiest approach would be to carry out the reaction in the dark. Another approach would be to use vessels made with intensely tinted glass. A more sophisticated method would be to use vessels made with intensely tinted glass of different colors.

8.47 (a) Shift to the left (reactants); (b) shift to the left (reactants); (c) shift to the left (reactants); (d) shift to the left (reactants); (e) no effect; (f) shift to the right (products)

8.49 An overall equilibrium constant for a sequence of reactions can be calculated only if a product of the first reaction is a reactant of the next reaction, and so on, for each succeeding reaction.

8.51 E_a of the uncatalyzed reaction is larger. Catalysts provide an alternate pathway for a reaction of lower E_a, which results in faster rate. ΔH is unchanged, because the reactants and products are the same.

8.53 $K = 0.0047$

Chapter 9

9.1 $K_W = 1.00 \times 10^{-14}$

9.3 A substance that can act as an acid or a base

9.5 A strong base is 100% dissociated in aqueous solution. Examples: NaOH and KOH.

9.7 $[H_3O^+] = [Cl^-] = 0.30\ M$

9.9 $[OH^-] = [Tl^+] = 0.25\ M$

9.11 $[OH^-] = 2.86 \times 10^{-14}\ M$

9.13 $[H_3O^+] = 2.5 \times 10^{-13}\ M$

9.15 The pH of pure water at 25°C is 7.00.

9.17 pH = 2.0

9.19 pH = 12.18; pOH = 1.82

9.21 (a) 2.48; (b) 1.60; (c) 3.14; (d) 1.13

9.23 A weak acid is one that is incompletely dissociated in aqueous solution. Examples: acetic and phosphoric acids.

9.25 Neutral

9.27

	Acid	Base
(a)	HNO_2	NO_2^-
	H_3O^+	H_2O
(b)	$H_2PO_4^-$	HPO_4^{2-}
	H_3O^+	H_2O

9.29 $K_b = 5.75 \times 10^{-10}$

9.31 $pK_a = 4.76$; $pK_b = 9.24$

9.33 $pK_a + pK_b = 14.00$

9.35 $H_2CO_3(aq) + H_2O(l) \rightleftharpoons H_3O^+(aq) + HCO_3^-(aq)$
$HCO_3^-(aq) + H_2O(l) \rightleftharpoons H_3O^+(aq) + CO_3^{2-}(aq)$

9.37 $K_{a1} = \dfrac{[H_3O^+][HCO_3^-]}{[H_2CO_3]} = 4.45 \times 10^{-7}$

$K_{a2} = \dfrac{[H_3O^+][CO_3^{2-}]}{[HCO_3^-]} = 4.72 \times 10^{-11}$

9.39 Minor significance

9.41 (a) Neutral; (b) basic; (c) acidic; (d) acidic; (e) neutral

9.43 (a) Neutral; (b) basic; (c) basic; (d) basic; (e) basic

9.45 The conjugate acid prevents shifts to basicity, and the basic form prevents shifts to acidity.

9.47 pH = 4.89

9.49 pH = 7.38

9.51 $[HCl] = 0.017\ M$

9.53 $[H_2SO_4] = 0.0170\ M$

9.55 Dissolve 16.3 g of H_3PO_4 in sufficient water to make 1.00 L of solution

9.57 0.072 N

9.59 (a) $Na^+ = 0.125\ M$; (b) $K^+ = 0.035\ M$; (c) $Ca^{2+} = 0.036\ M$; (d) $Mg^{2+} = 0.014\ M$

9.61 (a) 2.51; (b) 0.00; (c) 2.16; (d) 1.72; (e) 1.64

9.63 (a) 2.21 (b) 2.09; (c) 1.25; (d) 0.00; (e) 2.00

9.65 (a) Acidic; (b) neutral; (c) basic; (d) acidic; (e) basic

9.67 Molar ratio of $[HPO_4^{2-}/H_2PO_4^-] = 10/1$

9.69 0.0310 N

9.71 (a) pOH = 11.53, $[OH^-] = 3.0 \times 10^{-12}\ M$;
(b) pOH = 12.40, $[OH^-] = 4.0 \times 10^{-13}\ M$;
(c) pOH = 10.86, $[OH^-] = 1.4 \times 10^{-11}\ M$;
(d) pOH = 12.87, $[OH^-] = 1.4 \times 10^{-13}\ M$

9.73 Ammonium acetate in water will form a neutral solution.

9.75 A buffered solution contains approximately equal concentrations of a conjugate acid–base pair.

9.77 A conjugate acid–base pair consists of a weak acid and the basic anion resulting from its dissociation.

9.79 5.31 g

9.81 pH = 8.38

9.83 0.15 atm

9.85 The pH is equal to the pK_a.

9.87 $Na_3PO_4 > Na_2HPO_4 > NaH_2PO_4$

9.89 $CO_3^{2-} > HCO_3^-$; $PO_4^{3-} > HPO_4^{2-} > H_2PO_4^-$

9.91 The pH of solution A is 2, and the pH of solution B is 12.

Chapter 10

10.1 Alpha-radiation consists of helium nuclei, charge 2+, mass, 4 amu.

10.3 Gamma-radiation has no mass or charge and is highly penetrating.

10.5 ^4_2He

10.7 $^0_{-1}e$

10.9 Radioactivity can be induced by bombardment of nonradioactive elements with high-energy subatomic particles.

10.11 A radioactive decay series is a series of nuclear reactions that begins with an unstable nucleus and ends with the formation of a stable one.

10.13 1.20×10^4 years old

10.15 Beta-radiation

10.17 It becomes attenuated.

10.19 The ionization of water and the production of hydrated electrons

10.21 The secondary chemical processes

10.23 Radiation sickness is caused by nonlethal exposure to radiation and is characterized by nausea and a severe drop in white-blood-cell count.

10.25 She must double the distance from the source to reduce the intensity by 1/4—move from 4 feet away to 8 feet away from the radiation source.

10.27 The detection of ions caused by radiation

10.29 A rad is the amount of energy absorbed in matter when exposed to radiation without regard to the effects caused.

10.31 No. The RBE for α-rays is 10, and that for γ-rays is 1.0.

10.33 No. The radiation will be absorbed by tissue before it can be detected externally.

10.35 Yes. Cobalt-60 is an energetic source of γ-radiation that can penetrate deeply into tissue to cause radiation damage.

10.37 No. Positrons can be detected only when they interact with an electron to produce γ-rays, which are in turn detected indirectly by the presence of the ions that they produce as they pass through matter.

10.39 No. Beta-emission would be absorbed by tissue and could not be used for detection of disorders within the body.

10.41 A CAT scan produces images of internal structures of the body with the use of X-rays.

10.43 31.25 mg

10.45 2.33×10^8 disintegrations per second

10.47 22.1%

10.49 Radioactivity is the name of the process in which atomic nuclei spontaneously decompose.

10.51 No. Gamma-radiation does not affect atomic mass or number.

10.53 Rapidly dividing tissue—for example, intestinal epithelium

10.55 Both processes release enormous amounts of energy. In nuclear fission, unstable nuclei of high atomic mass are split by interaction with a high-energy particle into lower-atomic-mass nuclei. In nuclear fusion, low-atomic-mass nuclei fuse to form new nuclei of higher atomic mass.

10.57 X-rays are produced by electrons accelerated from an anode at high voltage and then colliding with a heavy metal cathode.

10.59 No. K-shell X-rays are the result of the ejection of an electron from an inner atomic electron shell. Bremsstrahlung refers to X-rays generated by close encounters of incoming electrons with the atomic nuclei of the target.

10.61 X-rays passing through tissue are therefore absorbed differentially. The emerging pattern reveals the underlying structure.

10.63 Stratospheric ozone is formed when high-energy cosmic radiation cleaves O_2 into free radicals that recombine into O_3 (ozone) molecules.

Glossary

accuracy *See* **error.**

acid *See* **Brønsted–Lowry acid.**

acid dissociation constant An equilibrium constant for the dissociation of a weak acid.

action potential After stimulation, an electrical depolarization that moves along a neuron's membrane.

activated complex A transitory molecular structure formed by the collision of two reacting molecules.

activation energy The minimum energy that colliding reactant molecules must possess to reach the activated complex and undergo chemical reaction to form new products.

active site A location on an enzyme's surface where catalysis takes place.

addition reaction A reaction in which all the elements of a reactant add to the double or triple bond of a compound.

alkali metal An element in Group I of the periodic table.

alkaline earth metal An element in Group II of the periodic table.

α-particle The nucleus of a helium atom emitted by a radioactive substance.

amphipathic molecule A molecule that contains both hydrophilic and hydrophobic groups.

amphoteric compound A compound that is both an acid and a base.

anhydrous substance A substance that does not contain water.

anion An ion with a negative charge.

anode The positive electrode in a battery or an electrophoretic or other apparatus.

antioxidant A substance that protects other substances against damage from oxidation.

aqueous solution A solution with water as the solvent.

atmosphere *See* **standard atmosphere.**

atomic symbol A one- or two-letter symbol for an element or an element's atoms.

basal metabolic rate The measurement of basal metabolic activity.

basal metabolism The minimal metabolic activity of a human at rest whose gastrointestinal tract is empty.

base *See* **Brønsted–Lowry base.**

base dissociation constant An equilibrium constant for the dissociation of a weak base.

base unit A fundamental unit of measurement for one of the base quantities in the SI system.

β-particle An electron emitted from the nucleus of a radioactive substance.

binary compound A compound consisting of two elements.

binding site The structural component of an enzyme's catalytic site where substrate is bound by secondary forces.

biochemistry The study of the structures and functions of living organisms at the molecular level.

biological membrane A membrane that surrounds a cell or organelle.

biomolecules The molecules of which living organisms are composed.

boiling point The temperature at which a substance boils when the atmospheric pressure is 760 torr.

bond angle The angle between two bonds that share a common atom.

bremsstrahlung X-rays used in medical imaging.

Brønsted–Lowry acid Any substance that can donate a proton.

Brønsted–Lowry base Any substance that can accept a proton.

buffer system An aqueous solution containing a Brønsted–Lowry acid with its conjugate base.

calorie The heat absorbed when the temperature of 1.0 g of water rises 1 Celsius degree between 14.5°C and 15.5°C.

catabolism Metabolic reactions in which molecules are degraded.

catalysis A process in which the rate of a chemical reaction is increased by the presence of a catalyst.

catalyst A substance that takes part in a chemical reaction and accelerates its rate but emerges unchanged at the reaction's conclusion.

cathode The negative electrode in a battery or an electrophoretic or other apparatus.

cation An ion with a positive charge.

Celsius scale A temperature scale, in degrees, that defines the freezing point of water at 0°C and the boiling point at 100°C.

centimeter A length equal to 1/100 of a meter.

chemical bond An electrical force or attraction strong enough to hold atoms together to form compounds.

chemical change A process through which substances lose their chemical identities and form new substances with new properties.

chemical equation A shorthand representation of a chemical reaction that uses formulas for reactants and products and numbers before components to represent their mole proportions.

chemical equilibrium A state in which the rate at which products form is equal to the rate at which reactants form.

chemical kinetics The study of the rate of a chemical reaction.

chemical property The ability of a pure substance to undergo chemical change.

colloidal particle A particle smaller than 1×10^{-4} cm.

combining power The number of bonds formed by an atom when the atom is present in a covalent compound.

combustion The burning of an element or a compound in air.

complementarity principle A principle that accounts for the selectivity of enzymes. The structure of the substrate must complement the structure of the enzyme's binding site. *See also* **lock-and-key theory.**

compound A pure substance composed of atoms of two or more elements present in a fixed and definite ratio.

concentration The quantity of a component of a mixture in a unit of mass or a unit of volume of the mixture.

condensation The conversion of the gaseous state into the liquid state.

conversion factor *See* **unit-conversion factor.**

cosmic radiation Ionizing radiation emanating from the sun and outer space and consisting mostly of protons.

covalent bond The attractive force holding two atoms together resulting from the sharing of a pair of electrons.

degree Celsius *See* **Celsius scale.**

degree Fahrenheit *See* **Fahrenheit scale.**

dehydrogenation A reaction that proceeds with the loss of two hydrogen atoms from a molecule.

density A derived unit defined as mass per unit volume.

derived quantity A unit that is a mathematical relation between two or more base quantities, such as centimeters squared (cm^2) or centimeters per second (cm/s).

derived unit of measurement *See* **derived quantity.**

diatomic molecule A molecule consisting of two atoms, such as O_2.

diffraction A wave property of light that allows it to bend around corners.

diffusion A reduction in a concentration gradient resulting from the random motion of particles.

dipole Any molecule that is electrically neutral overall but electrically asymmetrical—that is, containing separated partial and opposite electrical charges.

dipole–dipole force The secondary attractions between dipoles in different molecules or between dipoles in different portions of the same molecule.

dipole moment A measure of the size of a dipole.

diprotic acid An acid containing two dissociable hydrogen atoms.

dissolution The process of dissolving or of preparing a mixture that will form a solution.

dynamic equilibrium An equilibrium that is the result of two opposing processes both taking place at the same rate.

electrode An electrically conductive solid suspended in a conductive medium through which electricity can flow.

electrolyte Any substance that, when dissolved in water, will allow the solution to conduct electricity.

electromagnetic radiation Radiation, such as heat and light, that has its origin in the oscillation of charged particles.

electron configuration The complete description of the organization of the electrons of an atom.

electronegativity The ability of an atom covalently bonded to another atom to draw the bonding electrons toward itself.

electron shell An organization level of atomic electrons defined by a principal quantum number.

electron volt (eV) An energy unit used for radiation (1 eV = 96.5 kJ/mol).

element A substance in which all the atoms have the same atomic number and electron configuration.

elemental symbol A symbolic representation of the name of an element, such as He for helium.

empirical formula *See* **formula, empirical.**

emulsifying agent A substance that stabilizes a suspension of colloidal droplets of one liquid in a continuous phase of another liquid.

emulsion A stable suspension of colloidal droplets of one liquid in a continuous phase of another liquid.

endothermic reaction A reaction in which heat is absorbed.

end point In a titration, the experimentally determined point at which the unknown acid or base is completely neutralized.

energy The capability to cause a change that can be measured as work.

energy level An atomic energy state defined by a principal quantum number.

enzyme A molecule, usually a protein, that catalyzes a biochemical reaction.

enzyme–substrate complex A temporary combination of enzyme with its substrate before catalysis.

equilibrium *See* **chemical equilibrium.**

equilibrium concentration The concentration of a product or a reactant of a chemical reaction at equilibrium.

equilibrium constant A constant that is calculated from the relation between molar concentrations of products and reactants of a chemical reaction at equilibrium.

equilibrium expression The relation between molar concentrations of products and reactants of a chemical reaction at equilibrium.

equivalence point In a titration, the theoretically expected point at which the unknown acid or base should be completely neutralized.

equivalent mass The equivalent mass of an acid is the formula mass of the acid divided by the number of reacting H^+ ions per mole of acid.

error The difference between the value considered true or correct and the measured value.

erythrocyte A red blood cell.

evaporation The vaporization of a liquid into the atmosphere.

exact number A number that can be considered to have an infinite number of significant figures.

exothermic reaction A chemical reaction that evolves heat.

extensive property A property, such as mass or volume, that is directly proportional to the size of the sample.

Fahrenheit scale A temperature scale, in degrees, on which the freezing point of water is 32°F and the boiling point is 212°F.

formula, chemical A representation of how many atoms of each element are in a fundamental unit of a compound.

formula, empirical A representation that gives the smallest whole-number ratio of the atoms of the elements of a compound.

formula, molecular The formula that shows the numbers of the atoms of the elements in a molecule of a compound.

formula, structural The formula that shows how the various atoms in a molecule are bonded together.

formula mass The sum of the atomic masses of the atoms in a formula of a compound in atomic mass units (amu).

formula unit The smallest particle that has the composition of the chemical formula of the compound.

formula weight *See* **formula mass.**

freezing point The temperature at which a substance undergoes the transition from a liquid to a solid.

γ ray Radiation of energy higher than that of an X-ray.

gradient The change in value of a physical quantity with distance.

gram A mass equal to 1/1000 of a kilogram.

half-life The time required for a substance to lose one-half of its physical or chemical activity.

heat A form of energy that moves between two objects in contact that are at different temperatures.

heat of fusion *See* **molar heat of fusion.**

heat of reaction *See* **molar heat of reaction.**

heat of vaporization *See* **molar heat of vaporization.**

heterogeneous mixture A mixture in which there are visual discontinuities in composition.

homeostasis The steady-state physiological condition of the body.

homogeneous mixture A mixture in which there are no visual discontinuities in composition.

hydrate A compound in which water is present in a fixed molar proportion of the other constituents.

hydration A reaction that proceeds with the addition of water to a double or triple bond. In aqueous solutions, the association of water with ions by secondary forces.

hydrocarbon An organic compound that contains only carbon and hydrogen.

hydrogen bond The secondary attractions present in molecules that contain O—H, N—H, or H—F bonds.

hydrolysis reaction The reaction of an organic compound with water resulting in cleavage of the compound into two organic fragments each of which combines with a fragment (H^+ or OH^-) from water.

hydrophilic Refers to a molecule or a part of a molecule that is attracted to water.

hydrophobic Refers to a molecule or a part of a molecule that is repelled by water.

inorganic chemistry The study of compounds that contain elements other than carbon.

inorganic compound A compound that contains elements other than carbon.

intermolecular force A secondary force operating between molecules.

intermolecular process A process (physical or chemical) that takes place between molecules.

International System of Units *See* **SI units.**

intramolecular force A secondary force operating within a molecule.

intramolecular process A process (physical or chemical) that takes place within a molecule.

in vitro Refers to a substance or a process that is outside an organism.

in vivo Refers to a substance or a process that is inside an organism.

ionizing radiation Radiation that enters a medium and creates ions from the molecules therein.

joule The SI unit of energy (4.184 J = 1 cal).

kelvin The SI unit of temperature; its size is equal to $1/100$ of the temperature interval between the freezing point and the boiling point of water.

kilocalorie The quantity of heat equal to 1000 calories.

kilogram The SI unit of mass.

kilojoule The quantity of energy equal to 1000 joules.

kinetic energy The energy of a moving body.

kinetics *See* **chemical kinetics.**

like-dissolves-like rule A solute is soluble in a solvent only if the secondary attractive forces between molecules of solute are similar to the secondary attractive forces between molecules of solvent.

liter A volume equal to 1000 cm^3, or 1000 mL.

lock-and-key theory A theory to account for the selectivity of enzymes for their substrates. The substrate must fit into a binding site just as a key fits into a lock.

London force The weak secondary attractive forces possessed by all molecules, independent of the presence of permanent dipoles or hydrogen bonds.

macromolecule *See* **polymer.**

mass A measure of the quantity of matter relative to a reference standard.

mass number The sum of protons and neutrons in an isotope of an element.

matter Anything that has mass and occupies space.

measurement An instrumental determination of a physical quantity.

mechanism of reaction The molecular-level details of how reactants change into products.

melting point The temperature at which a solid is transformed into a liquid.

metabolism All the chemical reactions that take place in an organism.

metal An element or combination of elements that is shiny, conducts electricity, and, if solid, is malleable.

metalloid An element that has some of the properties of metals and nonmetals.

meter The SI unit of length (1 m = 100 cm).

metric system A decimal system of weights and measures in which base units are converted into smaller or larger multiples by movement of a decimal point; superseded by the International System of Units (SI system).

microgram A mass equal to $1/1000$ of a milligram.

microliter A volume equal to $1/1000$ of a milliliter.

milliliter A volume equal to $1/1000$ of a liter.

millimeter A unit of length equal to $1/1000$ of a meter.

mixture Matter consisting of two or more pure substances in varying proportions.

molar concentration A solution's concentration in units of moles of solute per liter of solution; molarity.

molar heat of fusion The amount of heat required to melt 1 mole of a pure solid.

molar heat of reaction The amount of heat generated or absorbed per mole of reactant when a chemical reaction takes place.

molar heat of vaporization The amount of heat required to vaporize 1 mole of a pure liquid.

molarity *See* **molar concentration.**

molar mass The mass of 1 mole of a compound (either molecular or ionic) in grams.

mole An Avogadro number of a substance's formula units.

molecular compound A compound whose atoms are joined by covalent bonds.

molecular formula *See* **formula, molecular.**

molecular mass The mass of one molecule of a molecular compound in atomic mass units (amu).

molecular weight *See* **molecular mass.**

molecule The smallest particle of a molecular compound.

monoprotic acid An acid containing one dissociable hydrogen atom.

neutralization reaction A reaction between an acid and a base.

noble gas An element in Group VIII of the periodic table.

nomenclature The naming of compounds.

nonbonding electron An electron located in the outer shell of an atom but not participating in bonding to other atoms.

nonelectrolyte Any substance that, when dissolved in water, will not allow the solution to conduct electricity.

nonmetal An element that is not shiny, cannot conduct electricity, and is not malleable.

orbital The region of space in which an electron resides.

orbital hybridization The excitation of an element's ground state to produce a different electronic configuration.

organic chemistry The study of compounds that contain carbon.

organic compound A compound that contains carbon.

oxidation of an inorganic compound An increase in the number of oxygen atoms or an increase in the positive charge of the metallic element or both.

oxidation of an organic compound An increase in the number of oxygen atoms or a decrease in the number of hydrogen atoms (or both) bonded to one or more of the carbon atoms of the compound.

oxidizing agent A substance that oxidizes another substance.

photon A unit of light energy.

physical change A change in a physical property.

physical property Any observable physical characteristic of a substance other than a chemical property.

physical quantity A property that is described by a quantity and a unit.

physical state The state of being either a gas, a liquid, or a solid.

physiological function A function of a living organism or of an individual cell, tissue, or organ of which the organism is composed.

physiological saline solution A solution of sodium chloride with the same osmolarity as that of blood.

plasma The liquid part of blood, in which blood cells and other substances are dissolved or suspended.

polar covalent bond A covalent bond in which the electron pair resides closer to one of the bond partners than the other.

polar molecule A molecule that has a dipole moment.

polyatomic ion An ion, such as OH^- or CO_3^{2-}, made up of two or more atoms.

polyatomic molecule A molecule, such as $C_6H_{12}O_6$ or H_2O, made up of two or more atoms.

polymer High-molecular-mass molecule produced by bonding together large numbers of smaller molecules.

polyprotic acid An acid that has two or more dissociable protons.

precipitate A solid that forms in a solution as the result of a chemical reaction.

precipitation The formation of a precipitate.

precision A measure of how close a series of measurements agree with one another.

pressure Force per unit area.

product A substance that is produced by a chemical reaction.

pure substance *See* **substance, pure.**

quantum A quantity of energy contained by a photon of light.

radical A species, such as ·OH, with an unpaired electron.

reactant A substance that undergoes chemical change in a reaction.

reaction, chemical *See* **chemical change.**

reducing agent A substance that reduces another substance.

reduction of an inorganic compound A decrease in the number of oxygen atoms or a decrease in the positive charge of the metallic element or both.

reduction of an organic compound A decrease in the number of oxygen atoms or an increase in the number of hydrogen atoms (or both) bonded to one or more of the carbon atoms of the compound.

representative element Any element that appears in an A group of the periodic table.

respiration The uptake of oxygen and release of carbon dioxide by either cells or the body.

reversible reaction A chemical reaction that can reach equilibrium by starting with either reactants or products.

salt The product (other than water) of the reaction between an acid and a base.

scientific notation A method of writing a number as a product of a number between 1 and 10 multiplied by 10^x where x can be either a positive or a negative number.

secondary force An attractive force between identical molecules (such as H_2O and H_2O) or between different molecules (such as formaldehyde and water) or between different parts of the same molecule (such as a polypeptide).

semipermeable membrane A membrane that will permit passage of particular substances and no others.

significant figure The number of digits in a numerical measurement or calculation that are known with certainty plus one additional digit.

single bond One bond between two atoms.

SI units Units that have the same names as those in the older, metric system but have new reference standards for the base units.

solute The substance present in smaller amount in a solution.

solution A homogeneous mixture of two or more substances that is visually uniform throughout.

solvent The substance present in larger amount in a solution.

specific gravity The ratio of the density of a test liquid to the density of a reference liquid.

specific heat The heat absorbed or lost per Celsius degree change in temperature per gram of substance.

spectroscopy Measurement of the results of the interaction of electromagnetic radiation with atoms and molecules.

standard, reference A base unit of measurement such as the meter or the kilogram.

standard atmosphere The pressure that supports a column of mercury 760 mm high at zero degrees Celsius.

standard conditions of temperature and pressure (STP) A temperature of zero degrees Celsius and one atmosphere of pressure.

standard solution A solution for which its concentration is known with accuracy.

state of matter A condition in which matter can exist: solid, liquid, or gas.

stoichiometry The calculation of the quantities of the elements or compounds taking part in a chemical reaction.

STP *See* **standard conditions of temperature and pressure.**

strong acid An acid that is completely dissociated in aqueous solution.

strong base A base that is completely dissociated in aqueous solution.

structural formula *See* **formula, structural.**

substance, pure An element or a compound; not a mixture.

substrate The substance that is the object of an enzyme's catalysis.

surface tension A force at the surface of a liquid that reduces the area of the surface.

suspension A mixture in which the solute particles are larger than colloidal in size.

temperature A measure of the hotness or coldness of an object.

tetrahedral bond angle The 109.5° bond angle.

tetrahedron A geometrical shape with four triangular faces of the same size.

theory A fundamental assumption that explains a large number of observations, facts, or hypotheses.

thermal expansion The increase in volume of a substance in response to an increase in its temperature.

thermal property The manner in which a substance responds to changes in its temperature.

transition element An element between Group IIA and Group IIIA and in the actinide or lanthanide family.

transition state Another name for the activated complex.

triatomic molecule A molecule containing three atoms, such as H_2O or CO_2.

trigonal bond angle The $120°$ bond angle.

triprotic acid An acid containing three dissociable hydrogen atoms.

uncertainty The estimate of the last number in a measurement.

unit-conversion factor A fraction, such as 2.54 cm/in., that allows the conversion of one unit into another unit.

unsaturated hydrocarbon A hydrocarbon with a multiple bond.

URL A Web site location, called a Uniform Resource Location.

vacuum An enclosed space containing no matter.

valence electron An electron in the outermost electron shell of an atom.

valence shell The outermost electron shell of an atom.

valence-shell electron-pair repulsion theory (VSEPR theory) A theory that accounts for the symmetrical distribution of atoms around a central atom in a covalent compound.

vaporization The conversion of a liquid into a gas.

vapor pressure The pressure of a vapor in equilibrium with its liquid.

volatile liquid A liquid with a high vapor pressure.

volume The capacity of an object to occupy space.

water of hydration Water contained in a pure solid in a specific ratio of moles of water to moles of substance.

weak acid An acid that undergoes incomplete dissociation in aqueous solution; one that has a small dissociation constant.

weak base A base that undergoes incomplete dissociation in aqueous solution; one that has a small dissociation constant.

weight The gravitational force on an object relative to some reference standard.

Index

Note: Page numbers followed by f, t, and b indicate illustrations, tables, and boxed material, respectively.